Rheinwerk Computing

The Rheinwerk Computing series offers new and established professionals comprehensive guidance to enrich their skillsets and enhance their career prospects. Our publications are written by the leading experts in their fields. Each book is detailed and hands-on to help readers develop essential, practical skills that they can apply to their daily work.

Explore more of the Rheinwerk Computing library!

Christian Wenz, Tobias Hauser
PHP and MySQL: The Comprehensive Guide
2025, 1005 pages, paperback and e-book
www.rheinwerk-computing.com/6022

Torsten T. Will
C++: The Comprehensive Guide
2025, 1089 pages, paperback and e-book
www.rheinwerk-computing.com/5927

Johannes Ernesti, Peter Kaiser
Python 3: The Comprehensive Guide
2022, 1036 pages, paperback and e-book
www.rheinwerk-computing.com/5566

Philip Ackermann
JavaScript: The Comprehensive Guide
2022, 982 pages, paperback and e-book
www.rheinwerk-computing.com/5554

Sebastian Springer
Node.js: The Comprehensive Guide
2022, 834 pages, paperback and e-book
www.rheinwerk-computing.com/5556

www.rheinwerk-computing.com

Nouman Azam

Rust

The Practical Guide

Editor Megan Fuerst
Acquisitions Editor Hareem Shafi
Copyeditor Yvette Chin
Cover Design Graham Geary
Photo Credits iStockphoto: 168329140/© ilbusca; Shutterstock: 1126625519/© Jackie Niam
Layout Design Vera Brauner
Production Hannah Lane
Typesetting SatzPro, Germany
Printed and bound in Canada, on paper from sustainable sources

ISBN **978-1-4932-2687-0**
1st edition 2025

Rheinwerk Publishing, Inc.
2 Heritage Drive, Suite 305
Quincy, MA 02171
USA
info@rheinwerk-publishing.com
+1.781.228.5070

Represented in the E.U. by:
Rheinwerk Verlag GmbH
Rheinwerkallee 4
53227 Bonn
Germany
service@rheinwerk-verlag.de
+49 (0) 228 42150-0

Library of Congress Cataloging-in-Publication Control Number: 2024061177

Contents at a Glance

Contents

PART II Intermediate Language Concepts

8 Flexibility and Abstraction with Generics and Traits 213

9 Functional Programming Aspects 259

Preface

Welcome to this practical guide for Rust programming. Since its inception, Rust has rapidly gained popularity due to its emphasis on memory safety, performance, and concurrency. As the demand for safer and more efficient systems programming grows, developers and organizations have increasingly turned to Rust to build reliable software. This book is designed to bridge the gap between Rust's powerful features and practical implementation, offering a structured and hands-on approach to mastering the language.

This book provides a deep dive into Rust's core concepts, covering everything from ownership and borrowing to advanced features like smart pointers, concurrency, and metaprogramming. We explore how Rust enables developers to write safe and efficient code without compromising performance, making it an excellent choice for a wide range of applications—from embedded systems to web services. Whether you're new to Rust or looking to refine your expertise, this book serves as a comprehensive resource, equipping you with knowledge and tools to build robust and high-performance software in Rust.

The Purpose of This Book

For many learning Rust, it quickly becomes apparent that while numerous resources exist, few provide in-depth examples that clarify complex concepts. Rust's steep learning curve, and particularly its unique ownership system and borrow checker, often makes it difficult for beginners to write even simple programs without encountering compiler errors. Understanding why the compiler rejects code and how to fix these issues can be overwhelming. Many existing resources are scattered or lack a structured approach, making it difficult to understand how to progress systematically through the language. This book addresses these gaps by organizing material into meaningful sections, arranged in a progressive manner to help you build understanding step by step. Through clear explanations, practical examples, and structured learning paths, this book aims to provide a smoother and more intuitive journey into Rust.

The Structure of This Book

This book is organized into three parts: Part I, Basic Programming with Rust; Part II, Intermediate Language Concepts; and Part III, Advanced Language Concepts.

Part I spans Chapter 1 to Chapter 7. This part of the book starts with an introduction to Rust programming, including how to install and set up the coding environment and some tips on how to take full advantage of the material in the book. Next, it introduces basic programming constructs such as variables, data types, functions, conditionals, and control flow. Following this, it delves into Rust's unique ownership model, covering borrowing and dereferencing. The last two chapters in this part cover custom types like structs and enums, useful library types such as `Option` and `Result`, and practical aspects of organizing and testing code.

Part II includes Chapter 8 to Chapter 12. This part of the book expands on generics and traits for flexibility and abstraction, followed by functional programming aspects like closures, functional pointers, and iterators. It also covers memory management features, including lifetimes and smart pointers, and guides you through implementing common data structures and useful patterns for handling structs.

Part III starts with Chapter 13 and ends at Chapter 18. This part tackles complex topics like size in Rust, concurrency with threads and asynchronous programming, and the powerful macro system. This part also addresses real-life problems requiring data structure-heavy solutions, web programming fundamentals, and text processing, file handling, and directory management.

Let's take a look at the topics covered in each chapter of this book:

- **Chapter 1: Introduction**
  This chapter lays the groundwork for Rust development by covering installation, setup, and compilation of your first Rust program. It ensures a smooth start by guiding you through configuring Visual Studio Code (VS Code) and making the most of the book's material.
- **Chapter 2: Variables, Data Types, and Functions**
  This chapter introduces variables and includes topics such as mutability, scope, and shadowing, alongside a deep dive into primitive and compound data types. It also explores functions and code blocks, establishing a strong foundation for program development.
- **Chapter 3: Conditionals and Control Flow**
  You'll explore Rust's control flow constructs, including `if-else` statements and `match`. The chapter also covers looping mechanisms such as `for`, `while`, and infinite loops, along with essential input/output handling techniques.
- **Chapter 4: Ownership**
  This chapter explains Rust's ownership system, focusing on how memory is managed through ownership, borrowing, and dereferencing. By mastering these core principles, you'll learn to write memory-safe and efficient Rust programs.
- **Chapter 5: Custom and Library-Provided Useful Types**
  This chapter covers defining custom types using structs and enums while also exploring key library types like `Option`, `Result`, `HashMap`, and `HashSet`. Understanding these types helps with building efficient and structured Rust applications.

- **Chapter 6: Organizing Your Code**
  You'll learn best practices for structuring Rust projects using modules, organizing code logically, and managing privacy. The chapter also covers re-exporting, incorporating external dependencies, and publishing crates. This ensures you can develop scalable and well-structured Rust applications.
- **Chapter 7: Testing Code**
  This chapter explores the fundamentals of testing in Rust, covering unit tests and integration testing. It also delves into controlling test execution and testing configurations. By the end, you'll understand how to write reliable and efficient Rust programs through rigorous testing.
- **Chapter 8: Flexibility and Abstraction with Generics and Traits**
  This chapter introduces generics for writing reusable code and traits for defining shared behavior across types. You'll also learn about trait bounds, supertraits, trait objects, and associated types to build flexible and abstract designs.
- **Chapter 9: Functional Programming Aspects**
  Rust's functional programming features are explored in this chapter, including closures, functional pointers, and iterators. The chapter also covers combinators and iterating through `Option` types for expressive and concise programming.
- **Chapter 10: Memory Management Features**
  This chapter covers lifetimes, lifetime elision, and smart pointers like `Box`, `Rc`, and `RefCell`. These concepts help ensure safe memory management and efficient resource allocation in Rust.
- **Chapter 11: Implementing Typical Data Structures**
  You'll learn to implement singly and doubly linked lists while understanding reference cycles and memory leaks. The chapter provides hands-on examples to build efficient and safe data structures.
- **Chapter 12: Useful Patterns for Handling Structs**
  This chapter explores struct initialization patterns, including the builder pattern and techniques for simplifying struct design. These approaches make Rust code more maintainable and flexible.
- **Chapter 13: Understanding Size in Rust**
  This chapter explores the distinction between sized and unsized types, including references to unsized types and their memory implications. It also covers unsized coercion, optionally sized traits, and zero-sized types like the never type, unit type, unit structs, and phantom data.
- **Chapter 14: Concurrency**
  You'll learn about concurrency in Rust, including threads, ownership in multithreaded contexts, and communication techniques like message passing and shared states. This chapter also introduces synchronization methods, async/await, Tokio tasks, and practical applications such as web scraping.

- **Chapter 15: Macros**
  This chapter introduces Rust's macro system, explaining how to create macros, capture types, and repeat patterns for code generation. You'll explore practical examples that enhance code flexibility and metaprogramming capabilities.
- **Chapter 16: Web Programming**
  Rust's web programming capabilities are covered in this chapter, starting with setting up a web server and handling HTTP requests. You'll also learn about managing multiple requests using threads to build efficient web applications.
- **Chapter 17: Text Processing, File Handling, and Directory Management**
  This chapter covers file and directory handling, including reading, writing, and managing paths. Regular expressions, string manipulation, and text processing techniques are also explored for efficient data handling.
- **Chapter 18: Practical Problems**
  In this chapter, you'll apply Rust's data structures to real-world scenarios like search results, product popularity, and stock price analysis. Advanced problems include efficient storage and retrieval techniques for handling large-scale data.

Target Audience

This book is designed for a wide range of readers, from beginners with no prior experience in Rust to experienced programmers looking to deepen their understanding of the language. Those new to Rust will benefit from its structured progression, which introduces fundamental concepts before advancing to more complex topics like ownership, borrowing, and lifetimes. Developers familiar with other languages such as C++, Python, or Java will find detailed explanations of Rust's unique features, helping them transition smoothly. The book is also valuable for system programmers, application developers, and those interested in safe concurrency. Whether you are a student, a professional developer, or a researcher, this book provides the necessary tools to master Rust efficiently.

Acknowledgments

Writing this book would not have been possible without the incredible Rust community, whose dedication and enthusiasm have made learning and mastering Rust an engaging journey. The discussions on Rust forums, blog posts, and insightful threads have been invaluable sources of knowledge and inspiration. I am particularly grateful for *The Rust Programming Language* (commonly known as the *Rust Book*), which serves as a cornerstone for understanding Rust's core principles.

I would also like to extend my sincere gratitude to the National University of Computer and Emerging Sciences, my employer, for providing the necessary infrastructure and support during the writing of this book. Their encouragement and resources played a crucial role in bringing this project to completion. Lastly, I appreciate all the contributors to Rust's open-source ecosystem, whose work has made Rust both a powerful and accessible language for developers worldwide.

PART I

Basic Programming with Rust

Chapter 1
Introduction

As with any journey, the first step is the most crucial. Let's embark on our exploration of Rust, where the foundations of programming await.

This chapter introduces you to setting up Rust and its web-based environment, guiding you step by step through the installation process. We'll cover how to run and compile your first Rust program, show you how to set up Visual Studio Code (VS Code) for Rust development, and offer valuable advice on maximizing the benefits of this book. This foundational chapter ensures a smooth start for anyone new to Rust.

1.1 Installing Rust and Its Web-Based Environment

Getting started with Rust begins with setting up its development environment. This section guides you through installing Rust using `rustup`, the official installer, and exploring Rust's web-based tools like the Rust Playground, a convenient platform for experimenting with code without requiring a local setup.

1.1.1 Installing Rust

To begin with installation, navigate to the official Rust website at *https://www.rust-lang.org*. This website serves as a hub for all things Rust and includes comprehensive documentation, installation guides, a supportive community for troubleshooting and advice, and information on upcoming events and conferences. The Rust community is known for being friendly and supportive, making it a fantastic resource for developers of all levels.

To install Rust, click the **Get Started** button on the homepage. Download the appropriate installer for your operating system:

- For Windows, select the 64-bit installer.
- For other systems, follow the instructions for your specific environment.

The Rust installation is free and does not require any payment. After downloading, run the installer and follow the prompts. This process will set up the Rust toolchain, including its package manager, called *Cargo*.

For an efficient development experience, we recommend using VS Code, which runs on macOS, Linux, and Windows. You can download it from *https://code.visualstudio.com* for free. Once downloaded, install it by following the installation prompts.

You should install Rust support for VS Code, as several options are available to enhance your coding experience. These extensions provide valuable features such as code completion, jump-to-definition capabilities, reference tracking, code formatting, and other helpful tools to make writing Rust code more efficient and effective. The most recommended extension is the rust-analyzer. To install this component, go to the **Extensions** menu pane on the left, enter "rust" in the search field, and then choose **rust-analyzer**, as shown in Figure 1.1.

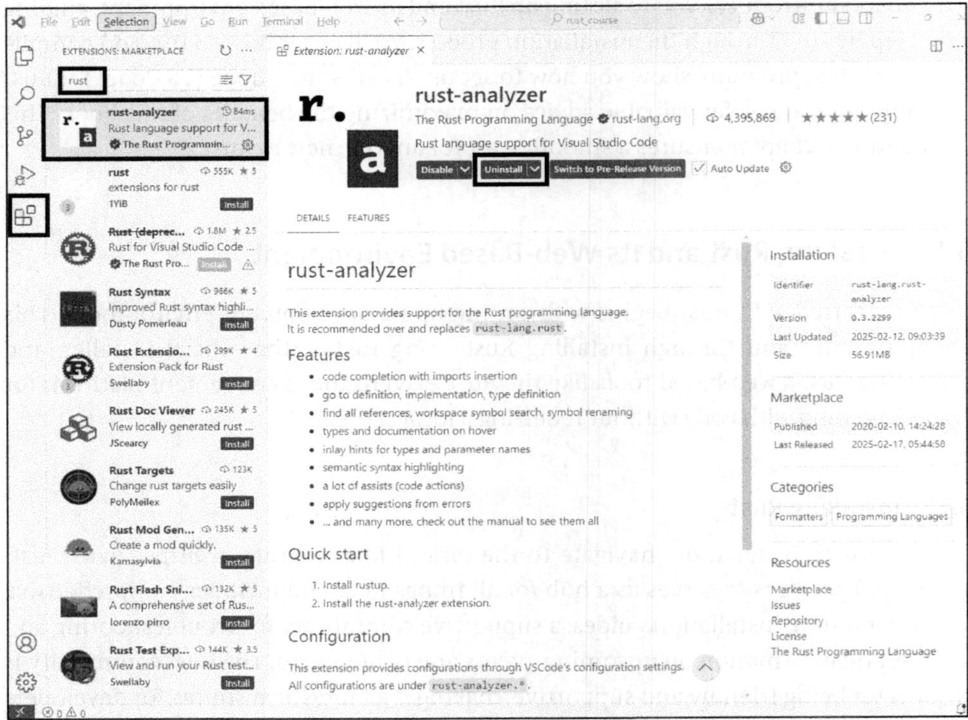

Figure 1.1 Installing the rust-analyzer Extension

To verify that Rust is installed correctly, open the terminal by pressing Ctrl + ` or go to **View • Terminal**. Then, enter the following command:

```
c:\> rustc --version
```

If this command fails, restart VS Code and the terminal to ensure that the PATH variable updates correctly.

Rust includes its own package manager, called *Cargo*. Just as other programming languages rely on package managers to handle package installation, dependency tracking, and related tasks, Rust uses Cargo to streamline these processes efficiently, much like Node Package Manager (npm) for Node.js, Composer for PHP, or pip for Python. To ensure that Cargo is installed correctly, run the following command:

```
c:\> cargo --version
```

Rust also includes the following command to update an installation:

```
c:\> rustup update
```

These simple steps are pretty much it for Rust installation.

1.1.2 Rust's Web-Based Compiler

The Rust Playground is ideal for quick experiments and sharing small code snippets. For larger projects, a full Rust installation is required so you can leverage advanced features and customize your development environment. The Rust Playground can be accessed at *https://play.rust-lang.org*. By default, the Rust Playground already contains a small Rust program. You can run this program without installing anything else by clicking the **Run** button at the top left.

Let's break down the code contained in the Rust Playground, as shown in Listing 1.1.

```
fn main() {
    println!("Hello, world!");
}
```

Listing 1.1 Hello World Program Already Contained in the Rust Playground

Every Rust program starts with the `main` function, which serves as the entry point. The `fn` keyword declares a function, and the empty parentheses `()` indicate that this function takes no input parameters. The code within the curly braces `{}` is the function body. In this case, the `println!` macro will print a message to the console. The text within the double quotes (`Hello, world!`) is the string we want to print.

This online version has many useful features. At the top left, clicking the **Debug** button lets you choose between two modes: **Debug** and **Release**. Debug mode is slower but easier for troubleshooting, while release mode runs more quickly by providing less debugging support. The difference, however, will not be evident in small programs. Typically, in the context of larger programs, you would select **Release**.

Next to the **Debug** button is the **Stable** button at the top left. The Rust Playground offers three compiler channels: stable, beta, and nightly. Stable is the most stable and widely used version, while beta provides early access to upcoming features. Nightly is the bleeding-edge version with the latest features and the highest potential for instability, often used for experimentation and testing.

Clicking the Rust Playground's **Share** button creates a GitHub gist of your code. This gist provides a permanent link to your code, making it easy to share with others or to reference later. Additionally, you can choose to include compiler output, such as warnings and errors, in the gist for more comprehensive information sharing.

The **Tool** button at the top right offers many useful features. For now, the most relevant ones are Rustfmt and Clippy. Rustfmt automatically formats your code, ensuring consistent indentation and spacing. Clippy goes a step further by analyzing your code for potential style issues, performance bottlenecks, and common mistakes. By using these tools, you can write cleaner, more efficient, and more maintainable Rust code.

1.2 Running and Compiling Your First Program

Let's make a new folder named `Rust Examples` for storing all our Rust code files. We'll add the *example1.rs* file to it containing the same code shown earlier in Listing 1.1. Let's open this file in VS Code by accessing **File • Open File**. You should be aware that the editor will not auto-save your program; be sure you save manually after making changes.

To compile a program, we'll write commands in the terminal. The terminal may be made visible via the **View** menu by selecting **Terminal**. Enter following command to compile the Rust code file:

```
c:\ Rust Examples> rustc example.rs
```

Program execution will take a few seconds, and if no errors arise, the program will return the control to the terminal without any additional messages. You may notice that it has generated executable file corresponding to the code file in the **Explorer** pane. Now, to run the executable file, enter the following command:

```
c:\ Rust Examples> ./example.rs
```

This approach is suitable for simple programs contained within a single file. However, in real-world scenarios, programs are usually complex, often involving multiple files and dependencies. In such cases, utilizing the Cargo utility is the ideal solution.

Cargo helps you manage Rust projects by efficiently organizing code, dependencies, and build processes. This tool is especially useful when working with multiple files, libraries, and external crates, by making the development process more streamlined and scalable. To create a project using Cargo, enter the following command:

```
c:\> cargo new Rust_Examples
```

This command sets up the directory structure for a new project named `Rust_Examples` inside the specified folder. The directory contains many files. The *Cargo.toml* file is the central configuration file for a Rust project managed by Cargo. This file contains metadata and settings that define the project's dependencies, build settings, and other relevant information. The *.gitignore* file contains the files or directories generated during the build process, temporary files, or sensitive data that should not be included in version control. The `target` directory contains build artifacts generated by Cargo, including compiled binaries, intermediate files, and dependencies. This directory organizes these files by building profiles (e.g., `debug`, `release`) and is automatically created when you build a Rust project.

Finally, the *src* folder contains all the Rust code source files with *.rs* extensions. By default, it contains a *hello.rs*, which is an autogenerated file and contains code similar to Listing 1.1. The `cargo run` command compiles your Rust project and immediately executes the resulting binary in one step. This command simplifies the development workflow by combining building and running. For instance, to execute our created project and the binary file residing inside it, enter the following command:

```
c:\Rust_Examples> cargo run
```

Note

Note for now that all files with *.rs* extensions are binary files with the exception of library (more on this topic in Chapter 6).

Make sure that the project `Rust_Examples` is in your current working directory. At this point, we've covered the basics of executing Rust programs.

The `cargo build` command is used in situations where you only want to build the program executable and are not interested in the resulting output of your program. This command simple compiles the Rust project into an executable or library, depending on the project configuration. By default, Cargo builds the project in debug mode. The resulting binaries are stored in the *target/debug* directory. This command is commonly used during development to quickly compile and test changes.

When building for production, you can use the `cargo build --release` command. This option may take some time, but once complete, you'll see a message like `Finished release [optimized]`, indicating the code is now optimized for production use. The optimized executable is created in the `target/release` folder within your project directory. This version is suitable for deployment, as the optimizations improve the performance of your Rust program, though they may increase its compilation time.

To summarize, the build process involves two primary scenarios:

1. In the development phase, quick and frequent builds are necessary since the code is still evolving. For this case, the default debug build is ideal.
2. Once the code is finalized for delivery, the focus shifts to optimization for speed, and the `--release` option is used to build the final executable.

For most of this book, we'll use the command cargo run for simplicity and for development purposes. Additionally, we'll primarily write our code in the *main.rs* file located in the *src* folder.

1.3 Visual Studio Code Settings

Efficiency in coding often starts with a well-configured editor. This section focuses on setting up VS Code, a popular choice for Rust developers, to optimize your development experience. These preferences reflect my own personal preferences, but you can adjust the configuration to suit your own needs and workflow.

To set up your preferences, navigate to **File • Preferences • Settings**, as shown in Figure 1.2. Let's make some changes in the **Settings** area.

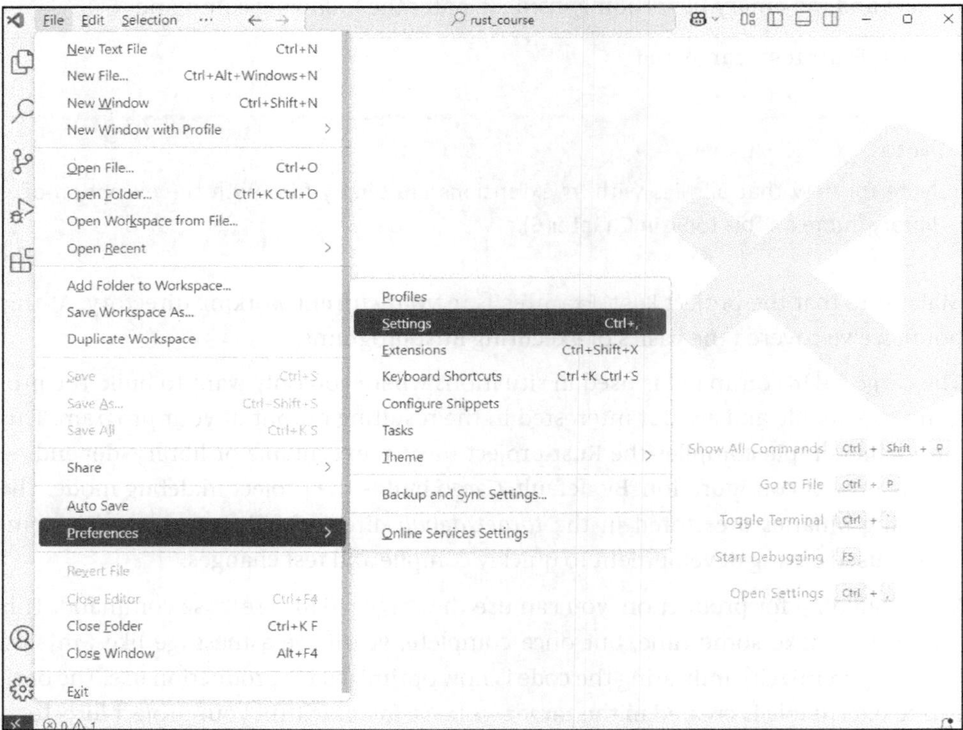

Figure 1.2 Setting Up Your Preferences

The first setting is **Format on Save**. To enable this feature, search for "format on save" in the search settings; then select the corresponding checkbox, as shown in Figure 1.3. This setting ensures that your code is automatically formatted whenever you save the file. For example, if we write some code in the main function and then save it, the code will be formatted automatically.

The blinking cursor can sometimes be distracting, so you might prefer a solid cursor instead. To adjust the cursor, search for "cursor blinking" in the search settings. By default, the cursor is set to blink, but you can change it to solid if that suits your preference better.

Next, we'll set up the font zoom functionality for the code. Search for "mouse wheel zoom" in the search setting. This feature allows you to zoom in and out of the code using the mouse wheel while holding the `Ctrl` key. Using this feature to adjust the text size quickly improves readability and reduces eye strain during long coding sessions.

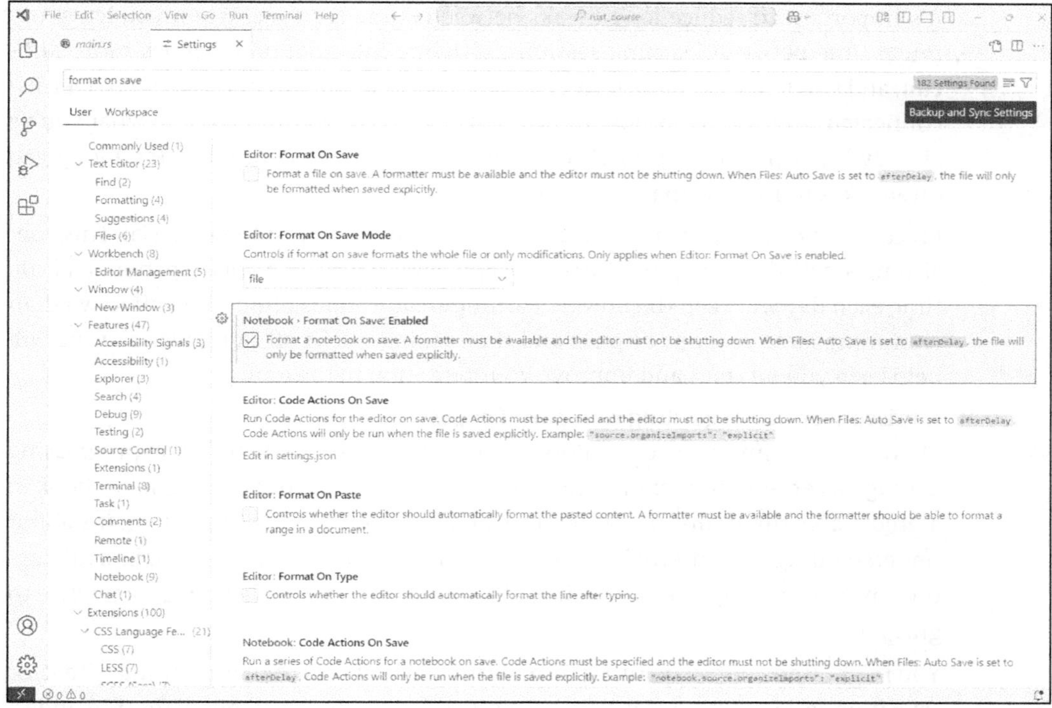

Figure 1.3 Setting Up the Format on Save Option

Next, let's set up some terminal-related settings. Search for "terminal font" in the search setting. Set the font size and font family according to your liking.

Finally, search for "windows zoom." The **Windows Zoom** field allows you to magnify the entire screen, so you can more easily focus on specific details and improve accessibility to overcome visual impairments. If the items are a bit smaller than your liking, try increasing the number from the default value of 1.

Now, let's quickly cover a few shortcuts that will be handy. To toggle the terminal, the default shortcut is `Ctrl` + `` ` ``. The shortcut for **Explorer** pane's visibility is `Ctrl` + `b`. To set different key bindings, navigate to **File • Preference • Keyboard Shortcuts**.

Then mention the shortcuts you want to update, for instance, write "toggle terminal," and then click and mention the new keys. Hold all the keys together and then press Enter. There are many additional shortcuts you may want to set up here.

1.4 Making the Most of This Book

Let's take a quick pause to prepare before we move on to the next chapter. Some advice that will help you as you embark on your Rust learning journey includes the following:

- **Reduce time gaps**

  It's important to reduce long breaks between study sessions. When you leave too much time between learning sessions, recalling information becomes more difficult, and you'll require more effort to get back on track. Content may start to feel disconnected, and you might lose momentum, making it harder to retain what you've already learned. The process of re-learning or refreshing concepts after a long gap often takes more time than studying a little bit each day.

  I recommend setting aside a bit of time daily to engage with the material. This consistent practice will help reinforce concepts and make them stick. Spending a little time each day will keep the flow of learning steady and is often more effective than cramming large sections at once and then taking weeks off. A regular routine will help keep you on track and improve your retention in the long run.

- **A misconception regarding number of pages**

  There is a common misconception among learners: the belief that progressing through a certain number of pages per day will allow a reader to complete a book in a predictable timeframe. However, learning is not merely about reading; it's about understanding and internalizing concepts. Rust's unique features, such as its approach to ownership and borrowing, often require deeper reflection and practice to grasp fully.

  You might find yourself pausing to analyze a piece of code, experimenting with it, or revisiting earlier sections for clarity. This iterative process is essential for true understanding but can make the learning journey longer than initially anticipated. Embrace this as part of the experience. Rust rewards persistence and patience, and the time you invest in truly learning the language will pay off when you start building safe, efficient, and powerful software.

- **Don't just copy the code**

  Simply copying code from this book or other examples without engaging deeply with the underlying concepts can limit your understanding. Instead of replicating the code as-is, take the opportunity to experiment. Try making small tweaks, introducing variations, or even challenging yourself to rewrite sections of code with a different approach. For instance, change variable types, modify function parameters, or restructure control flows.

This hands-on exploration will help you uncover nuances and possibilities that might not be immediately apparent. Through experimentation, not only can you solidify your understanding of Rust's syntax and its rules but also develop an intuition for how different features interact.

Additionally, breaking things intentionally, that is, causing errors or unexpected outcomes, can be an incredibly effective way to learn. By debugging and fixing the issues you encounter, you'll gain a deeper appreciation of Rust's error messages and safety mechanisms, which are invaluable in real-world programming.

Remember, the goal isn't just to write code that works but to truly *understand why it works* (or doesn't) and how it can be adapted for various scenarios. This active engagement will accelerate your journey to master Rust.

- **Don't just rely on a good start**
  For those with prior programming experience, starting with Rust might feel intuitive and straightforward. Your familiarity with fundamental programming concepts will give you an initial advantage, making the first few weeks feel like smooth sailing. However, don't let this promising start lead you to complacency. Rust is a unique language with features and paradigms that differ significantly from many others. Concepts like ownership, borrowing, lifetimes, and its strict safety guarantees are unlike what most programmers have encountered before.

  At some point, you'll probably face challenges that slow down your progress. These challenges might stem from understanding how Rust enforces memory safety, from learning to navigate its strict compile-time checks, or from adapting to its functional programming aspects. Stress is a common experience, and it's entirely normal.

  The key is to anticipate these difficulties and approach them with patience and persistence. Learning Rust requires not just knowledge but a shift in mindset, especially when dealing with concepts that break away from conventional programming norms. When the journey becomes tough, remind yourself that every hurdle overcome is a step closer to mastering a language that empowers you to write robust, efficient, and safe code. Stay resilient, embrace the challenges, and trust the process.

- **What to do when you're stuck**
  At some point during your learning journey, you'll inevitably encounter a roadblock. This happens to everyone, regardless of their experience level. When it does, the most important thing to remember is: Don't panic. It's natural to feel frustrated when you're stuck, but that frustration can cloud your ability to think clearly and to solve problems effectively.

  An effective strategy to become unstuck is to simply step away and take a break. I know it sounds like common advice, but I've seen firsthand how powerful it can be. Your brain needs time to process and internalize the new concepts you're learning. If you keep pushing yourself without giving your mind time to rest, you risk burning out and diminishing your capacity to absorb information.

Taking a short break, whether it's a walk, a cup of coffee, or even a few minutes of mindfulness, allows you to reset. When you return to the problem with a clear mind, you'll often find that the solution becomes much easier to spot. Sometimes, all it takes is a brief pause to give your brain the space it needs to make connections that aren't possible in the moment.

So, when you hit a wall, don't force it. Give yourself permission to take a step back and remember that a bit of distance can provide new perspectives.

- **Do not jump around the sections**
  You might be tempted, especially if you're an experienced programmer, to skip ahead to the parts that seem more relevant or interesting. However, in this book, the content is carefully structured to build upon each previous section. Each topic introduces foundational concepts that will support your understanding of more advanced material later. Skipping around can lead to gaps in your knowledge, which might make it harder to grasp more complex ideas down the road.

  Think of the learning process like building a house: You need a solid foundation before you can add the walls, roof, and finishing touches. If you skip the groundwork, the structure cannot stand firm, and it could collapse when you try to build further upon it.

  Of course, if you already have some experience with Rust, you might find certain sections easier and can choose to skim or even skip them. However, for most readers, following the progression of topics is crucial to ensure that each concept is thoroughly understood before moving on. Taking your time and working through the material in order will help reinforce each new idea and ensure a deeper, more comprehensive understanding of the language.

- **Be prepared for some hard work**
  Learning Rust can be challenging, and you should be ready for some hard work. The language's unique features, such as its concepts of ownership and borrowing and its strict compiler, can sometimes feel like obstacles. However, these features are what make Rust powerful and safe, and they require attention to detail.

  At times, concepts will seem tough, or sometimes, a problem just won't click. These challenges are completely normal and part of the learning process. Don't be discouraged—perseverance is key.

Learning Rust takes patience and dedication. The hard work you put in now will lead to greater mastery down the road, helping you write safer, more efficient code.

1.5 Summary

In this chapter, we laid the groundwork for your Rust journey, starting with setting up the Rust programming environment. This chapter covered the installation process, how to run and compile your first Rust program, and how to configure VS Code for optimal development. Additionally, we provided advice on how to make the most of this book's content and set the stage for a seamless and productive learning experience in Rust programming.

Now, we're ready to dive into the details of Rust, starting with core elements like variables and functions in the next chapter.

optimal development. Additionally, we provided advice on how to make the most of this book's content and set the stage for a seamless and productive learning experience in Rust programming.

Now, we're ready to dive into the details of Rust, starting with core elements like variables and functions in the next chapter.

Chapter 2
Variables, Data Types, and Functions

Understanding the building blocks of a language is like knowing the ingredients of a recipe. This chapter will equip you with the essentials to create your own programming masterpieces.

In this chapter, we'll explore how to define and work with variables, including their mutability, scope, and shadowing. We'll also cover constants and the different primitive and compound data types in Rust. You'll learn about integers, floats, chars, and Booleans, as well as strings, arrays, vectors, and tuples. Finally, we'll discuss type conversion, aliasing, functions, and code blocks. By the end, you'll have a solid understanding of these fundamental concepts in Rust programming.

2.1 Variables

Variables are a cornerstone of programming, serving as named storage for values used by your code. In this section, we'll explore the nuances of variables in Rust, including their definition, mutability, scope, and unique features like shadowing and constants.

2.1.1 Definition

Variables in Rust are defined using the `let` keyword. For example, the following line defines a variable and assign it a value of `10`:

```
let x = 10;
```

If the Visual Studio Code (VS Code) extension is properly installed (see Chapter 1, Section 1.1.1), the type of the variable will be automatically inferred. The variable along with its type will be shown in the editor. In particular, you'll see the following line in the editor:

```
let x: i32 = 10;
```

The type of variable is specified after the colon `:` that follows a variable's name. In this case, the VS Code displays the type as an `i32`, which means a 32-bit integer type.

Before going further, you need to understand the important term *binding*. Referring to associating a name with a value, binding is the process of creating a variable and assigning it a value, which is done using the `let` keyword. In the preceding example, `x` is a binding that holds the value `5`. The value is stored in memory, and the binding provides access to that value.

Changing a value may change the type. For instance, consider the following line:

```
let x = 10.0;
```

In this case, the editor will display the type of `x` to be `f64`, which represents a 64-bit float number. We can also explicitly add the type annotation:

```
Let x: i16 = 10;
```

In this case, `x` is now a `16` bit integer. This code works and compiles because `10` can be represented as a 16-bit integer. However, let's now change the value to `10.0`, with an explicitly mentioned type of `i16`, for example:

```
Let x: i16 = 10.0; // Error
```

Now, the compiler will throw an error when we save. The error is "mismatched types expected `i16`, found floating-point number."

Variables can be printed using a print statement, such as the following example:

```
println!("x is {x}");
```

This statement will print `x is 10` on the terminal when executed.

2.1.2 Mutability of Variables

In Rust, variables are immutable by default, meaning their values cannot be changed once assigned. For instance, the following code will not compile:

```
Let y = 5;
y = 10; // Error
```

The compiler throws an error like "cannot assign twice to immutable variable `y`." Variables are immutable by default. For an immutable variable, once a value is bound to the variable, we can't change that value. To fix this error, you can make the variable mutable by adding the `mut` keyword, as in the following example:

```
Let mut y = 5;
y = 10; // Error
```

The following code, however, will compile:

```
Let y;
y = 10;
```

Compiling is possible because immutable variables can be assigned a value once. After the assignment of a value, any subsequent attempt to update the value throws an error. For instance, the following code will throw an error:

```
Let y;
y = 10;
y = 5; // Error
```

2.1.3 Scope of Variables

In Rust, the scope of a variable refers to the part of the program where the variable is accessible. A variable is only valid within the block or function in which it is declared, and once a variable goes out of scope, it is no longer accessible. Understanding scope is essential for managing memory and preventing errors related to variable accessibility.

From a syntax perspective, *scope* means code lines that reside between an opening brace ({) and closing brace (}). The code shown in Listing 2.1 defines a variable inside a scope and then tries to access the same variable outside the scope.

```
fn main(){
   {
         let i = 50;
   }
let j = i; // Error
}
```

Listing 2.1 A Variable Defined within a Scope

Variable `i` is defined inside the scope. Accessing it outside the scope gives an error of "cannot find value `i` in this scope." The variable `j`, however, resides within `main` function scope, which starts with the opening braces after the `main` and ends at the ending braces, at the end of the `main`.

2.1.4 Shadowing

Shadowing in Rust allows you to reuse a variable name by declaring a new variable with the same name in a different scope. Unlike mutability, shadowing enables you to overwrite the value of a variable, creating a new binding without affecting the original one. This feature provides flexibility in variable management while maintaining clarity in your code.

Consider the code shown in Listing 2.2.

```
fn main() {
    let x = 10;
    let x = x + 10;
    println!("x is {x}");
}
```

Listing 2.2 Shadowing of Variable x

The code first binds x to a value of 10. Then, it creates a new variable x by adding 10 to the original value. In any subsequent code, the second instance (or redefinition) of x is what the compiler will recognize when you reference the variable. In other words, the second variable x shadows the first one. Therefore, the print statement will display the updated value of x, which is 20.

Shadowing has many useful use cases. For example, you might initially want your variable to be of a certain type, like an integer. However, after performing some operations, you may want to change the variable to a different type, such as a float.

Another common use case of shadowing is for nested scopes. Consider the code shown in Listing 2.3.

```
fn main() {
    let x = 30;
    {
       let x = 40;
        println!("inner x is: {x}");
    }
    println!("x is: {x}");
}
```

Listing 2.3 Shadowing in Scopes

The variable x in the inner scope will shadow the original variable x in the outer or main scope. The print statement in the inner scope will therefore print a value of 40. However, in the outer scope, its value will remain at a value of 30. This distinction is handy in situations where the same variable needs to be reused in the inner scope, possibly in a different context.

Mutability versus Shadowing

An important difference to understand is the difference between mutability and shadowing. *Mutability* allows you to modify the value of a single variable after it has been declared. In contrast, *shadowing* involves creating a new variable with the same name, effectively replacing the original variable in scope. The new variable shadows the previous one, but they are treated as two distinct variables, each with its own values.

2.1.5 Constants

Constants in Rust are immutable values that are bound to a name and remain fixed throughout the program's execution. They are declared using the `const` keyword instead of the `let` keyword and requires an explicit type annotation. Consider the following example of declaring a constant:

```
const MAX_VALUE: u32 = 100;
```

Once created, constants cannot be mutated. Rust's naming convention for constants is SCREAMING_SNAKE_CASE, meaning that all the letters are capitalized and underscore is used between words. The value of the constant must be known at compile time. Thus, we cannot declare a constant with a type but no value. The following line will therefore generate an error:

```
const MAX_VALUE: u32; // Error
```

2.1.6 Statics

Statics are similar to constants and are declared using the `static` keyword. Let's consider the following example of a static variable:

```
static WELCOME: &str = "Welcome to Rust";
```

Like constants, we must provide an explicit type annotation and use the naming convention of SCREAMING_SNAKE_CASE.

The essential difference between statics and constants is that constants are *inlined*. To understand what we mean, consider the code shown in Listing 2.4.

```
fn main() {
    static WELCOME: &str = "Welcome to Rust";
    const PI: f32 = 3.14;

    let a = PI;
    let b = PI;

    let c = WELCOME;
    let d = WELCOME;
}
```

Listing 2.4 Difference between Statics and Constants

Inlining of constants in Rust refers to how constant values (`const`) are directly embedded into the places they are used during compile time. As a result, the compiler replaces every occurrence of a constant with its actual value. The values `a` and `b` will both get the literal value `3.14` directly embedded into the generated binary, as `PI` is inlined. In contrast, static variables like WELCOME do not get inlined. Instead, WELCOME is stored in a fixed memory location, and both c and d refer to this same memory address.

If you're a beginner, we advise defaulting to constants, unless you have a good reason to switch to statics. Some reasons for using statics include when you want to refer to some large amount of data in memory or when you're interested in interior mutability, something that we'll explore later in Chapter 10, Section 10.2.3.

2.1.7 Unused Variables

By default, Rust uses snake_case for variable names, where all letters are lowercase and words are separated by underscores. This convention applies to regular variables, function parameters, and function names. For example, a variable holding the number of students might be named num_students.

An unused variable in Rust is a variable that is declared but not used anywhere in the program, which typically results in a compiler warning. To suppress the compiler warning, prefix the variable name with an underscore _. This prefix indicates that the variable is intentionally unused. Let's look at an example of this scenario:

```
let _n = 5;
```

This approach is particularly useful when destructuring a tuple where some values might not be needed, such as in the following example:

```
let (x, _) = (10, 20);
```

In this example, the second value is not required. Additionally, the unused variable warning can be silenced by adding the line #[allow(unused_variables)], at the start of the program. A best practice, however, is to use underscores for cleaner code.

Underscores can also improve the readability of larger numbers. For instance, the number in the following code is easier to read:

```
let x = 40_000_000;
```

2.2 Data Types

Data types in Rust define the kind of values a variable can hold, thus ensuring type safety and clarity in a program. In this section, we'll explore the essential data types required to perform typical programming tasks effectively.

2.2.1 Primitive Data Types

Primitive data types in Rust are the building blocks of all data manipulation, providing fundamental types like integers, floating-point numbers, Booleans, and characters.

These types are simple, efficient, and form the foundation for more complex data structures in Rust.

Integers and Floats

Integers are available in two types: unsigned and signed. Unsigned integers are positive integers and start with the letter u followed by the number of bits that could be stored in that specific type. Consider the following example of an unsigned integer:

```
let unsigned_num: u8 = 5
```

Other available unsigned types include `u16`, `u32`, `u64`, and `u128`. The more bits used in the type, the greater the range of numbers it can represent.

Signed integers follow a similar pattern. They start with the letter `i`. The type `i32` is the default type for all integers in Rust. Following is an example of an 8-bit signed integer:

```
let signed_num: i8 = 5;
```

Other signed integer types include `i16`, `i32`, `i64`, and `i128`.

Note that some platform-specific integer types exist; they aren't too important for now but still good to know:

- `usize` represents a pointer-sized, unsigned integer.
- `isize` represents a pointer-sized, signed integer.
- If you have a type of an indefinite size, we call those types *unsized types* in Rust or sometimes *dynamically sized types*. We'll cover these topics further in Chapter 13.

Char and Boolean

The `char` type can represent a single character. The character value must be enclosed in single quotes. Consider the following example of declaring a variable of the `char` type:

```
let char = 'a';
```

A Boolean variable can hold two values, either `true` or `false`. This kind of variable is commonly used in conditionals and is represented using the `bool` keyword. Consider the following example of declaring a variable of `bool` type:

```
let b = true;
```

Type Conversion and Aliasing

Type aliasing allows us to create a new name for an existing type using the `type` keyword. This feature is useful for improving code readability, simplifying complex types, or enabling easier refactoring. Type aliases do not create new types. They merely provide an alternate name for an existing one. Consider the following code:

```
type Age = u8;
let peter_age: Age = 42;
```

The code creates an alias named Age for the u8 type. The variable peter_age is now declared with the type Age.

Type conversion is frequently required when you need precision in computation or need to overcome compatibility issues between different parts of the code. Type conversion is done by writing the as keyword followed by the target type. Consider the following examples:

```
let a = 10;
let b = a as f64;
```

The variable a has a default type of i32. The variable b, which is assigned the value of a, will take on the type f64 as a result of type conversion.

2.2.2 Compound Data Types

Compound data types allow you to group multiple values into a single entity, enabling more complex and structured data management. Rust provides compound types of arrays, vectors, and tuples. In the upcoming sections, we'll cover each of these data types.

Arrays

Arrays hold multiple values of the same type. The following code initializes an array of integers:

```
let mut array_1: [i32; 5] = [4, 5, 6, 8, 9];
```

The array type is specified using brackets [] containing the type of the array and the number of elements. We can remove the type annotation, i.e., [i32; 5], and the editor will automatically show the type annotation. Once declared, the size of the array cannot change and is therefore of fixed size. As a result, we cannot remove or add elements to an array.

We can index values in an array using the variable name followed by brackets. The indexes start at 0 in Rust. The following line obtains the fourth element in the array:

```
Let num = array_1[3];
```

We can use the indexing into array to mutate a specific value. The following line will mutate the value at the third index:

```
array_1[2] = 10;
```

The following line prints the entire array to terminal:

```
println!("{:?}", array_1);
```

The {:?} format specifier in Rust is used to print values using the debug trait. While this format specifier can be used to print compound data types like arrays, tuples, or structs, you are not limited to compound types. This approach works with any type that implements the debug trait. We'll cover traits in more detail in Chapter 8, Section 8.2, so don't worry about them right now. For now, at this stage, just know that you can print compound types with this format specifier.

Sometimes, you want to create an array where all its elements have the same default value. You can achieve this goal using the following syntax:

```
let array_2 = [0; 10];
```

In this case, the syntax consists of the default value, followed by a semicolon ; and then the size of the array. In this example, an array of size 10 is created, where all elements are initialized to 0. By default, this statement creates an array of type i32.

Vectors

Vectors are growable arrays that allow you to store multiple values of the same type. Unlike fixed-size arrays, vectors provide the flexibility to dynamically resize, making them ideal for situations where the size of the collection is not known at compile time.

Vectors are created using the vec! macro. The following statement is an example of declaring a vector:

```
let vec_1: Vec<i32> = vec![4, 5, 6, 8, 9];
```

The type is annotated using the syntax of Vec followed by angled brackets <>, which specifies the type of values. Explicit type information is not required; however, we mention this information for the sake of comprehensiveness. In contrast to arrays, no size information exists for vectors. Moreover, all elements in a vector must be of the same type.

Just like arrays, a specific element in a vector can be accessed using indexing:

```
let num = vec_1[3];
```

In this code line, the value at the fourth position (index 3) of vector vec_1 is retrieved and assigned to the variable num. Note that indexing also starts at 0 in case of vectors.

Tuples and Empty Tuples

Unlike arrays, which can only store values of the same type, tuples can store values of different types within a single collection. Tuples are created using parentheses that contain multiple values. Consider the following example of a tuple:

```
let my_info = ("Salary", 40000, "Age", 40);
```

In this case, the `my_info` tuple contains a string slice `&str` (a type we'll cover Section 2.2.3), an integer, another string slice, and a second integer.

We can access elements of a tuple by specifying the variable name, followed by a dot (`.`), and the index number of the desired element. For instance, consider the following statement:

```
let salary_value = my_info.1;
```

This statement will assign the second value, which is an integer to the variable `salary_value`.

You can also destructure an entire tuple, allowing you to assign its individual values to specific variables, for example, with the following statement:

```
let (salary, salary_value, age, age_value) = my_info;
```

Finally, you may have an empty tuple, which is also known as the *unit type*. An empty tuple is simply a tuple created with nothing inside, for example, through the following statement:

```
let unit = ();
```

While not technically a compound type, the unit type is related to tuples, so we include it in our current discussion. The unit type is typically returned implicitly when a function does not return any meaningful value. For example, functions without a specific return value will implicitly return the unit type, which is zero-sized and consumes no memory. We'll revisit this concept in greater detail in Chapter 13, Section 13.5.

2.2.3 Text-Related Types

Rust has two common types for representing texts or strings. The first type is called a string slice, which is represented by an ampersand `&` followed by `str`. Consider the following example of declaring a `&str` type:

```
let fixed_str = "Fixed length string";
```

The second type is simply called `String`. The `String` type comes from the Rust standard library. You can create a `String` by calling the `from` function using the double colon syntax, as in the following example:

```
let mut flexible_str = String::from("This string will grow");
```

In Rust, the `::` syntax is used to access associated functions or methods of a type or a module. We'll explore this syntax further in Chapter 6, Section 6.2.3.

The difference between the two types is that a string slice is of fixed size. By fixed size, we mean that, once created, we cannot add or remove text from it. On the other hand, the String type allows text to be modified and can grow in size. For instance, we can add a single character to the existing text using the push function, as in the following example:

```
flexible_str.push('s');
```

Typically, a string slice is used for read-only access to string data. More differences between the two types will become clear later in the book.

2.3 Functions

At this point, you've already encountered functions in Rust, specifically the main function.

Functions in Rust are declared using the fn keyword, followed by the function name, parentheses (which can contain input parameters), and finally curly braces that enclose the function's body. The following code defines a function:

```
fn my_fn(s: &str) {
    println!("{s}");
}
```

The naming convention for functions in Rust is snake_case, meaning everything is written in lowercase with an underscore between words. Parameters are defined like variables, with the variable name followed by a type annotation. We can add more parameters by simply separating them with a comma and using the same syntax. In this case, the function takes a string slice as a parameter and prints its value.

To call the function in the main function, we simply mention the name of the function, followed by parentheses, and then pass the value as an argument. The following code shows how to use the function my_fn defined previously in main:

```
fn main() {
    my_fn("This is my function");
}
```

Instead of passing a concrete value, we can pass a variable:

```
fn main(){
    let str = "Function call with a variable";
    my_fn(str);
}
```

The variable type that we are passing to the function and function parameter type must match.

Functions can also return a value. For instance, consider the function shown in Listing 2.5.

```
fn multiplication(num1: i32, num2: i32) -> i32 {
    println!("Computing multiplication");
    num1 * num2
}
```

Listing 2.5 Function Returning a Value

The return value is specified using the arrow syntax -> followed by the data type. In this case, we are returning an `i32` value.

The final expression in a function becomes its return value.

Expressions versus Statements

Expressions are code lines that evaluate to a value, while *statements* are instructions that do not return any value. For example, the `println!` statement in the function is a statement, as it does not return anything. However, the last line in the function is an expression because it returns a value.

For a function to return the last value as the return value, we must omit the semicolon. If we add a semicolon, we'll encounter a type mismatch error.

When compiled, the code shown in Listing 2.6 will throw an error of "mismatched types expected `i32`, found `()`." The function was expecting an `i32` value; however, it found a unit value.

```
fn multiplication(num1: i32, num2: i32) -> i32 { // Error
    println!("Computing multiplication");
    num1 * num2;
}
```

Listing 2.6 Function Not Returning a Valid Value

Functions that don't return anything instead return a unit type. To properly return a valid value, let's remove the semicolon from the last code line in the function.

Note that you cannot have multiple returning expressions inside a function. One and only one returning expression can exist, and that expression must be the last expression of the function. If you want to return early (i.e., before the final expression), add the `return` keyword with a proper return value, matching the return type. For instance, Listing 2.7 shows how you can return early from the function defined in the code shown earlier in Listing 2.5.

```
fn multiplication(num1: i32, num2: i32) -> i32 {
    return 45;
    println!("Computing multiplication");
    num1 * num2
}
```

Listing 2.7 Function Returning Early

The return value from the function can be stored in a variable. To store the resulting value from the `multiplication` function shown earlier in Listing 2.5, use the following line of code:

```
let answer = multiplication(10, 15);
```

Functions can return multiple values. The code shown in Listing 2.8 defines a function that returns multiple values.

```
fn basic_math(num1: i32, num2: i32) -> (i32, i32, i32) {
    (num1 * num2, num1 + num2, num1 - num2)
}
```

Listing 2.8 Function Returning Multiple Values

Multiple returning values are mentioned using tuples. As shown in Listing 2.8, the syntax `(i32, i32, i32)` means that the function will return a tuple containing three `i32` values. The function returns the multiplication, addition, and subtraction of the two input parameters inside a tuple.

The returning tuple may be destructed into individual variables as follows:

```
let (multiplication, addition, subtraction) = basic_math(10, 15);
```

2.4 Code Blocks

A code block is a section of code enclosed in curly braces `{}`. You use code blocks in many contexts, including functions, loops, conditionals, and more. They group together multiple statements and expressions, and in Rust, the last expression in a code block is automatically returned as the value of that block.

We mentioned code blocks earlier in Section 2.1.3 when discussing the scope of a variable. Now, consider the code shown in Listing 2.9.

```
fn main() {
    let full_name = {
        let first_name = "John";
        let last_name = "Archer";
```

```
        format!("{first_name} {last_name}")
    };
}
```

Listing 2.9 An Example of a Code Block

The `format!` macro is used for `String` formatting. The placeholders inside the double quotes will be filled with the respective variable values. Like functions, the last expression without a semicolon is the returning value from the code block. In our example, we'll store the returning value in the variable `full_name`. The whole code block is an assignment statement in this case; therefore, we can add a semicolon at the end of the code block. The variables defined within a scope are limited to that scope.

Functions versus Code Blocks

Code blocks share some similarities with functions. Like functions, they have their own separate bodies, can return values, and may have variables limited in scope to their bodies. Some key differences exist between code blocks and functions, however. First, code blocks are not designed for reuse, whereas functions are. Thus, code blocks are intended for one-time execution, while functions can be called and executed any number of times. Second, code blocks do not have explicit parameters. All variables in the scope of the code block are visible to it. In contrast, functions have an explicit list of parameters and can only access variables that are either passed as parameters or locally defined within the function.

2.5 Practice Exercises

In this section, we'll provide some practice questions to help reinforce your learning. The exercise questions cover almost all the concepts introduced in the chapter. Solutions to each exercise will follow in Section 2.6.

1. **Correctly defining variables**
   Fix the following code:

   ```
   fn main(){
       my_age  = 40;
       println!("My age is: {}", my_age);  // do not change this line
   }
   ```

2. **Checking mutability**
   Correct the following code:

   ```
   fn main(){
       let x1 = 40;
       let x2 = x1;
   ```

```
    x2 = x1-2;     // do not change this
    println!("x1 is: {} and x2 is: {}", x1,x2); // do not change this
}
```


3. **Mutability of variables**
   Without executing the following code, determine if it will compile:

```
fn main() {
    let mut x1 = 40;
    let x2;
    x1 = x1 * 3;
    x2 = x1 - 2;
    println!("x1 is: {}, x2 is: {}", x1, x2);
}
```

4. **Correctly shadowing a variable**
   Fix the following code:

```
fn main() {
    let a = "three";  // don't change this line
    a = 10; // don't change the name of this variable
    println!("a is: {}", a);
}
```

5. **Correcting variable assignment for unsigned integer**
   Fix the following code by assigning a correct value to the variable:

```
fn main() {
    let x: u8; // Don't change this line!
    x = -1;
    println!("x is: {}", x);
}
```

6. **Adjusting variable type for floating-point assignment**
   Make the following program compile by replacing the variable type:

```
fn main() {
    let pi: i32;
    pi = 3.14159; // This value represents pi
    println!("pi is: {}", pi);
}
```

7. **Assigning appropriate data types to variables**
   Replace the placeholder DATA_TYPES_PLEASE with the appropriate data types in the following program:

```
fn main() {
        let a: DATA_TYPES_PLEASE = -15;
        let b: DATA_TYPES_PLEASE = 170;
```

```
        let name: DATA_TYPES_PLEASE = "Michael";
        println!("name is: {}, and the multiplication result is {}", name, a
* b);

}
```

8. **Defining a type alias for a tuple**
   Add a type alias for Book so that we are able to store the information:

```
fn main() {
    type Book = // Add your code here to this line
    let book1: Book = (
        String::from("Rust Programming Language"),
        String::from("RUST Community"),
        2010,
    );
    println!(
        "Book name: {}, Author: {}, Year {}",
        book1.0, book1.1, book1.2
    );
}
```

9. **Implementing basic arithmetic functions**
   In the following program, three functions are missing:
   - add_3(x): Adds three to the input x and returns the result.
   - add_5(x): Adds five to the input x and returns the result.
   - times(x, y): Multiplies the two input values x and y and returns the result.

   Your task is to define these functions so that the following program compiles:

```
fn main() {
    let x = 3;
    let y = 4;
    println!(
        "The result of x+3 times y+5 is {}",
        times(add_3(x), add_5(y))
    );
}
```

10. **Refactoring to remove intermediate variables**
    Refactor the code in main by taking rid of the variables x and y. In other words, rewrite the code in main to produce the same outcome as the original code, but without using any variables:

```
fn double(x: i32) -> i32 {
    x * 2
}
```

```
fn triple(x: i32) -> i32 {
    x * 3
}
fn main() {
    let x = triple(double(5));
    let y = triple(x);
    println!("Answer: {}", y);
}
```

11. **Correcting function call for tuple parameter**
    Fix the following code by making changes to the line in which a function call is made:

```
fn print_distance(point: (f32, f32)) -> f32 {
    let (x, y) = point;
    (x.powf(2.0) + y.powf(2.0)).sqrt() // Formula for computing distance
}
fn main() {
    println!(
        "The distance of the number the point from the origin is {}",
        print_distance(5.0, 4.0) // concentrate on the call to the function
    );
}
```

12. **Implementing the quadruple function using double**
    Add the definition of the following quadruple function by calling the double function twice inside the quadruple function:

```
fn double(x: i32) -> i32 {
    x * 2
}
fn quadruple(x: i32) -> i32 {
    // your code here //
}
fn main() {
    println!(
        "For 1: the expected value is 4 while the output is {}",
        quadruple(1)
    );
    println!(
        "For 2: the expected value is 8 while the output is {}",
        quadruple(2)
    );
    println!(
        "For 3: the expected value is 12 while the output is {}",
        quadruple(3)
```

```
        );
        println!(
            "For 4: the expected value is 16 while the output is {}",
            quadruple(4)
        );
    }
```

2.6 Solutions

This section provides the code solutions for the practice exercises in Section 2.5. The code is largely self-explanatory; however, we have included comments and additional explanations wherever necessary to enhance understanding.

1. **Correctly defining variables**

```
fn main(){
    let my_age  = 40;
    println!("My age is: {}", my_age);
}
```

2. **Checking mutability**

```
fn main(){
    let x1 = 40;
    let mut x2 = x1;
    x2 = x1-2;
    println!("x1 is: {} and x2 is: {}", x1,x2);
}
/* Note: There will be a warning when you execute that the value of x2 has
been immediately overwritten before being read. You may just ignore it
otherwise add the line #![ allow(unused)] at the top before the fn main() if
you do want to see the warning. */
```

3. **Mutability of variables**

```
fn main() {
    let mut x1 = 40;
    let x2;
    x1 = x1 * 3;
    x2 = x1 - 2;
    println!("x1 is: {}, x2 is: {}", x1, x2);
}
/* Explanation: Yes, it will compile. Although the variable x2 is not mutable
and we are trying to assign a value to this variable in the line "x2 = x1 -
2" but its ok since the immutable variables can be assigned values once.
```

```
Since we are not updating its value later on therefore the variable x2 is
assigned a value once which is consistent with the definition of immutable
variables, i.e., they can only be assigned value once.*/
```

4. **Correctly shadowing a variable**

```
fn main() {
    let a = "three";
    let a = 10;
    println!("a is: {}", a);
}
```

5. **Correcting variable assignment for unsigned integer**

```
fn main() {
    let x: u8;
    x = 1;
    println!("x is: {}", x);
}
```

6. **Adjusting variable type for floating-point assignment**

```
fn main() {
    let pi: f32;
    pi = 3.14159;
    println!("pi is: {}", pi);
}
```

7. **Assigning appropriate data types to variables**

```
fn main() {
        let a: i16 = -15;
        let b: i16 = 170;
        let name: &str = "Michael";
        println!("name is: {}, and the multiplication result is {}", name, a
* b);
}
```

8. **Defining a type alias for a tuple**

```
fn main() {
    type Book = (String, String, u32);
    let book1: Book = (
        String::from("Rust Programming Language"),
        String::from("RUST Community"),
        2020,
    );
    println!(
        "Book name: {}, Author: {}, Year {}",
```

```
        book1.0, book1.1, book1.2
    );
}
```

9. **Implementing basic arithmetic functions**

```
fn add_3(x: i32) -> i32 {
    x + 3
}
fn add_5(x: i32) -> i32 {
    x + 5
}
fn times(x: i32, y: i32) -> i32 {
    x * y
}
fn main() {
    let x = 3;
    let y = 4;
    println!(
        "The result of x+3 times y+5 is {}",
        times(add_3(x), add_5(y))
    );
}
```

10. **Refactoring to remove intermediate variables**

```
fn double(x: i32) -> i32 {
    x * 2
}
fn triple(x: i32) -> i32 {
    x * 3
}
fn main() {
    println!("Answer: {}", triple(triple(double(5))));
}
```

11. **Correcting function call for tuple parameter**

```
fn print_distance(point: (f32, f32)) -> f32 {
    let (x, y) = point;
    (x.powf(2.0) + y.powf(2.0)).sqrt()
}
fn main() {
    println!("The distance of the number the point from the origin is {}",
        print_distance((5.0, 4.0))); // The function needs a tuple as an
                                     // input and not two floats.
}
```

12. **Implementing the quadruple function using double**

```
fn double(x: i32) -> i32 {
    x * 2
}
fn quadruple(x: i32) -> i32 {
    double(double(x))
}
fn main() {
    println!("For 1: the expected value is 4 while the output is {}",
quadruple(1));
    println!("For 2: the expected value is 8 while the output is {}",
quadruple(2));
    println!("For 3: the expected value is 12 while the output is {}",
quadruple(3));
    println!("For 4: the expected value is 16 while the output is {}",
quadruple(4));
}
```

2.7 Summary

In this chapter, we explored the core building blocks of Rust programming, starting with variables and their definition and exploring mutability, scope, shadowing, and constants. We then delved into data types, discussing both primitive types such as integers, floats, chars, and Booleans and compound types like strings, arrays, vectors, and tuples. We introduced you to functions and code blocks to demonstrate how Rust structures its logic. By the end of this chapter, you should have a solid grasp of how Rust manages variables and types and of its basic program flow. This chapter concluded with practice exercises and solutions to reinforce the concepts we've covered.

In the next chapter, we'll build on the foundations set in this chapter by exploring conditional statements and control flow, enabling you to write programs that make decisions and respond to various scenarios dynamically.

Chapter 3

Conditionals and Control Flow

In programming, the choices we make determine the paths we take. Let's delve into how control structures guide our code through various scenarios.

3

In this chapter, we'll explore conditionals and control flow in Rust. You'll learn how to use `if else` statements and the `match` expression for decision-making. We'll also cover different looping constructs like simple loops, `for` loops, and `while` loops. Finally, we'll touch upon handling comments, print commands, and user input.

3.1 Conditionals

In programming, conditional statements are the decision-makers, allowing a program to execute different actions based on varying conditions. Rust's approach to conditionals is no exception, offering a syntax that's both expressive and type safe. Let's dive into using conditionals through `if else`, `if else if` ladder, and `match` statements in the upcoming sections.

3.1.1 If Else

The most basic conditional structure is the `if else` statement. Consider the code shown in Listing 3.1.

```
fn main() {
    let num = 40;
    if num < 50 {
        println!("The number is less than 50");
    } else {
        println!("The number is greater than or equal to 50");
    }
}
```

Listing 3.1 Basic if else Statement

The condition after the if statement checks whether the value is less than 50 or not. If the value is less than 50, then the code block after the if statement will be executed. In any other case, the code block after the else part will be executed.

The else part is not mandatory, and the code will compile if we delete it. Leaving out the else can be useful in situations where the code inside the if block ensures that the remaining part of the program will not need any further else logic to run. This omission helps make your code simpler and easier to read. For instance, the code shown in Listing 3.2 will compile perfectly fine.

```
fn main() {
    let num = 40;
    if num < 50 {
        println!("The number is less than 50");
    }
}
```

Listing 3.2 Code Containing Only the if Statement

The condition after the if statement must evaluate to a Boolean value, meaning either true or false. The compiler will complain if we change the condition to a non-Boolean value, as shown in Listing 3.3.

```
fn main() {
    let num = 40;
    if num { // Error
        println!("The number is less than 50");
    } else {
        println!("The number is greater than or equal to 50");
    }
}
```

Listing 3.3 Non-Boolean Condition after the if Statement

The compiler throws an error of "mismatch types, expected bool, found integer." This error arises because the expression after the if statement is not something that evaluates to a Boolean value. In particular, num is just an integer variable and does not represent a Boolean value. In Rust, if is not just a statement but also an expression. Thus, any if statement can return a value, which is especially useful when initializing variables. Consider the code shown in Listing 3.4.

```
fn main() {
    let number = 10;
    let result = if number % 2 == 0 {
        "even"
    } else {
```

```
        "odd"
    };
    println!("The number is {}", result);
}
```

Listing 3.4 Using if else as an Expression

The % operator in Rust calculates the remainder of the division of two numbers. The values returned from both the code blocks associated with the `if` and `else` part must be of the same type. Like functions and code blocks, the returning values do not end in semicolons. However, the returning value must follow the same rules as in the case of functions and code blocks covered earlier in Chapter 2, Section 2.4. For instance, if you have multiple statements in a code block corresponding to `if` or `else`, only the last statement without a semicolon will be the returning value. At the end of the `if else`, you must add the mandatory semicolon since the return value itself is then used as an assignment statement.

3.1.2 If Else If Ladder

With an `if else if` ladder, you can evaluate multiple conditions in sequence, making this conditional a powerful tool for branching logic in your program. The ladder evaluates conditions from top to bottom, executing the first block where the condition evaluates to `true`.

Consider the code shown in Listing 3.5.

```
fn main() {
    let marks = 95;
    let mut grade = 'N';
    if marks >= 90 {
        grade = 'A';
    } else if marks >= 80 {
        grade = 'B';
    } else if marks >= 70 {
        grade = 'C';
    } else {
        grade = 'F';
    };
}
```

Listing 3.5 Example of an if else if Ladder

The conditions will be checked in sequence. The final `else` in an `if else if` ladder acts as the default case. This final case executes when none of the preceding conditions are met and must be placed at the end of the `if else if` ladder. However, this final case is optional. If you delete it from the shown in Listing 3.5, the code will still compile. The else statement is useful when you need to handle all remaining cases that are not covered by previous if or else if conditions; however, if no additional cases arise to handle or if the logic does not require it, this else statement can be omitted entirely.

You can also convert an if else if ladder to an expression by assigning it to a variable, for instance, the variable grade, as shown in Listing 3.6.

```
fn main() {
    let marks = 95;
    let mut grade = if marks >= 90 {
        'A'
    } else if marks >= 80 {
        'B'
    } else if marks >= 70 {
        'C'
    } else {
        'F'
    };
}
```

Listing 3.6 The if else if Ladder as an Expression

Note again that the returning values from all the branches must have the same type. Moreover, the ladder must end with a semicolon since it is now considered as an assignment statement.

3.1.3 Match

The match statement allows you to compare a value against multiple patterns, thus enabling expressive and exhaustive decision-making. This versatile tool can handle everything from basic matching to complex pattern destructuring, all while ensuring safety and clarity in your code.

Consider the code shown in Listing 3.7, which is a revised version of the code shown in Listing 3.5, this time using match.

```
fn main() {
    let marks = 95;
    let mut grade = 'N';
    match marks {
        90..=100 => grade = 'A',
        80..=89 => grade = 'B',
        70..=79 => grade = 'C',
```

```
        _ => grade = 'F',
    }
}
```

Listing 3.7 Revision to Listing 3.5 Using the match Statement

Each matching pattern, along with its corresponding code block, is referred to as an *arm*. In the code example, `90..=100 => grade = 'A'` is the first arm. Each arm consists of two parts: a pattern to match on the left and the code block to execute if the pattern matches on the right. If the right-hand side contains multiple lines of code, these lines should be enclosed in curly braces.

The `..` syntax is used to specify a range of values in Rust. The range begins with the value on the left of the two dots and ends with the value on the right. The `=` in `..=` includes the endpoint in the range. To exclude the endpoint, use `..` without the `=`. The match is exhaustive. It requires that that every possible value or variant of the input type is covered in the `match` statement. To ensure this, the last arm, often called the default arm, is included. If none of the patterns match the input value, the code in the default arm will execute. In Rust, a pattern with just an underscore `_` acts as a catch-all and matches any value.

If we remove the default arm, the compiler will throw an error of "non-exhaustive patterns." This error arises because we are matching on an i32 variable, which can take values from -2^{31} to $2^{31} - 1$, and the arms only check for values between `60` to `100`. The remaining values in the range are thus ever considered, and therefore, the match arms are not exhaustive.

Beginners tend to make a few common mistakes. The first such mistake is an unreachable pattern. Consider the code shown in Listing 3.8.

```
fn main() {
    let marks = 95;
    let mut grade = 'N';
    match marks {
        90..=100 => grade = 'A',
        80..=89 => grade = 'B',
        _ => grade = 'F',
        70..=79 => grade = 'C',
    }
}
```

Listing 3.8 Code with an Unreachable Pattern

Although the code compiles, the compiler issues a warning about an "unreachable pattern." This problem occurs because the default arm matches all values, making any subsequent arms unreachable. Our example is just one example of an unreachable pattern. They generally occur when the compiler detects that a particular pattern cannot match because all possible values or variants of the input type have already been handled by earlier arms.

The second mistake is overlapping patterns. Consider the code shown in Listing 3.9.

```
fn main() {
    let marks = 95;
    let mut grade = 'N';
    match marks {
        90..=100 => grade = 'A',
        80..=90 => grade = 'B',
        70..=79 => grade = 'C',
        _ => grade = 'F',
    }
}
```

Listing 3.9 An Overlapping Pattern in the First and Second Arms

The value 90 is included in both the first and the second pattern. Recall that arms are executed in sequence one after the other. The value 90 is already covered in the first arm; therefore, the pattern for the second arm will never match on value 90. The compiler is telling us exactly this problem.

Finally, like `if`, `match` can be treated as an expression with a returning a value. Listing 3.10 shows the updated version of the code shown earlier in Listing 3.7. Now, the match returns a grade value in the variable `grade`.

```
fn main() {
    let marks = 95;
    let mut grade = match marks {
        90..=100 => 'A',
        80..=90 => 'B',
        70..=79 => 'C',
        _ => 'F',
    };
}
```

Listing 3.10 Treating a Match as an Expression

3.2 Control Flow

Control flow (or the flow of control) refers to the order in which individual statements, instructions are executed or evaluated in a program. In this section, we'll go over the fundamental control flow construct called a *loop* and its different types. You'll use loops to handle repetitive tasks efficiently.

3.2.1 Simple Loops

The first type of loop is simply a loop. The code after this loop construct executes forever. For instance, the following code will execute infinitely:

```
loop {
        println!("Simple loop");
}
```

This code will print Simple loop forever. To exit out of a loop, you can use a break statement, as in the following example:

```
loop {
        println!("Simple loop");
        break;
}
```

If you have nested loops (i.e., a loop containing another loop), the break will only exit out of the inner loop, not from the outer loop. To explicitly tell the compiler that to exit out of the outer loop, you'll use labeling. Listing 3.11 shows how to exit out of a loop using labeling.

```
fn main() {
    'outer: loop {
        'inner: loop {
            println!("Simple loop");
            break 'outer;
        }
    }
}
```

Listing 3.11 Exiting from a Loop Using Labeling

Labels start with the backtick (') followed by a colon (:). Inside the loop, after the break, you can specify the loop from which you want to break out.

The loops can also be treated like an expression returning a value. Consider the following code example:

```
let a = loop {
    break 5;
};
```

This code is admittedly a simplified example with little practical use. However, a similar pattern can be applied when you have a potentially failing operation. In such cases, you can keep attempting the operation until it succeeds and then assign the resulting value to your variable.

3.2.2 For and While Loops

A for loop allows you to loop through collections such as vectors and arrays. Consider the code shown in Listing 3.12.

```
fn main() {
    let vec = vec![45, 30, 85, 90, 41, 39];
    for i in vec {
        println!("{i}");
    }
}
```

Listing 3.12 Example of a for Loop Iterating over the Elements in a Vector

In this code, the for loop iterates over each element in the vector vec. For each element i, it prints the value to the console. Note that, in contrast to a simple loop, which runs indefinitely until explicitly broken, a for loop iterates over a specific range or collection, automatically handling the start, end, and increment without requiring manual control.

While the for loop provides a concise and efficient way to iterate over ranges or collections, the while loop offers more flexibility by allowing iteration based on a condition, making it suitable for situations where the number of iterations isn't known in advance. In particular, while loops will continue to execute while any given condition is true. To demonstrate, let's define a variable. Listing 3.13 shows a simplified example of the while loop.

```
fn main() {
    let mut num = 0;
    while num < 10 {
        num = num + 1;
    }
}
```

Listing 3.13 Example Using while Loop

In this program, the while loop repeatedly increments the num variable by 1 as long as its value is less than 10. The loop continues executing until num reaches 10, at which point the condition num < 10 becomes false.

3.3 Comments, Outputs, and Inputs

Comments, print commands, and input handling are essential tools for writing clear, interactive, and user-friendly code. Let's go through each of these tools in detail in this section.

3.3.1 Comments

Commented lines start with two back slashes (//), as in the following example:

```
// The current line is a comment line
```


Rust supports multiline comments using the /* */ syntax. These comments can span multiple lines and are useful for temporarily disabling sections of code or for writing longer explanations without using multiple // lines. For example:

```
/* This is a multi-line comment
   that spans multiple lines
   and can be used anywhere in the code.
*/
```

Other types of commenting styles include document comments, which are further divided into outer document comments and inner document comments. We'll cover these topics later in Chapter 6, Section 6.6.2.

3.3.2 Formatting Outputs with Escape Sequences and Arguments

The `println!` macro, introduced earlier in Chapter 2, is typically used for printing something to the terminal. When `print!` is used instead of `println!`, the output is displayed on the same line without a newline character at the end. For instance, consider the following lines:

```
print!("This is a print command");
print!("This is going to be printed on the same line");
```

These lines will produce the following output:

```
This is a print commandThis is going to be printed on the same line
```

Next, let's look at escape sequences inside a print statement. An *escape sequence* contains a backslash (\) symbol followed by one of the escape sequence characters.

Some of the most commonly used escape sequences in Rust include the following:

- **\n: Newline character**
  The newline character `\n` ensures that subsequent output appears on the next line.
- **\t: Tab space**
  The tab character `\t` inserts a tab space, creating a space equivalent to a tab.
- **\r: Carriage return**
  A carriage return `\r` moves the cursor back to the start of the current line. Anything that you print afterwards will delete any previous text that is present on the same line. For instance, consider the following code line:

```
println!("This will be overwritten \r This text will only appear on the
screen");
```

The first part of the text "This will be overwritten" in this case will be overwritten by the second part of the text, which is the text after \r.

- **\": Double quote**
  Double quotes cannot be included directly in a println! statement, as in the following example:

  ```
  println!("Prints double quotes " "); // Error
  ```

  This code results in a syntax error because double quotes are used to delimit text in a print command. The final double quotes are interpreted as marking the end of the text rather than being part of it, which is invalid in Rust syntax.

  To fix the double quote syntax error, you must use a backslash (\) to escape the double quotes, as in the following example:

  ```
  println!("Prints double quotes \" ");
  ```

- **\\: Backward slash**
  A backslash can be included in the output by escaping it with an additional backslash, resulting in two backslashes (\\).

Another important tool for formatting outputs in Rust is the *positional argument*, which allows you to reference values in a specific order within a formatted string. You use curly braces {} as placeholders, and the arguments provided after the string are substituted into these placeholders in the order they appear. Consider the following example:

```
println!("Hello, {}! You are {} years old.", "Alice", 30);
```

In this case, Alice replaces the first {} and 30 replaces the second {}, thus producing the following output:

```
Hello, Alice! You are 30 years old.
```

Named arguments in Rust provide a way to use descriptive labels instead of positional indexes in formatted strings. Using named arguments improves clarity, especially in complex print statements. Consider the following example:

```
println!("{name} scored {points} points!", name = "Alice", points = 50);
```

In this case, name and points are the named arguments, making the statement easier to read and maintain compared to relying on positional placeholders like {0} and {1}.

3.3.3 User Input

Like other programming languages, Rust uses the functions from a standard input/output (I/O) library to handle user inputs. The standard I/O functionality in Rust is provided by the `std::io` module. Specifically, the `read_line` function defined in the standard library can read user inputs in the `String` format. Let's look at the code in shown Listing 3.14 to illustrate how to handle user inputs.

```
fn main() {
    let mut n = String::new();
    std::io::stdin()
        .read_line(&mut n)
        .expect("failed to read input.");

    let n: f64 = n.trim().parse().expect("invalid input");
}
```

Listing 3.14 An Example of Taking User Input

In this example, first we defined a `String` variable `n`, which is then passed to the `read_line` function. To access the `read_line` from the standard library, we'll use the double colon syntax, `std::io::stdin().read_line`. Note that the `&mut` syntax will be covered in Chapter 4, Section 4.6. The return value from `read_line` is a `Result` enum type, which again will be covered in Chapter 5, Section 5.4. For now, just know that a `Result` will have two possible values, either `Ok` or `Err`. An `Ok` indicates that the operation was successful, and `Err` indicates that the operation failed. The `expect` function prints out a message, in case reading errors arise.

The final line of the code trims any whitespace from the input string, then attempts to parse it as a floating-point number (`f64`). If the input cannot be successfully parsed into an `f64`, the program will panic and display the message `"invalid input"`. This error handing is performed through the use of the `expect()` method, which ensures that any parsing error results in a clear and immediate program failure, thus making debugging easier.

3.4 Practice Exercises

In this section, we'll provide some practice questions to help reinforce your learning. The exercise questions cover almost all the concepts introduced in the chapter. Solutions to each exercise will follow in Section 3.5.

1. **Calculate the difference between square of sum and sum of squares**
   Complete the program to calculate the difference between the square of the sum and the sum of the squares for the first *N* natural numbers. The input *N* will be provided

by the user. Use the given code template and complete the missing logic. For example, if the user enters 5, the program should output 170 because:

- Square of sum = $(1 + 2 + 3 + 4 + 5)^2 = 225$
- Sum of squares = $1^2 + 2^2 + 3^2 + 4^2 + 5^2 = 55$
- Difference = 225 – 55 = 170

Your task is to compute the sum of the first *N* numbers and square it. Next, compute the sum of the squares of the first *N* numbers. Finally, calculate and print the difference between these two values. Fill in the code at the appropriate place in the following code template:

```
fn main() {
    let mut n = String::new();
    std::io::stdin()
        .read_line(&mut n)
        .expect("failed to read input.");
    let n: i32 = n.trim().parse().expect("invalid input");

    let mut square_of_sum = 0;
    let mut sum_of_squares = 0;

    /* Complete the code after this line */
}
```

2. **Sum of multiples of 3 or 5 below a given number**
   Complete the program to compute the sum of all natural numbers below *N* (provided by the user) that are multiples of either 3 or 5. Each multiple should only be counted once, even if it is divisible by both 3 and 5. For example, if N = 20, the multiples of 3 are 3, 6, 9, 12, 15, 18 and the multiples of 5 are 5, 10, 15. The sum is calculated as 3 + 5 + 6 + 9 + 10 + 12 + 15 + 18 = 78.

   Your task is to identify numbers below *N* that are divisible by 3 or 5. Next, calculate their sum, ensuring no number is counted more than once. Finally, print the resulting sum. Fill in the code at the appropriate place in the following code template to complete the tasks:

```
fn main() {
    let mut n = String::new();
    std::io::stdin()
        .read_line(&mut n)
        .expect("failed to read input.");
    let n: i32 = n.trim().parse().expect("invalid input");

    /* Add your code below this line */

}
```

3. **Calculating car production in an assembly line**

   This question involves writing code to analyze the production of an assembly line in a car factory. The assembly line has different speeds, ranging from 0 (off) to 10 (maximum). At the lowest speed of 1, the assembly line produces a total of 221 cars per hour. The production rate increases linearly with the speed, meaning that a speed of 4 produces 4 × 221 = 884 cars per hour.

   However, higher speeds increase the likelihood of producing faulty cars that need to be discarded. The success rate depends on the speed:

   - Speeds 1 to 4: 100% success rate.
   - Speeds 5 to 8: 90% success rate.
   - Speeds 9 and 10: 77% success rate.

   You need to write two functions:

   - First, the `total_production()` function calculates the total number of cars successfully produced without faults within a specified time given in hours. This function takes the number of hours and the speed as inputs and returns the number of cars successfully produced.
   - Second, the `cars_produced_per_minute()` function calculates the number of cars successfully produced per minute. This function takes the number of hours and the speed as inputs and returns the number of cars produced per minute.

   Write the code for both functions based on the provided specifications and using the following code template:

```
fn total_production(hours: u8, speed: u8) -> f32 {
    let success_rate: f32;
    /* Your code below this line*/
}
fn cars_produced_per_minutes(hours: u8, speed: u8) -> f32 {
    let success_rate: f32;
    /* Your code below this line*/
}
fn main() {
    println!("{}", total_production(6, 5) as i32); // to round the values we
use i32, which you may just ignore for now
    println!("{}", cars_produced_per_minutes(6, 5) as i32); // to round the
values we use i32, which you may just ignore for now
}
```

4. **Check if a string is a palindrome**

   A palindrome is a word, verse, or sentence that reads the same backward or forward, such as "Able was I ere I saw Elba" or a number like 1881.

   Write a function named `is_palindrome()` that checks whether a given string is a palindrome or not. The function should take a string as input and return a Boolean value indicating whether the string is a palindrome or not. Use the following code template for writing the code:

```
fn palindrome(input: String) -> bool {
    /* Your Code here */
}
fn main() {
    let input = String::from("1211");
    println!("It is {:?} that the given string is palindrome",
palindrome(input));
}
```

5. **Finding a Pythagorean triplet with a given sum**
   Complete the program to compute a Pythagorean triplet (a, b, c) such that the following statements are true:
   - $a < b < c$
   - $a^2 + b^2 = c^2$
   - $a + b + c = 1000$

   Your program should output the triplet values and verify the Pythagorean condition.

6. **Determine movie viewing eligibility**
   Write a function can_see_movie, which will check whether a person is old enough to see a movie. A person can see the movie if they are 17 years old or older, or if they are 13 or older and have a parent's permission. Thus, if you are 17 years old or older, you implicitly have permission.
   Following is the code template for doing the exercise:

```
fn can_see_movie(age: i32, permission: bool) -> bool {
    // Write your code here to implement the logic
    return false; // Remove 'return false' once you have written the code
}
fn main() {
    println!("John who is 18, can see the move: {}", can_see_movie(17,
true));
}
```

3.5 Solutions

This section provides the code solutions for the practice exercises in Section 3.4.

1. **Calculate the difference between square of sum and sum of squares**

```
fn main() {
    let mut n = String::new();
```

```
    std::io::stdin()
        .read_line(&mut n)
        .expect("failed to read input.");
    let n: i32 = n.trim().parse().expect("invalid input");

    let mut square_of_sum = 0;
    let mut sum_of_squares = 0;

    for i in 1..=n {
        square_of_sum = square_of_sum + i;
        sum_of_squares = sum_of_squares + i.pow(2);
    }

    let difference = square_of_sum.pow(2) - sum_of_squares;
    println!(
        "The difference of the square_of_sum and sum of Squares for N = {} is
{}",
        n, difference
    );
}
```

2. **Sum of multiples of 3 or 5 below a given number**

```
fn main() {
    let mut n = String::new();
    std::io::stdin()
        .read_line(&mut n)
        .expect("failed to read input.");
    let n: u32 = n.trim().parse().expect("invalid input");
    let mut sum: u32 = 0;
    for i in 1..n {
        if i % 3 == 0 || i % 5 == 0 {
            sum = sum + i;
        }
    }
    println!("\n\n The sum of the multiples are = {sum}");
}
```

3. **Calculating car production in an assembly line**

```
fn total_production(hours: u8, speed: u8) -> f32 {
    let success_rate: f32;
    if speed <= 4 {
        success_rate = 1.0;
    } else if speed >= 5 && speed <= 8 {
        success_rate = 0.9;
```

```
    } else {
        success_rate = 0.77;
    }
    hours as f32 * 221.0 * success_rate * speed as f32
}
fn cars_produced_per_minutes(hours: u8, speed: u8) -> f32 {
    let success_rate: f32;
    if speed <= 4 {
        success_rate = 1.0;
    } else if speed >= 5 && speed <= 8 {
        success_rate = 0.9;
    } else {
        success_rate = 0.77;
    }
    (hours as f32 * 221.0 * success_rate * speed as f32) / (60.0 * hours as
f32)
}
fn main() {
    println!("{}", total_production(6, 5) as i32); // to round the values we
use i32
    println!("{}", cars_produced_per_minutes(6, 5) as i32); // to round the
values we use i32
}
```

4. **Check if a string is a palindrome**

```
fn palindrome(input: String) -> bool {
    let mut is_palindrome = true;
    if input.len() == 0 {
        is_palindrome = true;
    } else {
        let mut last = input.len() - 1;
        let mut first = 0;
        let my_vec = input.as_bytes();
        while first < last {
            if my_vec[first] != my_vec[last] {
                is_palindrome = false;
                break;
            }
            first += 1;
            last -= 1;
        }
    }
    is_palindrome
}
```

```
fn main() {
    let input = String::from("1211");
    println!(
        "It is {:?} that the given string is palindrome",
        palindrome(input)
    );
}
```

5. **Finding a Pythagorean triplet with a given sum**

```
fn main() {
    let mut flag = true;
    for a in 1..=1000 {
        for b in a + 1..1000 {
            // this ensures that a < b
            for c in b + 1..1000 {
                // this ensure that b < c
                if a * a + b * b == c * c && a + b + c == 1000 {
                    println!(
                        "\n\n The required pathagorian triplet are ({}, {},
{}) \n\n",
                        a, b, c
                    );
                    flag = false;
                    break;
                }
            }
            if flag == false {
                break;
            }
        }
        if flag == false {
            break;
        }
    }
}
```

6. **Determine movie viewing eligibility**

```
fn can_see_movie(age: i32, permission: bool) -> bool {
    (age >= 17) || (age >= 13 && permission)
}
fn main() {
    println!("John who is 18, can see the move: {}", can_see_movie(17, true));
}
```

3.6 Summary

In this chapter, we delved into controlling the flow of Rust programs through conditionals and control flow constructs. We started by exploring how Rust handles decision-making using `if else` statements and `if else if` ladders, followed by the powerful `match` expression for pattern matching. Moving on to control flow, we examined different types of loops, including simple loops, `for` loops for iterating over collections, and `while` loops for conditional repetition. Additionally, we introduced comments, which improve code clarity, and discussed how to use print commands and handle user input for interactive programs. This combination of control structures and utility commands forms the backbone of Rust's program execution logic. To solidify these concepts, the chapter concluded with practical exercises and solutions, providing an opportunity for hands-on learning.

In the next chapter, we'll shift our focus to understanding ownership, a fundamental concept in Rust that governs memory management and ensures safety without the need for a garbage collector.

Chapter 4
Ownership

Mastering Rust begins with understanding ownership, a concept that ensures memory safety and reliability. As we journey into ownership, you'll uncover how Rust's unique model forms the backbone of efficient code.

This chapter dives into Rust's unique ownership model, a cornerstone of the language's memory safety. We'll start with the basics of ownership, exploring how it works and why it's crucial. This chapter explains how ownership is handled in functions in a way that ensures data is managed correctly across scopes. Borrowing is introduced next, and we'll show you how to reference data without taking ownership, along with some practical examples of functions. We'll also cover dereferencing, teaching you how to access the value behind a reference. Finally, we'll cover the distinctions between immutable and mutable bindings of references and provide a detailed discussion on the six types of references. By mastering these concepts, you'll learn to write efficient and safe Rust programs.

4.1 Ownership Basics

Ownership is one of the fundamental concepts that make Rust an unparalleled language. Ownership, although a common concept, carries profound significance in Rust. You must understand three simple rules of ownership:

1. Each value has a variable that serves as its "owner."
2. A value can have only one owner at a time.
3. If an owner goes out of scope, the value is cleaned up.

Let's explore these rules in more detail.

4.1.1 Values Must Have a Single Owner

Let's start with the first two rules, which go together naturally. Consider the following code:

```
let s1 = String::from("world");
let s2 = s1;
```

We create a `String` variable `s1` and then assign it to another variable `s2`. According to Rust's first and second ownership rules, each value must have one and only one owner. When we assign `s1` to `s2`, ownership of the `String` `s1` is moved to `s2`. As a result, `s1` is no longer valid, and any attempt to use `s1` after the move will result in a compile-time error. For instance, the code shown in Listing 4.1 will not compile.

```
let s1 = String::from("world");
let s2 = s1;
println!("s1 is: {s_2}"); // Error!
```

Listing 4.1 Accessing Variable after Ownership Change

You'll see an error, "borrow of moved value: s1, value borrowed here after move." To better understand this error, you must understand what is going under the hood, in the memory.

Generally speaking, memory can be divided into two categories:

- **Non-volatile memory**
  Also known as permanent memory, this kind of memory includes storage mediums like hard drives and solid-state drives (SSDs). These devices are typically slower but offer abundant storage capacity. Data in non-volatile memory persists even when the power is turned off, such as files stored on a hard drive.
- **Volatile memory**
  Also known as temporary memory, this kind of memory includes random access memory (RAM) and caches, which are fast but limited in quantity, i.e., not abundant. Volatile memory is used during program execution to manage a program's memory requirements. However, the data in volatile memory is lost when the power is turned off.

Non-volatile memory is managed by the operating system and is not directly relevant during the execution of a program. What we're interested in is volatile memory and how a program uses and interacts with it during execution.

Primarily, three separate memory regions are available in volatile memory for program execution, namely, static, stack, and heap.

The *static* region stores a program's binary instructions and static variables. This region is populated with relevant program data when your program starts up and is destroyed when your program ends. The cleanup of values from the static region is automatic.

The *stack* deals with data that has a fixed known size at compile time. Since the size is known, the values are stored in order, using the last in, first out (LIFO) strategy. The management of the stack is easy and fast since everything is predictable and no special computations are required.

In contrast, the *heap* deals with data of unknown size at compile time. Since the size is not known, the storage of this data cannot follow an exact order and therefore the data is stored all over the place, where a suitable fit for the data can be found in memory. The management of this part is therefore typically slow and requires a lot of management. During program execution, both the stack and the heap can grow in size.

Let's see what these different parts have to do with ownership, coming back to our program shown earlier in Listing 4.1. First, we create the variable s1. Figure 4.1 shows how the variable s1 will be laid down in memory. The variable s1, which is a string, is made up of a pointer to some allocated space in heap, containing the contents of the string, alongside other data such as its length and its capacity. The length indicates the length the string is occupying in memory, and the capacity indicates the number of bytes assigned by the allocator or manager of the heap to the string. The pointer, the length, and the capacity are of fixed sizes; therefore, they reside on the stack. The pointer size depends on the underlying hardware and is typically 32 or 64 bits. The value, however, is stored on the heap since it can grow in size and, therefore, does not have a fixed size. This whole makes up the string s1.

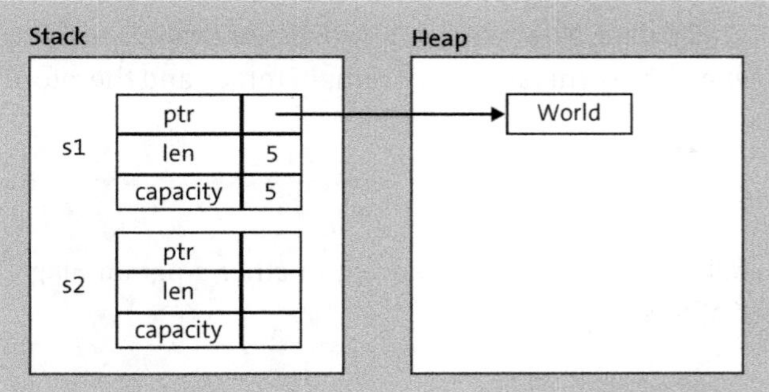

Figure 4.1 String in Memory

At this point, s1 is the owner of the data. Now, what happens when we assign the value of s1 to s2, as shown in line 2 of Listing 4.1? Since, according to the second ownership rule, a value can only have one owner, the value is being moved into s2. What happens inside the memory is shown in Figure 4.2.

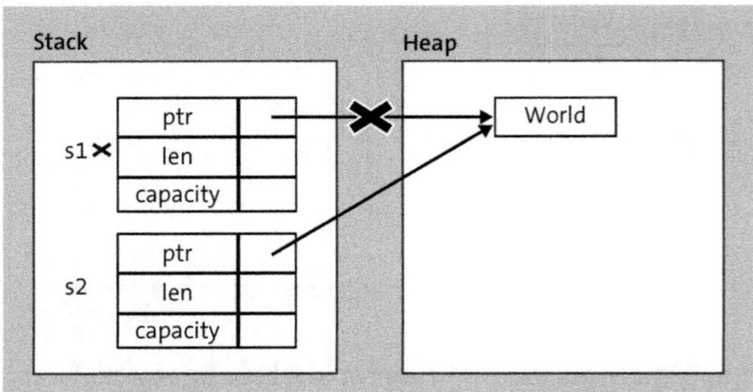

Figure 4.2 Ownership Change from s1 to s2

The pointer, the length, and the capacity are all copied inside the stack from s1 and pushed as new values on the stack as part of s2. To ensure there is a single owner, Rust will immediately invalidate s1. As a result, s1 will no longer be available for use and will be cleared from memory. This deletion is indicated by having a small cross after the variable s1, as shown in Figure 4.2. In other words, both the variable and its pointer will both be deleted and cleared from memory. Any attempt to access s1 after the move will throw an error. The third line in Listing 4.1, therefore, ends in an error.

Now, what if we didn't want to move the value, but copy it instead, in such a way that we can use s1 afterwards? You can use the clone method for this purpose, in the following way:

```
let s1 = String::from("world");
let s2 = s1.clone();
println!("s1 is: {s_2}");
```

This code now compiles. The clone method will not only make a copy of the stack data of s1 but also will make a copy its heap data. Now, you have two distinct values residing in the heap, each with a single owner. The ownership rules remain intact, and therefore, the compiler will have no issues.

Clone and Deep Copy

The clone method in Rust is similar to the deep copy concept in other programming languages, like Python, Java, and others.

Figure 4.3 shows exactly what happens when we call the clone method.

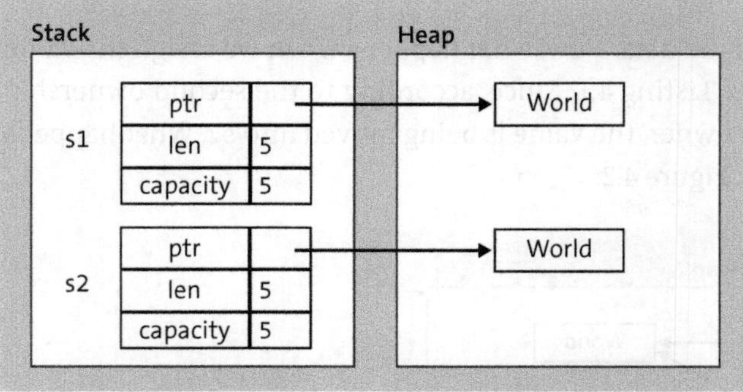

Figure 4.3 Cloning s1

Move versus Copy

Rust uses the term *copy* when only the stack data is being copied. Let's briefly explore what this term means.

When we assign the value of one variable to another, the value is being moved, thus leading to a change of ownership. However, this change is not true for some primitive types. Consider the following code:

```
let x = 15;
let y = x;
println!("x is: {x}");
```

You may expect an error from the `print` line. The value of variable `x` should have moved on the line above variable `y`, thereby making `x` invalid. This code works and compiles because, in Rust, primitives (i.e., integers, floats, Booleans, and characters) are entirely stored on the stack, with no reference to the heap. As a result, they are copied rather than moved, and each copy has its own separate owner. Types that behave in this more independent manner are often referred to as *copy types* or *stack-allocated types*.

4.1.2 When Owner Goes Out of Scope, the Value Is Cleaned Up

Now, let's talk about the third rule: When the owner goes out of scope, the value is dropped.

By default, the variables live within the `main` function scope. When the `main` exits, the variables will be dropped. Consider the code shown in Listing 4.2.

```
let s1 = String::from("world");
{
    let s2 = s1.clone();
}
println!("s1 is: {s_2}"); // Error
```

Listing 4.2 s2 with Limited Scope

Variable `s2` will be dropped when the scope ends. When we try to access `s2` after the scope in the print statement, we get an error, "cannot find `s2` in this scope." Variable `s2` is defined in the inner scope and is no longer dropped at the end of the `main` function; instead, the variable is dropped at the end of the scope. When the inner scope ends, it will be popped out of the stack, and the memory it is pointing to will be immediately freed up. This process prevents the following issues from arising:

- *Dangling pointers*, that is, pointers that point to invalid memory
- *Memory leaks*, that is, not releasing memory that is no longer required

All this cleanup is possible at compile time, while we are writing code. Furthermore, with the ownership rules in place, this cleanup incurs zero runtime costs.

4.2 Ownership in Functions

Ownership in Rust becomes particularly interesting when functions are involved. Functions can be categorized into three groups in the context of ownership: those that take ownership, those that give ownership, and those that take and then return ownership. Let's examine each of these types one by one.

4.2.1 Functions Taking Ownership

Consider a case where you need a function in which the caller transfers ownership of a value to the function. This transfer ensures the function has exclusive access to the value and prevents further use of the original value. Such functions are particularly useful in contexts involving resource management. Consider the code shown in Listing 4.3.

```
fn main(){
    let vec_1 = vec![1, 2, 3];
    takes_ownership(vec_1);
    println!("vec 1 is: {:?}", vec_1); // Error
}
fn takes_ownership(vec: Vec<i32>) {
    println!("vec is: {:?}", vec);
}
```

Listing 4.3 Function Taking Ownership

In main, we create a vector. Like strings, vectors are also heap allocated, with their pointers, lengths, and capacities stored on the stack. In the next line, we call the function takes_ownership passing in vec_1. The function takes_ownership accepts a vector and prints it.

The last line in main, which prints the vec_1 variable, throws the error message "borrowed of a moved value." Passing a variable to a function has the same effect as assigning a variable to another variable. Just as ownership would have moved when assigning vec_1 to vec_2, ownership will also move from vec_1 to vec when vec_1 is passed in the call to the function takes_ownership. The function name itself refers exactly to this behavior. Now, since vec_1 ownership has already been moved to vec, the variable will be invalidated and therefore cannot be printed.

The vec variable lives within the function scope. At the end of the function, the variable will be dropped, meaning that it will be cleaned up from the heap. One way to fix this code would be to clone vec_1, as shown in Listing 4.4.

```
fn main(){
    let vec_1 = vec![1, 2, 3];
    takes_ownership(vec_1.clone());
```

```
    println!("vec 1 is: {:?}", vec_1);
}
fn takes_ownership(vec: Vec<i32>) {
    println!("vec is: {:?}", vec);
}
```


Listing 4.4 Fixing the Code by Passing in a Clone to the Function

This code will work because we are not passing the ownership of vec_1. Instead, we are first cloning vec_1 and then sending out the cloned copy of vec_1 to the function. The cloned copy's ownership is moved to the variable vec, which will be dropped at the end of the function, resulting also in the dropping of the cloned copy itself. The original variable remains valid in main, and therefore, the print statement will not cause any error.

4.2.2 Function Returning Ownership

Now, let's explore how ownership can be transferred out of a function through code. Such functions can be useful when a function is responsible for initializing or generating resources. Consider the code shown in Listing 4.5.

```
fn main() {
    let vec_2 = gives_ownership();
    println!("vec 2 is: {:?}", vec_2);
}
fn gives_ownership() -> Vec<i32> {
    vec![4, 5, 6]
}
```

Listing 4.5 Function Giving Ownership

In this code, the gives_ownership function creates a new vector [4, 5, 6] and returns it. When gives_ownership is called in main, the ownership of the vector is transferred to the variable vec_2. The vector is now owned by vec_2 in the main function, and its contents are printed. The vector will remain valid until the end of the main function, at which point it will be dropped and cleaned up from memory.

4.2.3 Function Taking and Returning Ownership

Finally, let's consider another function that combines the two cases of taking ownership and giving it back to main. This type of functions is used in resource management or when applying specific data transformations before data is handed back for further use. Consider the code shown in Listing 4.6.

```
fn main() {
    let vec_1 = vec![1, 2, 3];
    let vec_2 = takes_and_gives_ownership(vec_1);
    println!("vec 2 is: {:?}", vec_2);
}
fn takes_and_gives_ownership(mut vec: Vec<i32>) -> Vec<i32> {
    vec.push(10);
    vec
}
```

Listing 4.6 Function Taking Ownership and Giving It Back

In this code, the `takes_and_gives_ownership` function accepts a vector as its parameter, modifies it by adding the value 10, and then returns the modified vector. When `vec_1` is passed to this function in `main`, its ownership is transferred to the parameter `vec`. After modifying the vector, the function returns it, transferring ownership to `vec_2` in `main`. The modified vector is then printed, showing the added value. `vec_2` will now be dropped at the end of the `main`. This example demonstrates how a function can take ownership of a value, modify it, and then return ownership back to the caller.

Moving a variable by ownership to inside the function usually means that the function is expected to consume the value. So, returning the value back is not usually the pattern we want. In most cases, we want to use references (also known as borrowing), which we'll cover shortly in Section 4.3.

Stack-Only Data Types and Functions

Stack-only data types or copy types, such as primitives, are not moved to inside the function. Consider Listing 4.7.

```
fn main() {
    let x = 10;
    stack_function(x);
    println!("In main, x is: {x}");
}
fn stack_function(mut var: i32) {
    var = 56;
    println!("In func, var is: {var}");
}
```

Listing 4.7 Stack-Only Data and Ownership

In this case, x is an integer, a stack-only data type. When we pass x to `stack_function`, a copy of x is made, which is assigned to the variable `var`, and the function operates on this copy. Stack-only data types, such as `integers`, `floats`, `bools`, and `chars`, are copied and not moved. As a result, the original x in `main` is unchanged and is still accessible after the function call. The two variables are now two distinct values residing in the stack. An update to the value of variable `var` inside the function should therefore not affect the value of the variable `x` in `main`. This separate behavior contrasts with heap-allocated types like vectors, where ownership is transferred instead of being copied. If we execute this code, notice how the value of the variable `x` in the `main` function is `10`, while the value of its copy in the function is `56`.

4.3 Borrowing Basics

A fundamental concept in Rust's ownership system, *borrowing* allows multiple parts of a program to interact with data safely and efficiently. In simple terms, borrowing involves creating a reference to a value. A reference is similar to a pointer but comes with specific rules and limitations. Unlike ownership, references do not take ownership of the values they point to, which is why the process is called borrowing; a reference temporarily borrows a value without claiming ownership.

Borrowing involves many details and explicit rules that you must follow. In the following sections, we'll explore several aspects of borrowing, starting with its rationale.

4.3.1 Why Borrowing?

First, let's discuss why borrowing is necessary. One key reason is the efficient use of memory, which leads to performance improvements. For instance, if a function only needs to read data, providing a reference to the data is more efficient than passing a clone or transferring ownership. Cloning or transferring ownership can be expensive, especially with data types that consume a significant amount of memory. In contrast, providing a reference is much cheaper and more efficient.

The second reason occurs when ownership is not required. Consider the function `takes_ownership`, shown earlier in Listing 4.3. This function simply prints the value of the vector that is passed to it. In this scenario, you probably don't want the function to take ownership of the vector, as that ownership would dictate when the vector is freed from memory. Instead, a better approach is to temporarily borrow the vector without taking ownership of it.

4.3.2 Borrowing Rules

Now, that we understand the motivation behind borrowing, let's go over the two rules for borrowing rules in Rust:

- At any time, you can have either one mutable reference or many immutable references.
- References must always be valid.

These rules solve two problems: data races and dangling references. Let's first understand the rules, and then we'll explain how enforcing these rules solves these two problems.

Either One Mutable Reference or Many Immutable References

Let's start with the first rule in this section. Consider the code shown in Listing 4.8.

```
fn main() {
    let mut vec_1 = vec![4, 5, 6];
    let ref1 = &mut vec_1;
    let ref2 = &mut vec_1; // Error
    println!("ref1: {:?}, ref2: {:?}", ref1, ref2);
}
```

Listing 4.8 Violation of First Borrowing Rule

We first created mutable vector of `vec_1`. Next, we created a mutable reference of `ref1` to the vector. The ampersand (&) is used to create a reference. References can be either immutable or mutable. The `mut` keyword after the `&` indicates that a reference is mutable, allowing you to modify the borrowed data. An immutable reference, on the other hand, allows you to borrow the data without making any changes. Both types of references do not take ownership of the data; instead, they temporarily borrow it. In the next line, we created another mutable reference to the data using `ref2`. Finally, we access the two references in a print statement.

The compiler does not like it, giving us an error, which states "cannot borrow `vec_1` as mutable more than once at a time." This issue arises because we are violating the first rule: You can have either one mutable reference or many immutable references at any given time, but not both simultaneously. Two mutable references to the same data at the same time are not allowed. This rule ensures that mutation occurs in a controlled manner, preventing multiple mutable references to the same data. Many beginners find this challenging because, in most other languages, mutation is generally unrestricted.

If we remove the print statement for the code shown in Listing 4.8, the code will compile, as shown in Listing 4.9.

```
fn main() {
    let mut vec_1 = vec![4, 5, 6];
    let ref1 = &mut vec_1;
    let ref2 = &mut vec_1;
}
```

Listing 4.9 Multiple Mutable References with No Violation of Rule 1

The issue with the code shown in Listing 4.8 arises because the Rust compiler tracks the active period of a reference (also known as its scope) from the line where it is introduced or defined until the last line where the reference is used. As shown in Listing 4.9, the scope of ref1 is limited to one line (that is, the line in which it is defined), and the scope of ref2 is limited to another line (again the line in which it is defined), with no overlap of code lines between the references. Thus, at any given time, only one mutable reference exists, so rule 1 is not violated. However, if we consider the code shown in Listing 4.9 with the print statement, printing ref1, an error will occur because, within the scope of ref1, no other mutable references to the same data should exist, which is violated by ref2.

According to rule 1, you can only have many immutable references to the same data. Consider the code shown in Listing 4.10.

```
fn main() {
    let mut vec_1 = vec![4, 5, 6];
    let ref1 = &vec_1;
    let ref2 = &vec_1;
    println!("ref1: {:?}, ref2: {:?}", ref1, ref2);
}
```

Listing 4.10 Multiple Immutable References

Immutable references are indicated by ampersand (&), and they cannot mutate the data. Listing 4.10 compiles since multiple immutable references are allowed by rule 1.

Let's consider one more case with regard to rule 1, as shown in Listing 4.11.

```
fn main() {
    let mut vec_1 = vec![4, 5, 6];
    let ref1 = &vec_1;
    let ref2 = &vec_1;
    let ref3 = &mut vec_1;  // Error
    println!("ref1: {:?}, ref2: {:?}, ref3: {:?}", ref1, ref2, ref3);
}
```

Listing 4.11 Violation of Rule 1 Due to Coexistence of Immutable and Mutable References

The compiler throws an error: "cannot borrow vec_1 as mutable because it is also borrowed as immutable." This error occurs because we are violating the rule that permits either one mutable reference or multiple immutable references at any given time, but not both simultaneously.

The code can be slightly modified to fix the issue, as shown in Listing 4.12.

```
fn main() {
    let mut vec_1 = vec![4, 5, 6];
    let ref1 = &vec_1;
```

```
    let ref2 = &vec_1;
    println!("ref1: {:?}, ref2: {:?}", ref1, ref2);
    let ref3 = &mut vec_1;
}
```

Listing 4.12 Immutable and Mutable References with No Coexistence

The code now compiles because the scopes of ref1 and ref2 end on the print line (their last usage), and therefore, immutable and mutable references do not coexist at any given time.

By enforcing this rule, Rust prevents data races at compile time. A *data race* occurs when multiple references to the same data exist, with at least one reference modifying the data, and there's no mechanism to synchronize access. Rust's borrowing rules ensure that data can be either read through immutable references or modified through a single mutable reference. This approach prevents data races and allows different parts of your code to interact with data safely.

References Must Always Be Valid

Now, let's look at the second rule, which states that references must always be valid. To illustrate this rule, let's explore what happens when we attempt to return a reference to a value created within an inner scope, as shown in Listing 4.13.

```
fn main() {
    let vec_1 = {
        let vec_2 = vec![1, 2, 3];
        &vec_2
    };
}
```

Listing 4.13 Attempting to Return a Reference to a Value Created in an Inner Scope

In this example, we've defined a vector, vec_2, inside an inner scope. We then attempt to return a reference to vec_2 from that scope and store it in a variable, vec_1. However, when we try to compile this code, the Rust compiler raises an error "vec_2 does not live long enough."

This error occurs because we've created what's known as a *dangling reference*. Inside the inner scope, vec_2 is created and takes ownership of the vector [1, 2, 3]. We then return a reference to vec_2, intending to use it outside the scope in which it was defined. However, when the inner scope ends, vec_2 is dropped, and the memory allocated for it is cleaned up. As a result, the reference to vec_2 that we tried to return is no longer pointing to valid data; it's now a dangling reference, which Rust's ownership rules strictly prohibit. For more information on dangling references, see Chapter 10, Section 10.1.1.

Rust's compiler prevents this situation by enforcing the rule that references must always be valid. This rule ensures that you can never accidentally reference memory that has already been freed, thereby preventing a class of bugs that can lead to undefined behavior in other programming languages.

4.3.3 Copying of References

Regardless of whether the data is stack allocated or heap allocated, a mutable reference to data can be copied only once. Consider the code shown in Listing 4.14.

```
fn main() {
    let mut vec_1 = vec![4, 5, 6];
    let ref1 = &mut vec_1;
    let ref2 = ref1;
    let ref3 = ref1; // Error
}
```

Listing 4.14 Trying to Create Multiple Copies of a Mutable Reference

The compiler does not allow the creation of multiple copies of a mutable reference and throws an error, "use of moved value: `ref1`." The compiler further explains that "the move occurs because `ref1` does not implement the `Copy` trait." We'll discuss traits in more detail in Chapter 8, but for now, this message simply means that mutable references cannot be copied; they can only be moved.

Note that references typically reside on the stack. In Section 4.1.1, you learned that stack-allocated data is copied, not moved, when assigned. However, mutable references, even though they are stored on the stack, are an exception. They are moved and not copied.

This behavior is enforced to maintain Rust's borrowing rules. According to the first borrowing rule, you cannot have multiple mutable references to the same data at the same time. If mutable references were allowed to be copied, instead of being moved, Rust's borrowing rules, which only permit a single mutable reference at any given time, would be violated.

Immutable references are, however, allowed to be copied many times, as shown in Listing 4.15.

```
fn main() {
    let mut vec_1 = vec![4, 5, 6];
    let ref1 = &vec_1;
    let ref2 = ref1;
    let ref3 = ref1;
```

```
    let ref4 = ref1;
}
```

Listing 4.15 Multiple Copies of an Immutable Reference

In this case, each assignment of ref1 creates a copy. Creating copies of immutable references does not violate any borrowing rules because the borrowing rules allow multiple immutable references. In summary, mutable references are moved when assigned, while immutable references are copied.

4.4 Borrowing in Functions

Let's revisit the function that takes ownership in Listing 4.16. The issue with this code is that it consumes `vec_1`, and therefore, the value is no longer accessible in `main`.

```
fn main(){
    let vec_1 = vec![1, 2, 3];
    takes_ownership(vec_1);
}
fn takes_ownership(vec: Vec<i32>) {
    println!("vec is: {:?}", vec);
}
```

Listing 4.16 Function That Takes Ownership

In Section 4.2.1, we mentioned that this issue can be fixed by cloning `vec_1`. However, cloning creates a new heap allocation, which is inefficient. Moreover, if you examine the function, notice how it only prints the vector passed to it, so it doesn't need to take ownership of the vector. The function shouldn't be responsible for deciding when the vector should be cleaned up.

A more sensible approach is to pass in a reference in this case. Update the code to pass in a reference instead, as shown in Listing 4.17.

```
fn main(){
    let vec_1 = vec![1, 2, 3];
    let ref1 = &vec_1;
    takes_ownership(ref1);
    println!("vec 1 is: {:?}", vec_1);
}
fn takes_ownership(vec: &Vec<i32>) {
    println!("vec is: {:?}", vec);
}
```

Listing 4.17 Passing in a Reference to the Function

In this example, ref1 is an immutable reference to vec_1, which is passed to the function takes_ownership. Notice that, in the function signature, the type of vec has been updated from Vec<i32>, which represents a type containing an owned value, to that of &Vec<i32>, which represents a type containing a borrowed value. Passing in the references in this case makes more sense because the function does not need ownership. Passing a reference is also cheaper than cloning because you aren't making any new heap allocations. Because the function is not taking ownership, vec_1 is available in the main function after the function call in the print statement.

Let's update the function name to something more accurate. The updated code is shown in Listing 4.18.

```
fn main(){
    let vec_1 = vec![1, 2, 3];
    let ref1 = &vec_1;
    borrows_vec(ref1);
    println!("vec 1 is: {:?}", vec_1);
}
fn borrows_vec(vec: &Vec<i32>) {
    println!("vec is: {:?}", vec);
}
```

Listing 4.18 Updated Code from Listing 4.17

The borrowing rules remain in place when we are dealing with functions that borrow values. Now, let's modify the code shown in Listing 4.18 and add a mutable reference to vec_1. The updated code is shown in Listing 4.19.

```
fn main(){
    let mut vec_1 = vec![1, 2, 3];     // vec_1 has to be mutable, in order to
                                       // create a mutable reference to it

    let ref1 = &vec_1;
    let ref2 = &mut vec_1;             // Error
    borrows_vec(ref1);
    println!("vec 1 is: {:?}", vec_1);
}
fn borrows_vec(vec: &Vec<i32>) {
    println!("vec is: {:?}", vec);
}
```

Listing 4.19 Violation of Borrowing Rule 1

This update code throws an error, "cannot borrow vec_1 as mutable because it is also borrowed as immutable." The reference ref1 scope extends from the line in which it is defined until the line in which it is passed to the function borrows_vec. During this period, you should not make any mutable references to it since mutable and immutable references cannot coexist. You are, however, free to make a mutable reference to the vector once the scope of the immutable reference ends. Consider the code shown in Listing 4.20.

```
fn main(){
    let mut vec_1 = vec![1, 2, 3];
    let ref1 = &vec_1;
    borrows_vec(ref1);
    let ref2 = &mut vec_1;
    println!("vec 1 is: {:?}", vec_1);
}
fn borrows_vec(vec: &Vec<i32>) {
    println!("vec is: {:?}", vec);
}
```

Listing 4.20 Mutable and Immutable Do Not Coexist Now

This code now compiles since mutable and immutable do not coexist within the same scope.

Let's remove the mutable reference and add one more function from Section 4.3 to the code shown in Listing 4.20. The updated code is shown in Listing 4.21.

```
fn main(){
    let mut vec_1 = vec![1, 2, 3];
    let ref1 = &vec_1;
    borrows_vec(ref1);
    takes_and_gives_ownership(vec_1);
    println!("vec 1 is: {:?}", vec_1);  // Error
}
fn borrows_vec(vec: &Vec<i32>) {
    println!("vec is: {:?}", vec);
}
fn takes_and_gives_ownership(mut vec: Vec<i32>) -> Vec<i32> {
    vec.push(10);
    vec
}
```

Listing 4.21 Code from Listing 4.20 with an Added Function

This code adds the `takes_and_gives_ownership` function, shown earlier in Listing 4.6. Additionally, in `main`, the code calls the added function with `vec_1` before printing `vec_1`. This line with the print statement results in an error on the line since the ownership of `vec_1` has been moved to inside the function and the returning owned value is not being captured in any variable. You can solve this problem by using variable shadowing (refer to Chapter 2, Section 2.1.4), as shown in Listing 4.22.

```
fn main(){
    let mut vec_1 = vec![1, 2, 3];
    let ref1 = &vec_1;
    borrows_vec(ref1);
    let vec_1 = takes_and_gives_ownership(vec_1);
    println!("vec 1 is: {:?}", vec_1);
}
fn borrows_vec(vec: &Vec<i32>) {
    println!("vec is: {:?}", vec);
}
fn takes_and_gives_ownership(mut vec: Vec<i32>) -> Vec<i32> {
    vec.push(10);
    vec
}
```

Listing 4.22 Code from Listing 4.21 Fixed Using Shadowing

The ownership of vec_1 is first transferred to the function takes_and_gives_ownership and then returned back in vec_1, which shadows the vec_1 defined in the first line of the main function. This code compiles but is inefficient because passing ownership of vec_1 to the function takes_and_gives_ownership involves moving the entire vector's data to the function's scope, which incurs the overhead of transferring ownership. After the function modifies and returns the vector, it is moved back to the original scope, resulting in unnecessary data movement. You can avoid these costs by using a mutable reference, which would allow the function to modify the vector in place without the overhead of ownership transfers.

Let's update the code shown in Listing 4.22 and pass in a mutable reference instead, as shown in Listing 4.23.

```
fn main(){
    let mut vec_1 = vec![1, 2, 3];
    let ref1 = &vec_1;
    borrows_vec(ref1);
    let ref2 = &mut vec_1;
    mutably_borrows_vec(ref2);
    println!("vec 1 is: {:?}", vec_1);
}
fn borrows_vec(vec: &Vec<i32>) {
    println!("vec is: {:?}", vec);
}
fn mutably_borrows_vec(vec: &mut Vec<i32>) { // name and signature updated
    vec.push(10);
}
```

Listing 4.23 Code from Listing 4.22 Updated by Passing in a Mutable Reference

The function name is updated to a more accurate name of `mutably_borrows_vec`. We created a mutable reference `ref2`, which is next passed into the function. For the function to accept a mutable reference, we updated the type of the variable `vec` from `mut vec: Vec<i32>` to that of `vec: &mut Vec<i32>`. The function does not need to return an updated vector anymore, since it will be updated through a mutable reference. The returning values and the return type from the signature are therefore removed. The function is not returning anything explicitly; therefore, we don't need to store the returning value of the function in `main`. Overall, the code shown in Listing 4.23 is now more meaningful than transferring the ownership and receiving it back.

Let's consider one final function from Section 4.2.2, the `gives_ownership` function, and understand its implications from the borrowing perspective. The function is shown in Listing 4.24.

```
fn main() {}
fn gives_onwership() -> Vec<i32> {
    vec![4, 5, 6]
}
```

Listing 4.24 Function gives_ownership from Section 4.2.2

Now, let's see what happens if, instead of returning the vector, we return a reference to the vector. Consider the code in Listing 4.25.

```
fn main(){}
fn gives_ownership() -> &Vec<i32> { // Error
    let vec = vec![4, 5, 6];
    &vec
}
```

Listing 4.25 Updated Function gives_ownership with an Immutable Reference

We encounter an error, "missing lifetime specifier, this function's return type contains a borrowed value, but there is no value for it to be borrowed from." Although we won't explore lifetime specifiers in this chapter, we'll discuss them in detail later in Chapter 10. In this particular case, the error occurs because we're violating borrowing rule 2, which states that references must always remain valid. The issue arises because, inside the function, we create the vector `vec`, which holds ownership. We then return a reference to this vector. However, once the function ends, the vector is dropped and cleaned up, resulting in a dangling reference, meaning the reference will no longer point to valid memory.

As a general rule of thumb, when you create a value within a function and intend to return it, you must transfer ownership of that value. Returning a reference is not advisable because the owned value will be automatically cleaned up at the end of the function, rendering the reference invalid. For the other two types of functions (i.e., ones which take ownership but do not require it or take ownership and then return it back), references can be updated with the help of immutable and mutable borrows.

4.5 Dereferencing

Dereferencing is the process of accessing the value pointed to by a reference or pointer. This concept is closely related to borrowing because when you borrow a reference to data, you often need to dereference it to manipulate or access the underlying value. Let's cover the basics of dereferencing through the example shown in Listing 4.26.

```
fn main() {
    let mut some_data = 42;
    let ref1 = &mut some_data;
    let deref_copy = *ref1;
    *ref1 = 13;
    println!("some_data is: {some_data}, deref_copy is: {deref_copy}");
}
```

Listing 4.26 Dereferencing Stack-Allocated Data behind a Mutable Reference

In this case, we have a mutable reference `ref1` to `some_data`. The line `let deref_copy = *ref1` essentially does two things. First, it dereferences the value pointed to by `ref1` on its right-hand side. Dereferencing is performed with the help of the * operator. What this operator does under the hood is access the actual value pointed to by `ref1`, in this case, the value of `42`. Second, the value is stored in a separate variable of `deref_copy` due to the assignment statement on the third line of the code in `main` (i.e., `let deref_copy = *ref1`). Recall that the assignment of stack-allocated data creates a copy and not a move. At this stage, `some_data` and `deref_copy` are two distinct values stored in the memory. Updating the value pointed by `ref1`, which is the value `42` held by variable `some_data`, will not affect the value of the `deref_copy`. Finally, the line `*ref1 = 13` will update the data pointed to by `ref1`, which is the value of `42` to a value of `13`. Since the value is being borrowed, ownership remains with `some_data`. As a result, `some_data` will be updated to a value of `42`. The `deref_copy` is a separate copy in stack; therefore, its value will not be updated. The print statement therefore prints the values of `13` and `42` for `some_data` and `deref_copy`, respectively.

Heap-allocated data, however, behaves differently. Consider the code shown in Listing 4.27.

```
fn main() {
    let mut heap_data = vec![5, 6, 7];
    let ref1 = &mut heap_data;
```

```
    let deref_copy = *ref1; // Error
}
```

Listing 4.27 Dereferencing Heap-Allocated Data behind a Mutable Reference

The code shown in Listing 4.27 is similar to the code shown in Listing 4.26. We first created some heap-allocated data and then created a reference. Finally, we dereferenced the value and assigned it to a variable `deref_copy`, which throws an error, "cannot move out of `*ref_1` which is behind a mutable reference."

Recall from Section 4.3.3 that assigning stack-allocated data results in a copy, while assigning heap-allocated data results in a change of ownership. As shown in Listing 4.26, the line `let deref_copy = *ref1` did not result in an error because a copy was made. However, as shown in Listing 4.27, the same line `let deref_copy = *ref1` caused an error, due to a move of ownership.

There are two potential issues with moving or changing ownership in the line `let deref_copy = *ref1` in Listing 4.27. First, `ref1` is a reference and does not own the data, so it isn't appropriate to use `ref1` to transfer ownership. Second, moving a value out of a mutable reference could leave the reference in an invalid state. To fix this error, use the `clone()` method, as shown in Listing 4.28.

```
fn main() {
    let mut heap_data = vec![5, 6, 7];
    let ref1 = &mut heap_data;
    let deref_copy = ref1.clone();
}
```

Listing 4.28 Making a Copy of Heap-Allocated Data behind a Reference Using a Clone

Calling `clone` on a reference, whether mutable or immutable, creates an owned copy of the value to which the reference points.

4.6 Mutable and Immutable Binding of References

The references we've created so far have been immutable. For instance, consider the code shown in Listing 4.29.

```
fn main() {
    let mut vec_1 = vec![1, 2, 3];
    let mut vec_2 = vec![4, 5, 6];
    let reference = &vec_1;
    reference = &vec_2; // Error
}
```

Listing 4.29 Immutable Binding of a Reference

In this case, the variable reference itself is immutable, and we cannot update it to point to some other vector. Updating the reference to point to vec_2, therefore, throws an error. This error arises because we have an immutable binding of an immutable reference. To enable mutation of a reference itself, you can make the reference mutable, as shown in Listing 4.30.

```
fn main() {
    let mut vec_1 = vec![1, 2, 3];
    let mut vec_2 = vec![4, 5, 6];
    let mut reference = &vec_1;
    reference = &vec_2;
}
```

Listing 4.30 Making reference Mutable to Enable Its Mutation

The code now works because reference is mutable, and we can therefore update it to point to some other vector. In the same way, you can have a mutable reference to vec_1, as shown in Listing 4.31.

```
fn main() {
    let mut vec_1 = vec![1, 2, 3];
    let mut vec_2 = vec![4, 5, 6];
    let reference = &mut vec_1;
    reference.push(10);
    reference = &mut vec_2; // Error
}
```

Listing 4.31 Immutable Binding of a Mutable Reference

In this case, reference is an immutable binding of a mutable reference. You can use the reference to update the vector; however, you cannot update it to point to some other vector. As in the previous case, we can fix the error by making the reference mutable, as shown in Listing 4.32.

```
fn main() {
    let mut vec_1 = vec![1, 2, 3];
    let mut vec_2 = vec![4, 5, 6];
    let mut reference = &mut vec_1;
    reference.push(10);
    reference = &mut vec_2;
}
```

Listing 4.32 A Mutable Binding of a Mutable Reference

In addition to immutable and mutable references, Rust allows a mutable reference to an immutable reference, as shown in Listing 4.33.

```
fn main() {
    let mut vec_1 = vec![1, 2, 3];
    let mut vec_2 = vec![4, 5, 6];
    let reference = &mut &vec_1;
    *reference = &mut &vec_2;
}
```

Listing 4.33 An Immutable Binding of a Mutable Reference to an Immutable Reference

The syntax `&mut &vec_1` may seem complex at first glance, but this topic will become clearer with an explanation. First, it's important to understand that this type of reference is different from both simple immutable and mutable references. The type accurately conveys its semantics: It's a mutable reference (indicated by `&mut` at the start), meaning you can modify what it points to. However, the type it points to is an immutable reference, i.e., `&vec`, which means the data itself cannot be changed, but the reference can be redirected to point to a different immutable reference.

In simple terms, this allows you to mutate the reference itself to point to a different vector (as done in the next line of `*reference = &mut &vec_2`), even though the reference is to an immutable type. Notice that, even though the reference variable is not mutable, you can still mutate the reference it holds. This behavior is not possible with other types of references. However, if we make the reference shown in Listing 4.33 mutable, we get the updated code shown in Listing 4.34.

```
fn main() {
    let mut vec_1 = vec![1, 2, 3];
    let mut vec_2 = vec![4, 5, 6];
    let mut reference = &mut &vec_1;
    reference = &mut &vec_2;
}
```

Listing 4.34 A Mutable Binding of a Mutable Reference to an Immutable Reference

In this case, we can mutate the reference without dereferencing through this mutable binding between a mutable reference and an immutable reference.

To summarize this discussion, six types of references exist, as shown in Table 4.1.

Reference Type	Meaning
`ref: &T`	Immutable binding of an immutable reference
`mut ref: &T`	Mutable binding of an immutable reference
`ref: &mut T`	Immutable binding of a mutable reference
`mut ref: &mut T`	Mutable binding of a mutable reference

Table 4.1 Types of References, with T Representing a Generic Type

Reference Type	Meaning
`ref: &mut &T;`	Immutable binding of a mutable reference to an immutable reference
`mut ref: &mut &T`	Mutable binding of a mutable reference to an immutable reference

Table 4.1 Types of References, with T Representing a Generic Type (Cont.)

The first four types of references are generally sufficient for routine programming. The last two types are used rarely but are good to know about.

Converting a Mutable Reference to an Immutable Reference

As a final note, in some situations, you might start with a mutable reference to modify data, but later, you only need to read the data without further modifications. In such cases, converting a mutable reference to an immutable one allows you to safely pass the reference around without worrying about violating Rust's borrowing rules. Listing 4.35 shows how to convert a mutable reference to an immutable reference.

```
fn main() {
    let mut x = 45;
    let z = &mut x;
    let y = &*z;
}
```

Listing 4.35 Conversion of a Mutable Reference to an Immutable Reference

In the code, `z` is a mutable reference to `x`. In the next line, the mutable reference `z` is first dereferenced using the `*z`, which gives access to the actual value, in this case, the value `45`. Next, we created a reference to the value using the syntax `&*z`. Thus, `&*z` is now an immutable reference to `x`, which is stored in variable `y`. This conversion allows `y` to safely access the value without permitting further modifications.

4.7 Practice Exercises

In this section, we'll provide some practice questions to help reinforce your learning. The exercise questions cover almost all the concepts introduced in the chapter. Solutions to each exercise will follow in Section 4.8.

1. **Fix the compilation error**
   Consider the following code snippet:

```
fn main() {
    let s1: String = String::from("this is me, ");
    let s2: &str = "Nouman";
    some_function(s1, s2); // Something is wrong here
```

```
    println!("{} {}", s1, s2);
}
fn some_function(a1: String, a2: &str) {
    println!("{} {}", a1, a2);
}
```

Identify and correct the issue causing the compilation error. Ensure that the program compiles and runs successfully while maintaining proper ownership and borrowing rules.

2. **Ownership in a loop**
   Review the following code:

```
fn main() {
    let mut my_vec = vec![1, 2, 3, 4, 5];
    let mut temp;
    while !my_vec.is_empty() {
        temp = my_vec; // Something wrong on this line
        println!("Elements in temporary vector are: {:?}", temp);
        if let Some(last_element) = my_vec.pop() { // pop() is used to remove
                                                    // an element from the vec
            println!("Popped element: {}", last_element);
        }
    }
}
```

   Fix the code to properly handle ownership within the loop. Explain how your solution addresses the issue with ownership and data transfer in this context.

3. **Correcting ownership transfer**
   Analyze the following code snippet:

```
fn main() {
    {
        let str1 = generate_string();
    }
    let str2 = str1;    // Something wrong with this line
}
fn generate_string() -> String {
    let some_string = String::from("I will generate a string");
    some_string
}
```

   Identify the issue with ownership and fix the code so that it compiles correctly. Ensure that the program correctly handles the return value and ownership of the string.

4. **Fix the borrowing issue**

   Examine the following code:

```
fn main() {
    let mut some_vec = vec![1, 2, 3];
    let first = get_first_element(&some_vec);
    some_vec.push(4);
    println!("The first number is: {}", first);
}
fn get_first_element(num_vec: &Vec<i32>) -> &i32 {
    &num_vec[0]
}
```

   Identify and correct the issue causing a borrowing problem. Ensure that the program correctly handles borrowing and mutable access to the vector.

5. **Correcting reference assignment**

   Review the following code snippet:

```
fn main() {
    let mut vec_1 = vec![1, 2, 3];
    let vec_2 = vec![4, 5, 6];
    let mut vec_ptr: &Vec<i32>;
    vec_ptr = vec_1;
    println!("vec ptr is pointing to vec_1");
    vec_ptr = vec_2;
    println!("vec ptr is updated and now pointing to vec_2");
}
```

   Fix the code to ensure it compiles correctly, considering the proper handling of references and their assignment.

6. **Resolve the mutable reference conflict**

   Analyze the following code:

```
fn main() {
    let first_num = 42;
    let second_num = 64;
    let ref1 = &mut first_num;
    let mut ref2 = &mut second_num;
    *ref1 = 15;
    *ref2 = 10;
    ref2 = ref1;
    println!("Updated first number: {ref2}");
}
```

   Identify and correct the issue related to mutable references and their assignments. Ensure the code compiles and correctly handles the mutable references.

4.8 Solutions

This section provides the code solutions for the practice exercises in Section 4.7. The code is largely self-explanatory; however, we have included comments and additional explanations wherever necessary to enhance understanding.

1. **Fix the compilation error**

```
fn main() {
    let s1: String = String::from("this is me, ");
    let s2: &str = "Nouman";
    some_function(&s1, s2);
    println!("{} {}", s1, s2);
}

fn some_function(a1: &String, a2: &str) { // update the first input to a
                                           // reference String
    println!("{} {}", a1, a2);
}
```

2. **Ownership in a loop**

```
fn main() {
    let mut my_vec = vec![1, 2, 3, 4, 5];
    let mut temp;

    while !my_vec.is_empty() {
        temp = my_vec.clone(); /* during the first iteration, the transfer of
                                  ownership occurs from my_vec to temp, which
                                  makes it impossible to access the variable
                                  my_vec in subsequent iterations */
        println!("Elements in temporary vector are: {:?}", temp);
        if let Some(last_element) = my_vec.pop() {
            println!("Popped element: {}", last_element);
        }
    }
}
```

3. **Correcting ownership transfer**

```
fn main() {
    let str1 = {
        let str1 = generate_string();
        str1
    };
```

```
    // An alternate solution would be to move the statement
    // let str1 = generate_string(); out of the scope

    let str2 = str1;   // Something wrong with this line
}

fn generate_string() -> String {
    let some_string = String::from("I will generate a string");
    some_string
}
```

4. **Fix the borrowing issue**

```
fn main() {
    let mut some_vec = vec![1, 2, 3];
    let first = get_first_element(&some_vec);
    //some_vec.push(4); // cannot borrow `some_vec` as mutable because it is
                        // also borrowed as immutable
    println!("The first number is: {}", first);
    some_vec.push(4);

    /*
    The problem with borrowing arises when we attempt to modify the some_vec
    vector after obtaining an immutable reference to its first element.
    This violates Rust's borrowing rules, according to which we cannot modify
    a variable if immutable references to it are still in scope.

    This rule ensures the safety and integrity of data in Rust,
    preventing potential conflicts and data races.
    */
}

fn get_first_element(num_vec: &Vec<i32>) -> &i32 {
    &num_vec[0]
}
```

5. **Correcting reference assignment**

```
fn main() {
    let mut vec_1 = vec![1, 2, 3];
    let vec_2 = vec![4, 5, 6];
    let mut vec_ptr: &Vec<i32>;
    vec_ptr = &vec_1; // The type of vec_ptr is a reference to a vector so we
                      // should borrow and not take ownership.
    println!("vec ptr is pointing to vec_1");
```

```
    vec_ptr = &vec_2; // We need to borrow using a reference and not take
                      // ownership.
    println!("vec ptr is updated and now pointing to vec_2");
}
```

6. **Resolve the mutable reference conflict**

```
fn main() {
    let mut first_num = 42;          // We are using mutable references to it
                                     // so the variable must be mutable
    let mut second_num = 64;         // We are using mutable references to it
                                     // so the variable must be mutable
    let ref1 = &mut first_num;
    let mut ref2 = &mut second_num;  // A mutable references means that the
                                     // reference can be updated to point
                                     // to some other variable


    *ref1 = 15;
    *ref2 = 10;
    ref2 = ref1;
    println!("Updated first number: {ref2}");
}
```

4.9 Summary

In this chapter, we provided a comprehensive overview of Rust's ownership model, which is fundamental to ensuring memory safety at compile time. We started by examining the basic rules of ownership, including the critical distinction between moving and copying data, and exploring how ownership is managed within functions. Our discussion included functions that take ownership, return ownership, and transfer ownership temporarily as well as how to handle stack-only data types. We then delved into borrowing, explaining its rules and how it allows safe, non-owning access to data. This chapter also covered dereferencing, which connects ownership and borrowing by allowing interactions with underlying data. Additionally, we explored mutable and immutable bindings of references. Mastering these concepts equips you with the skills to write efficient and reliable Rust code.

In the next chapter, we'll continue with custom types, like structs and enums, and explore some important library types such as `Option` and `Result`.

Chapter 5

Custom and Library-Provided Useful Types

5

In Rust, types do more than define data; they shape how you think about and manipulate data. From custom structures to powerful library types like Option, Result, and collections, this chapter explores the tools that bring safety and expressiveness to your code.

In this chapter, we'll explore how to define custom types with structs and enums, adding functionality to them to create robust data structures. The chapter also introduces important library types like `Option` and `Result`, which are essential for gracefully handling null values and managing errors. With `HashMaps`, you'll learn how to store and retrieve data efficiently using key-value pairs. Understanding these types enables you to build more complex and more functional applications.

5.1 Structs

In Rust, a structure (or struct) is a custom data type that allows you to group together related data under one name. Each piece of data within the structure is called a *field*, and each field can have its own type. With structs, you can create complex data types that model real-world entities, thus making your code more organized and easier to understand.

In the following sections, we cover the definition of structs, the instantiation of struct instances, different types of structs, how to add functionality to structs, and more.

5.1.1 Defining Structs

Let's create a `Car` struct to store car-related information in one place. Each car has an owner, make, model, year, fuel level, and price. The code shown in Listing 5.1 creates such a structure.

```
struct Car {
    owner: String,
    year: u32,
```

```
    fuel_level: f32,
    price: u32,
}
```

Listing 5.1 A Car Struct

You create structures using the struct keyword followed by the name of the struct, which in this case is Car. Inside the curly braces, we add the list of fields. Each field has a name and a type. The first field in this struct is owner, with the type String. The second field is year_of_make, with the type u32. Next, we have fuel_level, with the type f32, and finally, we have price, with the type u32.

5.1.2 Instantiating Struct Instances

When instantiating struct instances in Rust, several options are available to create and initialize them. You can provide values for all fields at once, use shorthand syntax when variable names match field names, or even leverage Rust's update syntax to create new instances by copying fields from an existing one.

Let's start by looking at the fundamental way of creating a struct instance, that is, by providing values for all fields at once. Consider the code shown in Listing 5.2.

```
struct Car {
    owner: String,
    year: u32,
    fuel_level: f32,
    price: u32,
}
fn main() {
    let mut my_car = Car {
        owner: String::from("ABC"),
        year: 2010,
        fuel_level: 0.0,
        price: 5_000,
    };
}
```

Listing 5.2 Creating a Struct Instance by Providing Values for All Fields

Each struct defined in the code is treated as a type in Rust. The variable my_car is of type Car. The individual fields are instantiated after the name of the field followed by a colon. For instance, the field year is initialized from the value of 2010. While creating an instance of the struct, all the fields must be initialized from some values. If we never mention the price field in the code shown in Listing 5.2, the compiler will throw an error, "missing structure fields: price."

To access a specific field within a struct, you can use the dot notation, as in the following example:

```
let car_year = my_car.year;
```

In this statement, the year field has been assigned to the variable car_year. The same syntax (i.e., the dot notation) can be used to mutate the value of a struct field, as in the following example:


```
my_car.fuel_level = 30.0;
```

The same statement would have thrown error if my_car (shown in Listing 5.2) was not mutable. This error arises because, when a variable representing an instance of the struct is immutable, we cannot mutate its fields.

Rust also provides a convenient way to create a new struct instance by reusing most of the fields from an existing instance. This reuse is achieved using the struct update syntax, which allows you to specify values for some fields while copying the remaining fields from another instance. Let's look at an example of this feature:

```
let another_car = Car {
    owner: "new_name".to_string(),
    ..my_car
};
```

The variable is now another_car instance with a new owner. The ..my_car syntax copies the remaining fields (year, fuel_level, and price) from the existing my_car instance, as defined in Listing 5.2.

Rust offers a shorthand syntax for initializing struct fields when the variable names match the field names. This syntax can make your code more concise and thus easier to read, as shown in Listing 5.3.

```
fn main() {
    let owner = String::from("John Doe");
    let year = 2021;
    let fuel_level = 45.5;
    let price = 10_000;

    let my_car = Car {
        owner,
        year,
        fuel_level,
        price,
    };
}
```

Listing 5.3 Shorthand Syntax for Struct Initialization When a Variable Name Matches a Field Name

The field names in the struct (i.e., owner, year, fuel_level, and price) match the names of the variables; therefore, instead of writing owner: owner, year: year, and so on, you can simply write owner, year, fuel_level, and price. Rust automatically assumes that the value for each field should come from the variable with the matching name.

5.1.3 Ownership Considerations

Rust's ownership rules apply to structs as well. For example, if a field is of a heap-allocated type like String, assigning it to another variable transfers ownership, making the original field inaccessible, which is known as a *partial* move. Let's see what happens when we add the following lines to Listing 5.2:

```
let extracted_owner = my_car.owner;
println!("Owner is: {}", my_car.owner); // Error
```

The compiler will generate an error on the print line, namely, "borrow of moved value: my_car.owner." To avoid losing access to the original field, you can clone the value in the following way:

```
let extracted_owner = my_car.owner.clone();
println!("Owner is: {}", my_car.owner);
```

While assigning a field of the struct, you must pay attention to heap-allocated types because their ownership may change.

5.1.4 Tuple Structs

Covered earlier in Chapter 2, Section 2.2.2, like structs, tuples can group data of different types together. Tuple structs are especially useful in scenarios where you need a lightweight, temporary grouping of values without needing to define a full struct. Consider the example shown in Listing 5.4.

```
let point1 = (1, 3);
let point2 = (4, 10, 13);
```

Listing 5.4 2D and 3D Points

point1 is tuple with two integers, and point2 is a tuple with three integers. However, the meaning of these numbers is not exactly clear. Let's change the names of these variables, as in the following example:

```
let point_2D = (1, 3);
let point_3D = (4, 10, 13);
```

The first tuple is a 2D point with the numbers inside the tuple, representing the point's coordinate values. Likewise, the second tuple is a 3D point with the numbers representing that point's coordinate values.

Although renaming variables is beneficial, some issues still exist. First, when passing these values to a function, the implementer of the function has the freedom to alter the variable names, which could lead to ambiguities. Second, there are no limitations on the number of elements or on the types of elements within a tuple. Tuple structs can help overcome these issues, as in the following example:

```
struct Point_2D(i32, i32);
struct Point_3D(i32, i32, i32);
```

Notice that in the field names we use parentheses instead of curly braces to define its body. Unlike regular structs, which require you to define named fields, tuples allow you to create anonymous groups of data. Rust will now ensure that our type has the correct values; that is, we cannot use a string as an element for type `Point_2D`. Moreover, we cannot add extra elements. For instance, attempting to create an instance of `Point_2D` with a string field or containing extra elements will lead to errors, as in the following examples:

```
let point1 = Point_2D(1, "2");  // Error
let point2 = Point_2D(1, 2, 3); // Error
```

5.1.5 Unit Structs

After covering the standard struct (a struct with field names) and tuple struct (a struct with no field names), the third type of struct is called a *unit struct*. They just have a name with no fields, as follows:

```
struct ABC;
```

The common use case for unit structs is to use a marker or flag to convey some information or behavior without needing to store any data. We'll come back to this topic in Chapter 13, Section 13.5.3.

5.1.6 Adding Functionality to Structs

Suppose we want to add the ability to print car information. Let's add a function, called `display_car_info`, to the code shown earlier in Listing 5.2. The updated code is shown in Listing 5.5.

```
struct Car {
    owner: String,
    year: u32,
    fuel_level: f32,
    price: u32,
}
```

```
fn display_car_info(car: &car) {
    println!(
        "owner: {}, Year: {}, Price: {}",
        car.owner, car.year, car.price);
}
fn main() {
    let mut my_car = Car {
       owner: String::from("ABC"),
       year: 2010,
       fuel_level: 0.0,
       price: 5_000,
    };
}
```

Listing 5.5 Code from Listing 5.2 with the Added Function display_car_info

The function prints the car information passed in as a reference. The function is not supposed to change the car's details; therefore, we'll pass in an immutable reference.

This code compiles but could be improved. The `display_car_info` function is completely separate from the type `Car`, having no dependency from a code perspective. Ideally, the `display_car_info` function should be defined on the `Car` type itself, instead of being defined independently. Rust provides *implementation blocks*, which solve exactly this problem. With implementation blocks, you can organize the functionalities that are defined on the type. Consider the example shown in Listing 5.6.

```
...
impl Car {
    fn display_car_info(&self) { // car: &Car changed to &self
        println!(
            "owner: {}, Year: {}, Price: {}",
            self.owner, self.year, self.price);
        }
    }
...
}
```

Listing 5.6 Code from Listing 5.5 Updated by Introducing an impl Block

The implementation block is created using the `impl` keyword, followed by the name of the type (in this case, the `Car` type) and then followed by curly braces. Notice that, instead of taking `Car` as an argument, the function now takes a reference to `self`. Inside an `impl` block, `self` has a special meaning: it refers to an instance of the implementing type (the type `Car` in this case) on which this function was being called.

Creating Methods

When a function inside an impl block uses any form of self (a reference or an owned type) as a parameter, that function is called a *method*. For a function to be considered a method, it must meet two requirements:

- First, the function must be located inside the implementation block of the type on which it is defined.
- Second, the first parameter must be self.

The following code adapted from Listing 5.6 defines the general syntax for creating a method corresponding to ABC:

```
struct ABC {}
impl ABC {
    fn method(&self) {}
    fn method_2(&mut self) {}
}
```

The call to a method uses a slightly different syntax compared to a call to a function:

```
my_car.display_car_info();
```

You may have noticed that we are not passing any arguments when the method expects self as a parameter. This is because, when calling methods on an instance of a type, the instance is automatically passed as the self parameter, so you don't need to explicitly provide the instance as a parameter in the method call. As a result, we no longer call display_car_info independently, and an error will arise when we try to call the method like a regular function:

```
display_car_info(&my_car); // Error
```

A method can take one of three forms of self. The first form is an immutable reference to self, as shown in Listing 5.6. This option is useful when you are only reading the data of self, not modifying it.

The second form is a mutable reference to self. Let's add one more method to the impl block, one that will refuel this car, as shown in Listing 5.7.

```
impl Car {
    fn display_car_info(&self) { // First form: &self
        println!(
            "owner: {}, Year: {}, Price: {}",
                self.owner, self.year, self.price);
        }
        fn refuel(&mut self, gallons: f32) { // Second form: &mut self
            self.fuel_level += gallons;
```

```
        }
}
```

Listing 5.7 Method refuel Added to the impl Block

The method `refuel` takes a mutable reference to `self` and updates the `fuel_level` based on the `gallons` passed in.

The last form of `self` a method can take is the *owned form* of `self`. Let's add one more method of `sell` to the `impl` block, as shown in Listing 5.8.

```
impl Car {
    fn display_car_info(&self) { // First form: &self
        println!(
            "owner: {}, Year: {}, Price: {}",
            self.owner, self.year, self.price);
        }
        fn refuel(&mut self, gallons: f32) {  // Second form: &mut self
            self.fuel_level += gallons;
        }
        fn sell(self) -> Self { // Third form: self
            self
        }
}
```

Listing 5.8 One More Method of sell Added to the impl Block

In an implementation block, `Self` with a capital "S" refers to the type for which the method is defined, in this case, the `Car` type. We could have used the type name as the returning type (i.e., `Car`). However, using `Self` is preferred and more idiomatic. This `sell` method will take ownership from the caller of the method and will return it to the new instance of the car. For instance, consider the following line in `main`:

```
let new_owner = my_car.sell();
```

This line will transfer the ownership from `my_car` to that of the `new_owner`. Since the method needs ownership, the input parameter is `self`, without a reference (`&`). The method will return ownership back, which is assigned a new instance of `Car` type. This type of method is typically used when you want to convert one type into another while ensuring the original instance remains inaccessible to the caller. For instance, the code in `main` shown in Listing 5.9 will generate an error.

```
fn main() {
    let mut my_car = Car {
        owner: String::from("ABC"),
        year: 2010,
        fuel_level: 0.0,
```

```
        price: 5_000,
    };
    let new_owner = my_car.sell();
    my_car.refuel(10.5); // Error
}
```

Listing 5.9 Trying to Access my_car after Selling

The compiler throws an error of "borrow of moved value." The sell method has already transferred ownership; after the car is sold, we don't want anyone to mess with this data in the future.

5.1.7 Associated Functions

Associated functions, also referred to as "static functions" in other languages, are tied to the type itself but do not operate on instances of that type. In other words, these methods are defined within the implementation block of a type but do not operate on or use an instance of the type. Nevertheless, they still maintain a relationship with the type. Consider the code shown in Listing 5.10.

```
impl Car {
    fn monthly_insurance () -> u32 {
        123
    }
...
}
```

Listing 5.10 Code from Listing 5.8 Updated with an Associated Function monthly_insurance

The associated function monthly_insurance simply returns fixed costs of insurance that you need to pay monthly. Notice that the associated function does not take self as an input parameter and is not called using the dot syntax.

Let's add one more method, called selling_price, to the impl block shown in Listing 5.11.

```
impl Car {
    fn monthly_insurance () -> u32 {
        123
    }
    fn selling_price(&self) -> u32 {
        self.price + Car::monthly_insurance() // calling associated functions
    }
...
}
```

Listing 5.11 Method selling_price Added to the impl Block from Listing 5.10

The double colon syntax (::) after the type calls associated functions.

A frequently observed pattern in Rust is for types to include an associated function, named new, that serves as a *constructor function*. Consider the code shown in Listing 5.12.

```
impl Car {
    fn new(name: String, year: u32) -> Self {
        Self {
            owner: name,
            year: year,
            fuel_level: 0.0,
            price: 0,
        }
    }
    ...
}
```

Listing 5.12 A New Constructor Function Added to Listing 5.11

The new function creates a new instance of the car with the owner and year values initialized from the values passed in, while the remaining values are set to some defaults. The returning type of Self (with a capital "S") indicates the Car type.

Constructor Functions

A *constructor function* in Rust is a special method for creating and initializing instances of a type. This kind of method typically returns a new instance of the type, often using the Self keyword to refer to the type being constructed, and is commonly defined within an impl block. Constructors help ensure proper initialization of a type by setting up its fields with appropriate values. We'll cover constructor functions in more detail in Chapter 12, Section 12.1.

You can create new instances more efficiently with less code in main, as shown in Listing 5.13.

```
fn main() {
    // Instead of using syntax below
    let mut my_car = Car {
        owner: String::from("ABC"),
        year: 2010,
        fuel_level: 0.0,
        price: 5_000,
    };
```

```
    // easier way to create new instances
    let new_car = Car::new("ABC".to_string(), 2010);
}
```

Listing 5.13 Creating Instances with a New Constructor Function

Instead of manually initializing each field, constructor functions help you initialize an instance with a single line of code. By calling the `new` constructor function in the code with the relevant inputs, the fields of the struct are automatically initialized.

5.2 Enums

Building upon the previous section on structs, let's now turn our attention to enums, another powerful feature in Rust for defining types that can represent one of several possible variants. In this section, we'll explore why enums are useful, show you how to define and enhance them with implementation blocks, and then consider adding data to enum variants to create even more flexible and expressive types.

5.2.1 Why Use Enums?

Consider a situation where you must define a mutable variable to store day-of-the-week information. You could simply create a variable and assign it the desired day as a string, as follows:

```
let mut day = "Monday".to_string();
```

While this approach works, it's far from ideal. The set of weekdays is limited, and you don't want to allow random strings. Additionally, there's a risk of misspelling a day name, leading to errors.

One potential solution might be to declare a vector of strings, with each string representing a day of the week, as shown in Listing 5.14.

```
let week_day = vec![
    "Monday".to_string(),
    "Tuesday".to_string(),
    "Wednesday".to_string(),
    "Thursday".to_string(),
    "Friday".to_string(),
    "Saturday".to_string(),
    "Sunday".to_string(),
];
day = week_day[1]
```

Listing 5.14 Creating an Array for String Day Information

You could then set the day variable to a specific day using an index. However, this approach has its own issues. For instance, attempting to move a value out of a vector, will result in an error. This is because, unlike structs, vectors do not support the partial move of elements. Although you could use the clone function to work around this issue, there are remaining unnecessary complexity, such as requiring you to memorize index values and their corresponding days.

5.2.2 Defining Enums

A much better solution is to use enums to define a type by enumerating all its possible variants, as shown in Listing 5.15, through the `enum` keyword.

```
enum WeekDay {
    Monday,
    Tuesday,
    Wednesday,
    Thursday,
    Friday,
    Saturday,
    Sunday,
}
```

Listing 5.15 A WeekDay Enum with Variants of Days

To define an enum, use the `enum` keyword, followed by a name for the enum. In our example, the days are called the *variants* of the enum. Enums are somewhat similar to structs. A key difference, however, is that structs associate a type with each field, while enums do not define a type for their variants. Instead, each variant is simply a member of the enum.

In the `main` function, now create a variable and set it as equal to the desired day variant:

```
fn main() {
    let day = WeekDay::Saturday;
    let day = WeekDay::Satrday; // Error
}
```

To create an instance of the enum, we must specify one of its variants using the double colon syntax (`::`). Using the enum, misspelling the day of the week cannot happen now. For instance, the second line of code will throw an error. Additionally, with enums, you don't need to remember the indexes that correspond to each day.

5.2.3 Implementation Blocks for Enums

Let's consider a scenario where a company has organized an event with many participants. Some participants traveled by car, some by train, and others by airplane. The company plans to reimburse travel expenses based on the mode of transportation. For this task, the company first must determine the participant's travel type and then calculate the allowance based on that type.

We can model this situation using an enum called `TravelType`, with three variants of `Car`, `Train`, and `Airplane`, as shown in Listing 5.16.

```
enum TravelType {
    Car,
    Train,
    Airplane,
}
```

Listing 5.16 Enum Determining the Travel Type

You can group functionality related to an enum using `impl` blocks, just like with structs. Let's add a method called `travel_allowance` to the `TravelType` enum. The updated code is shown in Listing 5.17.

```
impl TravelType {
    fn travel_allowance(&self, miles: f32) -> f32 {
        let allowance = match self {
            TravelType::Car => miles * 2.0,
            TravelType::Train => miles * 3.0,
            TravelType::Airplane => miles * 5.0,
        };
        allowance
    }
}
```

Listing 5.17 Method travel_allowance Added to the impl Block of Enum

This method takes a reference to `self` and the number of `miles` covered as parameters. The output is the computed allowance based on the travel type, which is comprehensively checked inside the `match` arms. We can now use the enum and the variant in the `main` function, as shown in Listing 5.18.

```
fn main() {
    let my_trip = TravelType::Car;
    let allowance = my_trip.travel_allowance(60.0);
    println!("Travel allowance: ${}", allowance);
}
```

Listing 5.18 Enum and the Method travel_allowance Used in main

Running this code will output the travel allowance based on the distance traveled. In this case, the result will be 120.

5.2.4 Adding Data to Enum Variants

Enums in Rust are powerful and can have data associated with them. Instead of passing the number of miles as a separate argument, for instance, you can associate this data directly with the enum variants, as shown in Listing 5.19.

```
enum TravelType {
    Car(f32),
    Train(f32),
    Airplane(f32),
}
```

Listing 5.19 TravelType Enum with Three Variants

Now, when declaring an instance of TravelType, you can specify the miles as associated data held by the variant in the following way:

```
fn main() {
    let my_trip = TravelType::Car(60.0);
}
```

You also need to update the travel_allowance method to work with this new type of enum. The updated code is shown in Listing 5.20.

```
impl TravelType {
    fn travel_allowance(&self) -> f32 { // parameter miles: f32 is
                                        // no longer required
        let allowance = match self {
            TravelType::Car(miles) => miles * 2.0,
            TravelType::Train(miles) => miles * 3.0,
            TravelType::Airplane(miles) => miles * 5.0,
        }
       allowance
    }
}
```

Listing 5.20 Method travel_allowance from Listing 5.17 Updated

You don't need to explicitly provide the miles information to the travel_allowance method. The miles information is already contained by the enum variants passed into the method. Notice that, in the match arms, we mentioned a variable in parentheses, which holds the associated data with the variants.

In main, we can now call the method with no inputs, as shown in Listing 5.21.

```
fn main() {
    let my_trip = TravelType::Car(60.0);
    let allowance = my_trip.travel_allowance(); // no inputs
    println!("Travel allowance: ${}", allowance);
}
```


Listing 5.21 Unlike Listing 5.18, This Method Is Called with No Inputs

5.3 Option

Following our discussion on enums, we now shift our focus to the Option type, a powerful enum provided by Rust that represents the possibility of a value being present or absent. In this section, we'll explore why the Option type is useful, examine its definition, and dive into pattern matching with Option. We'll also show you how to use if let for more concise handling of optional values.

5.3.1 Why Use Option?

Let's imagine we are implementing the code shown in Listing 5.22.

```
struct Student {
    name: String,
    grade: u32,
}
fn main() {
    let student_db = vec![
        Student {
            name: String::from("Alice"),
            grade: Some(95),
        },
        Student {
            name: String::from("Bob"),
            grade: Some(87),
        },
    ];
}
```

Listing 5.22 Program for Simulating a Student Database

In main, we are simulating a student database called student_db, containing several student records. Each student in the database is an instance of the Student struct, with two fields: name and grade. To add a student to the database, both the name and grade fields must be provided, but perhaps grades are not available yet, as they may not have been finalized by the teacher. The current implementation, however, does not allow for a missing or empty grade field. If we add an instance with no grades, the compiler will complain since Rust does not allow empty struct fields. We could set it equal to some default value, let's say, a value of 0, as shown in Listing 5.23.

```
fn main() {
    let student_db = vec![
        ...
        Student {
            name: string::from("Charlie"),
        grade: 0,
},
    ];
}
```

Listing 5.23 Student with an Empty Grade Field

This code works but is not efficient. Someone may interpret this value as a valid grade.

What we ideally want in this situation is to make it clear that there is no value. In some languages, we might use something like `null` in this case. However, Rust does not define `null`; the compiler cannot interpret `null` because there is no such concept in Rust. Fortunately, Rust provides the `Option` enum to handle such cases.

5.3.2 Defining the Option Enum

The `Option` enum is defined in the standard library as in the following example:

```
enum Option<T> {
    None,
    Some(T),
}
```

The `Option` enum has two variants of `Some`, which holds a generic value, and `None`, which signifies the absence of a value. You'll learn more about generics in Chapter 8, Section 8.1.

Let's change the type of the grade from `u32` to `Option<u32>`, as shown in Listing 5.24.

```
struct Student {
    name: String,
    grade: Option<u32>,
}
```

Listing 5.24 Updated Student Definition with grade Field Changed from u32 to Option<u32>

Note

One important aspect to note is that the Option enum is extensively used in Rust programs and is included in the Rust prelude. The prelude is a collection of items that Rust automatically imports into every Rust program. Thus, you can use Option without needing to write any additional import statements.

The student_db vector created earlier in Listing 5.22, along with the student having no grades, will now be redefined based on the new definition of student from Listing 5.24, as shown in Listing 5.25.

```
fn main() {
    let student_db = vec![
        Student {
            name: String::from("Alice"),
            grade: Some(95),
        },
        Student {
            name: String::from("Bob"),
            grade: Some(87),
        },
        Student {
            name: String::from("Charlie"),
            grade: None,
        },
    ];
}
```

Listing 5.25 Code from Listing 5.22 Updated Based on New Student Definition in Listing 5.24

Notice that when a student has grades, that information is wrapped inside the Some variant, and when there is no grade, we represent this lack using the None variant.

5.3.3 Matching on Option

Using match with the Option enum allows you to explicitly handle both possible cases: Some for when a value is present and None for when it is absent. This capability ensures exhaustive pattern matching, helping you write safe and clear code when working with optional values.

Let's add some functionality to the code shown in Listing 5.25. We'll add a function that, when given a student's name, will search for that name in the student database and will return his grade. The code is shown in Listing 5.26.

```
fn get_grade(student_name: &String, student_db: &Vec<Student>) -> Option<u32> {
    for student in student_db {
        if student.name == *student_name {
            return student.grade;
        }
    }
    None
}
```

Listing 5.26 Checking Whether a Student Is in the Database and Returning Their Grade

The input to the function is a student_name and student_db, and the returning value is an optional u32 value representing the respective grade. Inside the function, we iterate through the student_db and check if the student's name matches the provided student_name parameter. Since the student_name is behind a reference, we use the dereference character (*) on the if line statement. When a match is found, the function exits the loop, returning the respective student grade. In there is no match, the function returns None.

Let's use this function in the main function:

```
let student_name = String::from("Bob");
let student_grade = get_grade(&student_name, &student_db);
```

If you check the type of the variable student_grade in Visual Studio Code (VS Code), notice that the variable is not simply a u32 type, but instead, it is an option<u32> type. To properly handle the value, we'll check if it has some valid value or no value. A match expression is typically used for this purpose.

```
match student_grade {
     Some(grade) => println!("Grade is: {grade}"),
     None => {}
}
```

We matched on the student_grade, which contains the returning value from the get_grade and is of type Option<u32>. The two variants of the option are handled using the match arm. Moreover, in the case of Some variant, we bind the value inside the Some variant to the variable grade.

5.3.4 Use of If Let with Option

The previous code for the match performs as intended but can be improved. The arm related to None seems irrelevant since, in this scenario, we are only concerned about a specific grade. When you're interested in handling just one variant, while disregarding all others, you can utilize the if let syntax in the following way:

```
if let Some(grade) = student_grade {
    println!("Grade is: {grade}");
}
```

The if let syntax is a convenient way to match a single pattern and execute code based on that pattern while ignoring other possibilities. Instead of using a Boolean condition, you can define a variable with the let keyword and use a pattern, such as the Some variant, to match against. If student_grade matches the Some pattern, the value within the Some variant is assigned to the grade variable, which can then be used within the if statement. The if let syntax is more concise than a match expression and is best used when you only care about a single variant and want to ignore the others.

5.4 Result

Just like an Option enum, which provides the ability to handle situations where you might have some value or no value, Rust has a built-in enum called Result. This enum handles situations where an operation can be successful, thereby returning a valid value, or may fail, thereby resulting in an error.

In the following sections, we'll explain the motivation behind Result, examine the results of matching, and explore the related ? operator.

5.4.1 Why Use Result?

Consider the function shown in Listing 5.26 once more. This function lacks a clear mechanism to check if a student exists in the database. Currently, if no student name matches, the function implicitly assumes the student doesn't exist and returns None. However, within the function's context, None is interpreted as the absence of a grade and not as the absence of a student. Ideally, checking for a grade for a student whose record we don't have should be more correctly considered as an error. A more logical approach, therefore, would be to first check whether the student's name exists in the database. If it does, the function can proceed to retrieve the student's grade. Otherwise, the function should return an error. To achieve this functionality, we can use the Result enum.

5.4.2 Defining the Result Enum

Let's look at how this enum is defined in the standard library:

```
enum Result<T, E> {
    Ok(T),
    Err(E),
}
```

The Result enum has two variants: Ok, which holds a generic value, and Err, which holds another generic value representing the error details.

Note

Due to the importance and frequent usage of the Result enum, it is also included in the Rust prelude.

Let's consider adding a new function, check_student, using the Result enum, as shown in Listing 5.27.

```
fn check_student(student_name: &String, student_db: &Vec<Student>) ->
Result<(), String> {
    for student in student_db {
        if student.name == *student_name {
            return Ok(());
        }
    }
    Err(String::from("Student not found"))
}
```

Listing 5.27 Function for Checking the Presence of a Student in student_db

This function is quite similar to get_grade. However, instead of returning an Option, we now return a Result. This function might succeed in finding the student in the database, in which case, we'll return a unit type (since we are not interested in the name), or it might fail to find the student's name, in which case, we return a String message explaining the cause of the error. Note that the generic types T and E in the Result<T, E> are replaced by unit types and string types, respectively. The body of the function contains the same logic as shown earlier in Listing 5.26. The function iterates through the database entries, matching each student name with the provided name. If a match is found, we return a unit value. If no match is found, we return the Err variant with an error message.

5.4.3 Matching on Result

Like the Option enum, we can use the match expression to process the Result enum. Consider the code shown in Listing 5.28.

```
fn main() {
    let student_db = vec![
        ...
    ];
    let student_name = String::from("Bob");
```

```
    let student_status = check_student(&student_name, &student_db);

    match student_status {
            Ok(_) => {
                Let student_grade = get_grade(&student_name, &student_db);
                if let Some(grade) = student_grade {
                    println!("Grade is: {grade}");
                }
            }
            Err(error_msg) => println!("{error_msg}"),
        }
}
```

Listing 5.28 Using Match to Process the Result Enum

Note that the type of the student_status in your editor is a Result<(), String> type. We are matching on the student_status. The match, as expected, has two arms, one corresponding to each variant. The underscore in the arm corresponding to Ok(_), means that we do not care about the variable associated with the Ok variant. Inside the body corresponding to this arm, we are obtaining the grade using the get_grade function, defined earlier in Listing 5.26. Next, the if let syntax checks the grades and prints it. The second arm corresponds to an Err variant, in which case, we print the respective error_msg. The variable error_msg is bound to the String that is associated with the Err variant.

We have some code redundancy in the definitions of the get_grade function in Listing 5.26 and check_student function in Listing 5.27. Both functions iterate through the same database. We can improve our code by introducing a single function rather than defining two distinct functions. For instance, consider the code shown in Listing 5.29.

```
struct Student {
    name: String,
    grade: Option<u32>,
}

fn check_student_get_grade(student_name: &String,student_db: &Vec<Student>) ->
Result<Option<u32>, String> {
    for student in student_db {
        if student.name == *student_name {
            return Ok(student.grade);
        }
    }
    Err(String::from("Student not found"))
}
```

```
fn main() {
    let student_db = vec![
        ...
    ];

    let student_name = String::from("Bob");
    let student_status = check_student_get_grade(&student_name, &student_db);

    match student_status {
        Ok(option_grade) => {
            if let Some(grade) = option_grade {
                println!("Grade is: {grade}");
            }
        }
        Err(error_msg) => println!("{error_msg}"),
    }
}
```

Listing 5.29 Revised Code Based on the Function check_student_get_grades

The function `check_student_get_grade` returns a `Result`, with an `Ok` variant containing an `Option<u32>` value instead of unit value. This capability is important because, if a student exists, it's still not guaranteed that their grades are present, and they may be missing or `None`. As we iterate though the database, when a student name is found, we return their grades wrapped inside an `Ok` variant. In any other case, the function returns an `Err` variant. The `main` function performs a `match` on the `student_status`. The `Ok` arm now contains an `option_grade` value, which is bound to the `Option<u32>` value contained inside the `Ok` variant. The `if let` syntax ensures that, if we have `Some grade`, then it will be printed.

5.4.4 The ? Operator

The `Result` type can sometimes be quite verbose, especially in situations when dealing with multiple operations that may fail. The ? operator provides a concise way to handle these cases.

Consider the function shown in Listing 5.30, which performs some division.

```
fn division(dividend: f64, divisor: f64) -> Result<f64, String> {
    let result = match divisor {
        0.0 => Err(String::from("Error: Division by zero")),
        _ => Ok(dividend / divisor),
    };

    let quotient = result?;
```

```
    println!("The quotient is: {:?}", quotient);
    Ok(quotient)
}

fn main() {
    println!("Result: {:?}", division(9.0, 3.0));
    println!("Result: {:?}", division(4.0, 0.0));
    println!("Result: {:?}", division(0.0, 2.0));
}
```


Listing 5.30 Using ? Operator to Handle Errors in a division Function

The ? operator simplifies error handling by automatically propagating errors. In this function, the division function first checks whether the divisor is zero. If so, the variable result is assigned an appropriate error message. If the divisor is a non-zero value, the division proceeds, and the result is wrapped in Ok.

Next, in the line let quotient = result?, applying the ? operator on result will first unwrap the result returned by the division. If the result is Ok, the value is assigned to quotient. If the result is an Err, the error is immediately returned from the function, and the remaining lines of code are skipped. This behavior allows the function to focus on its main logic without explicit error handling code. For instance, if the divisor is zero, the ? operator will cause the function to return the error message "Error: Division by zero" immediately, as is the case in the following line:

```
println!("Result: {:?}", division(4.0, 0.0));
```

In this case, the print statement within the division function will not execute because the error is propagated directly to the calling function. The ? operator streamlines error handling, making your code cleaner and easier to maintain.

5.5 HashMaps

Let's say you want to store information about some words along with their frequencies or counts. One way to store this information is by using two vectors. The first vector will contain the words, and the second will contain their counts, as follows:

```
let words = vec!["Hello", "world", "Rust", "Programming"];
let counts = vec![5, 2, 15, 5];
```

This example is another scenario that works but is not efficient. To find the frequency of a particular word, you would need to write logic that first searches for the word in the vector. If found, we then retrieve the count using the correct index. Additionally, managing two separate vectors can be cumbersome.

A better solution is to use a single vector where each element is a tuple containing a string slice and an i32 value, as in the following example:

```
let word_counts = vec![("Hello", 5), ("world", 2), ("Rust", 15),
("Programming", 5)];
```

This approach offers better organization since now we're only working with a single vector. Additionally, the words and their counts are stored at the same index within the vector as a tuple. This approach eliminates the need to worry about maintaining the correct index correspondence between specific words and their counts, as was necessary in the previous solution. However, finding the frequency of a particular word may still require additional logic, and there's a risk of accidentally entering duplicate entries. For instance, to find a particular word in the tuple and retrieve its count, you might write something similar to the code shown in Listing 5.31.

```
let target_word = "Hello";
let mut found = false;
for (word, _) in &word_counts {
    if word.to_lowercase() == target_word.to_lowercase() {
        found = true;
        break;
    }
}
```

Listing 5.31 Sample Code for Finding a Particular Word and Retrieving Its word_count Vector

HashMaps are a common data structure that provides better organization of data in such cases. These data structures are based on key-value pairs, which offer efficient retrieval. Listing 5.32 shows the syntax for using a `HashMap`.

```
let mut word_counts: HashMap<&str, u8> = HashMap::new();
word_counts.insert("Hello", 5);
word_counts.insert("world", 2);
word_counts.insert("Rust", 15);
word_counts.insert("Programming", 5);
```

Listing 5.32 HashMaps for Storing the Same Information

You can use the `new` constructor function to create a `HashMap`. The `insert` function adds an entry to the `HashMap`. Each entry in the `HashMap` is a key-value pair. For instance, in the statement `word_counts.insert("Hello", 5)`, `'Hello'` is the key, and the value, `5`.

Note

`HashMaps` are not included by default; you must include them manually using the `use` statement, as in `use std::collections::HashMap;`.

An important property of HashMaps is that the keys are unique. To demonstrate, let's add the following duplicate key entry in the code:

```
word_counts.insert("Programming", 15);
println!("The word counts are {:?}", word_counts);
```


Executing the code will show a count of 15 for the key Programming. Moreover, the word "programming" appears once, not twice. This code eliminates the chances of duplicate entries.

Searching for a key in a HashMap is extremely fast and does not require complex logic. The contains_key function checks for the presence of a specific key, as follows:

```
let target_word = word_counts.contains_key("Programming");
```

The function returns an Option wrapping the value inside Some if the key exists, and None, otherwise. As a result, we no longer require complicated logic for retrieving the counts of the words and checking for the presence of the words.

The entry method on HashMap provides a more flexible way of accessing or modifying the values associated with keys, as in the following example:

```
word_counts.entry("Hello");
```

This line returns an instance of the Entry enum, which has two variants of Occupied and Vacant. The Occupied variant indicates that the entry is already present, while the Vacant variant means that the entry is not present.

Following the entry method, you can perform operations such as insertion and modification. For instance, the or_insert method allows you to insert a value if the key does not exist:

```
word_counts.entry("Hello").or_insert(0);
```

In this case, since the key of "Hello" already exists, the value will not be inserted.

5.6 HashSets

In addition to HashMaps, Rust also has HashSets. The key difference between a HashSet and a HashMap is that a HashSet stores only unique keys without any associated values, while a HashMap stores key-value pairs, allowing each key to have an associated value. Consider the example HashSet shown in Listing 5.33.

```
let mut words: HashSet<&str> = HashSet::new();
words.insert("Hello");
words.insert("world");
```

```
words.insert("Rust");
words.insert("Programming");
```

Listing 5.33 Storing Unique Words Using a HashSet

Note

Like HashMap, HashSet is also not included by default, and you must include it manually using the use statement, as in use std::collections::HashSet;.

Using a HashSet offers an organized and efficient for managing collections of unique items. The HashSet ensures that no duplicate entries are stored, simplifying the process of maintaining uniqueness. For example, let's attempt to insert a duplicate entry with the following lines:

```
words.insert("Hello");
println!("The words are {:?}", words);
```

When we execute this code, it will show that "Hello" appears only once in the HashSet, demonstrating that the code does not allow duplicate entries. The HashSet automatically handles duplicates for us.

Searching for an item in a HashSet is amazingly fast and does not require complex logic. The contains method checks for the presence of a specific item, as follows:

```
let target_word = words.contains("world");
```

The contains method returns a bool, indicating whether the item is present or not. This approach simplifies the process of checking for the existence of an item.

The HashSet provides additional methods for modifying and accessing its contents. For example, the take method allows for removing an item while retrieving it:

```
if let Some(word) = words.take("world") {
    println!("Removed: {}", word);
}
```

The take method removes the item from the HashSet and returns it if it exists, allowing for both removal and retrieval in one operation. The take method is particularly useful and operates in different contexts with different uses. We'll see more examples of its usage in Chapter 11, Section 11.1.4.

Overall, HashSet offers a more efficient and straightforward way to manage collections of unique items, eliminating the need for manual checks and ensuring that duplicates are automatically handled.

5.7 Practice Exercises

In this section, we'll provide some practice questions to help reinforce your learning. The exercise questions cover almost all the concepts introduced in the chapter. Solutions to each exercise will follow in Section 5.8.

1. **Storing different types in a vector**
   In Rust, vectors are typically used to store elements of the same type. However, in some cases, you may want to store multiple data types in a single vector. The following code aims to store both integers and floats in a vector:

```
#[derive(Debug)]
enum Value {
    // Add code here
}

fn main() {
    let some_val = vec![Value::Integer(12), Value::Float(15.5)];
    for i in some_val {
        match i {
            Value::Integer(num) => println!("Integer: {} ", num),
            Value::Float(num) => println!("Float: {}", num),
        }
    }
}
```

   Define the Value enum with two variants:
   - Integer(i32) for storing integers
   - Float(f64) for storing floating-point numbers

2. **Implementing a library management system**
   You're tasked with designing a library management system in Rust that can handle both books and magazines. Follow these steps to complete the task.

   Define an Item struct with the following fields:
   - id: An integer representing the unique identifier of the item
   - title: A string representing the title of the item
   - year: An integer representing the publication year of the item
   - item_type: An enumeration that specifies whether the item is a book or a magazine

   Create an ItemType enum with the following variants:
   - Book: Represents a book
   - Magazine: Represents a magazine

Implement a function `display_item_info()` that takes an `Item` as input and prints the `Item ID`, `Title`, `Publication` year, and whether the item is a `book` or a `magazine`.

3. **Retrieving the first character of a vector**
   You're provided with the following code that does not compile. Identify and fix the issues to ensure it works as expected.

```
fn first_character(chars: &Vec<char>) -> Option<char> {
    if chars.len() > 0 {
        Some(chars[0])
    } else {
        None
    }
}

fn main() {
    let my_chars = vec!['a', 'b', 'c', 'd'];
    match first_character(&my_chars) {
        Some => println!("First character: {character}"),
        None => println!("Empty array"),
    }
}
```

4. **Checking for fruit in a basket**
   You're provided with the following code that does not compile. Identify and fix the issues to ensure it works correctly.

```
fn check_fruit(input_fruit: String) -> Option<String> {
    let fruit_basket = vec![
        String::from("mango"),
        String::from("apple"),
        String::from("banana"),
    ];
    for fruit in fruit_basket {
        if input_fruit == fruit {
            return Some(fruit);
        }
    }
}

fn main() {
    let user_fruit = String::from("apple");
    if let Some(fruit) = check_fruit(user_fruit) {
        println!("User's name: {fruit}");
    }
}
```

5. **Calculating area and perimeter of shapes**

   In the following code, there is an error in the main function. Fix it so that it will compile.

```
enum Measurement {
    CircleArea(f64),
    RectangleArea(f64, f64),
    TriangleArea(f64, f64),
    Perimeter(Vec<f64>),
}

impl Measurement {
    fn calculate(self) -> Result<f64, String> {
        match self {
            Self::CircleArea(radius) => {
                if radius < 0.0 {
                    Err(String::from("Radius cannot be negative"))
                } else {
                    Ok(std::f64::consts::PI * radius * radius)
                }
            }
            Self::RectangleArea(length, width) => {
                if length < 0.0 || width < 0.0 {
                    Err(String::from("Length and width cannot be negative"))
                } else {
                    Ok(length * width)
                }
            }
            Self::TriangleArea(base, height) => {
                if base < 0.0 || height < 0.0 {
                    Err(String::from("Base and height cannot be negative"))
                } else {
                    Ok(0.5 * base * height)
                }
            }
            Self::Perimeter(sides) => {
                if sides.len() < 3 {
                    Err(String::from("A polygon must have at least 3 sides"))
                } else {
                    Ok(sides.iter().sum())
                }
            }
        }
    }
}
```

```
fn main() {
    let user_input = Measurement::TriangleArea(5.0, 8.0);
    match user_input.calculate() {
        => println!("Result: {res}"),
        => println!("Error: {e}"),
    }
}
```

6. **Calculating the square of a number**
   Complete the following function signature:

```
fn calculate_square(num: i32) ->  {
    if num >= 0 {
        let result = num * num;
        println!("The square of {} is: {}", num, result);
        Ok(result)
    } else {
        Err("Negative number provided".to_string())
    }
}

fn main() {
    let number = 7;
    if let Err(e) = calculate_square(number) {
        println!("Error: {e}");
    }
}
```

7. **Implementing a student management system**
   In this exercise, you'll create a student management system using Rust. The system should store and retrieve student information based on their unique IDs. You're provided with a `Student` struct that contains fields for ID, name, and grade.

   Next, create a `StudentManager` structure that contains a field of students, which will be a `HashMap` where the key is an integer representing the student's unique ID, and the value is the complete details of the student (i.e., an instance of the `Student` structure).

   The `StudentManager` should implement the following methods:

   - `new() -> Self`: A constructor that initializes an empty student manager.
   - `add_student(&mut self, student: Student) -> Result<(), String>`: Adds a student to the manager. If the student's ID already exists, return an error message. Otherwise, add the student and return `Ok`.
   - `get_student(&self, id: i32) -> Option<&Student>`: Retrieves a student from the manager based on their ID. If the student is found, return `Some(student)`. Otherwise, return `None`.

Your task is to implement the StudentManager structure along with these methods. Additionally, provide sample usage of the system by adding a few students and retrieving their information using the get_student() method.

5.8 Solutions

This section provides the code solutions for the practice exercises in Section 5.7. The code is largely self-explanatory; however, we have included comments and additional explanations wherever necessary to enhance understanding.

1. **Storing different types in a vector**

```
#[derive(Debug)]
enum Value {
    Integer(i32),
    Float(f32),
}

fn main() {
    let some_val = vec![Value::Integer(12), Value::Float(15.5)];

    for i in some_val {
        match i {
            Value::Integer(num) => println!("Integer: {} ", num),
            Value::Float(num) => println!("Float: {}", num),
        }
    }
}
```

2. **Implementing a library management system**

```
#[derive(Debug)]
struct Item {
    id: i32,
    title: String,
    year: i32,
    type_: ItemType,
}

#[derive(Debug)]
enum ItemType {
    Book,
    Magazine,
}
```

```
fn display_item_info(item: &Item) {
    println!("Item ID: {:?}", item.id);
    println!("Title: {:?}", item.title);
    println!("Publication Year: {:?}", item.year);
    println!("Publication Type: {:?}", item.type_);
}

fn main() {
    let rust_book = Item {
        id: 1,
        title: String::from("The Rust Programming Language Book"),
        year: 2021,
        type_: ItemType::Book,
    };

    let rust_magazine = Item {
        id: 2,
        title: String::from("Rust Magazine"),
        year: 2022,
        type_: ItemType::Magazine,
    };

    display_item_info(&rust_book);
    display_item_info(&rust_magazine);
}
```

3. **Retrieving the first character of a vector**

```
fn first_character(chars: &Vec<char>) -> Option<char> {
    if chars.len() > 0 {
        Some(chars[0])
    } else {
        None
    }
}

fn main() {
    let my_chars = vec!['a', 'b', 'c', 'd'];
    match first_character(&my_chars) {
        Some(character) => println!("First character: {character}"),
        None => println!("Empty array"),
    }
}
```

4. **Checking for fruit in a basket**

```
fn check_fruit(input_fruit: String) -> Option<String> {
    let fruit_basket = vec![
        String::from("mango"),
        String::from("apple"),
        String::from("banana"),
    ];
    for fruit in fruit_basket {
        if input_fruit == fruit {
            return Some(fruit);
        }
    }
    None /* In case the if statement is not successful for any fruit in
            the basket, then the function will return a unit type while it
            is expecting an Option enum. Therefore we need to explicitly
            return the None variant */
}

fn main() {
    let user_fruit = String::from("apple");
    if let Some(fruit) = check_fruit(user_fruit) {
        println!("User's name: {fruit}");
    }
}
```

5. **Calculating area and perimeter of shapes**

```
enum Measurement {
    CircleArea(f64),
    RectangleArea(f64, f64),
    TriangleArea(f64, f64),
    Perimeter(Vec<f64>),
}

impl Measurement {
    fn calculate(self) -> Result<f64, String> {
        match self {
            Self::CircleArea(radius) => {
                if radius < 0.0 {
                    Err(String::from("Radius cannot be negative"))
                } else {
                    Ok(std::f64::consts::PI * radius * radius)
                }
            }
            Self::RectangleArea(length, width) => {
```

```
                if length < 0.0 || width < 0.0 {
                    Err(String::from("Length and width cannot be negative"))
                } else {
                    Ok(length * width)
                }
            }
            Self::TriangleArea(base, height) => {
                if base < 0.0 || height < 0.0 {
                    Err(String::from("Base and height cannot be negative"))
                } else {
                    Ok(0.5 * base * height)
                }
            }
            Self::Perimeter(sides) => {
                if sides.len() < 3 {
                    Err(String::from("A polygon must have at least 3 sides"))
                } else {
                    Ok(sides.iter().sum())
                }
            }
        }
    }
}

fn main() {
    let user_input = Measurement::TriangleArea(5.0, 8.0);
    match user_input.calculate() {
        Ok(res) => println!("Result: {res}"),
        Err(e) => println!("Error: {e}"),
    }
}
```

6. **Calculating the square of a number**

```
fn calculate_square(num: i32) -> Result<i32, String> {
    if num >= 0 {
        let result = num * num;
        println!("The square of {} is: {}", num, result);
        Ok(result)
    } else {
        Err("Negative number provided".to_string())
    }
}

fn main() {
    let number = 7;
```

```
    if let Err(e) = calculate_square(number) {
        println!("Error: {e}");
    }
}
```

7. **Implementing a student management system**

```
use std::collections::HashMap;
struct Student {
    id: i32,
    name: String,
    grade: String,
}

struct StudentManager {
    students: HashMap<i32, Student>,
}

impl StudentManager {
    fn new() -> Self {
        StudentManager {
            students: HashMap::new(),
        }
    }

    fn add_student(&mut self, student: Student) -> Result<(), String> {
        if self.students.contains_key(&student.id) {
            Err(format!("Student with ID {} already exists", student.id))
        } else {
            self.students.insert(student.id, student);
            Ok(())
        }
    }

    fn get_student(&self, id: i32) -> Option<&Student> {
        self.students.get(&id)
    }
}

fn main() {
    let mut manager = StudentManager::new();

    let student1 = Student {
        id: 1,
        name: String::from("Alice"),
        grade: String::from("A"),
```

```
    };
    let student2 = Student {
        id: 2,
        name: String::from("Bob"),
        grade: String::from("B"),
    };

    manager.add_student(student1).unwrap();
    manager.add_student(student2).unwrap();

    // Retrieve and print student information
    if let Some(student) = manager.get_student(1) {
        println!("Student ID: {}", student.id);
        println!("Student Name: {}", student.name);
        println!("Student Grade: {}", student.grade);
    }
    if let Some(student) = manager.get_student(2) {
        println!("Student ID: {}", student.id);
        println!("Student Name: {}", student.name);
        println!("Student Grade: {}", student.grade);
    }
}
```

5.9 Summary

In this chapter, we explored essential concepts in Rust, focusing on foundational data structures and their practical applications. We began with an in-depth look at structs, covering the basics of defining, instantiating, and utilizing different types of structs, including tuple and unit structs. Following this, we delved into how to add functionality to structs, enhancing their utility in real-world scenarios. We then transitioned to enums, discussing how to add data to enum variants and extend their functionality, making them versatile tools for handling different states or conditions in your programs. The chapter also provided a thorough understanding of the `Option` and `Result` types, essential for managing the presence or absence of values and handling errors gracefully. We covered the use of pattern matching with `Option`, the `if let` construct, and ownership considerations, as well as the powerful `?` operator for simplifying error handling with `Result`. Additionally, we explored `HashMaps` and `HashSets`, highlighting their roles in efficiently managing collections of key-value pairs and unique values, respectively. This chapter equips you with a solid understanding of these core Rust features, laying the groundwork for writing more robust and efficient code.

In the upcoming chapter, we'll examine best practices for organizing your Rust code, focusing on modularity and code structure.

Chapter 6
Organizing Your Code

Effective code organization is the foundation of maintainable software. In this chapter, we'll explore how Rust's module system and other organizational tools not only structure your project but also bring clarity and control to even the most complex codebases.

This chapter focuses on best practices for structuring Rust projects. You'll learn about the modular system in Rust, including how to create and use modules to organize code logically. This chapter discusses the visualization and organization of modules, the re-exporting of items, and the management of privacy. We'll also cover how you can incorporate external dependencies into projects and publish a crate to the Rust ecosystem. With this knowledge, you can create scalable and maintainable codebases.

6.1 Code Organization

Up until now, we've written all our code in a single file. However, as our projects grow, we'll need to organize our code effectively. This section will guide you through structuring and organizing code within a Rust project, focusing on three key components: packages, crates, and modules.

At the highest level of organization, we have packages. A *package* is created using the `cargo new` command and may contain one or more crates. For instance, to create a new package named `rust_book`, you'll enter the following command in terminal:

```
c:\> cargo new rust_book
```

The package includes a *Cargo.toml* file, which serves as the central configuration hub that manages metadata, dependencies, build instructions, and optional features. Packages enable you to build, test, and share your code, all managed through Cargo commands like `cargo build` and `cargo test`. Packages are the highest level of code organization and contain lower-level organizational units called crates.

A *crate* is a compilation unit that houses a set of modules and their associated items such as functions and structures. From a code organization perspective, a crate is essentially a tree of modules, representing a hierarchical structure. A library crate is a piece of code that produces a library, which you can share with other crates. The library crate is for sharing purposes and is not for execution. A binary crate, meanwhile, produces an executable (i.e., file that you can execute). A binary crate may contain code from a library crate.

Finally, *modules* allow you to group items at an exceptionally fine level. They also control matters of scope and privacy. We'll talk about scope and privacy in more detail throughout this chapter.

Packages must follow certain rules, including that packages must contain at least one crate. A package can have any number of binary crates, but at most only one library crate. Figure 6.1 shows the relationship between the packages, crates, and modules.

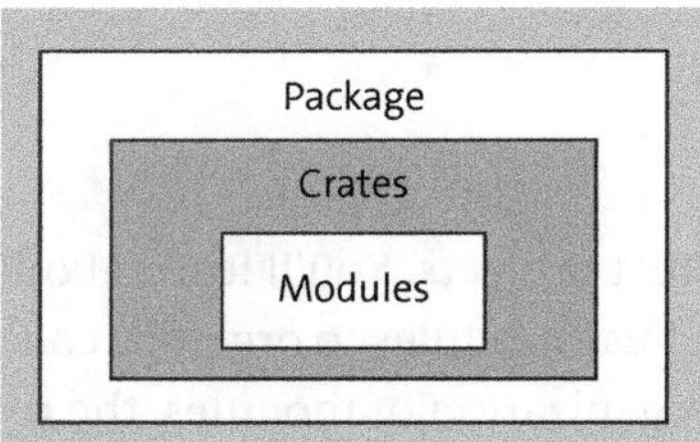

Figure 6.1 Relationship between Packages, Crates, and Modules

At the highest level, we have packages containing crates. Each crate, at a finer level, contains modules.

Figure 6.2 shows the structure of a typical Rust package. In this case, we have one binary crate and one library crate, which is typically named *lib.rs*. These crates contain modules. Each crate has a root module, which may contain further modules. The binary crate root module contains four submodules. We are free to create further levels of hierarchy by creating submodules of existing modules.

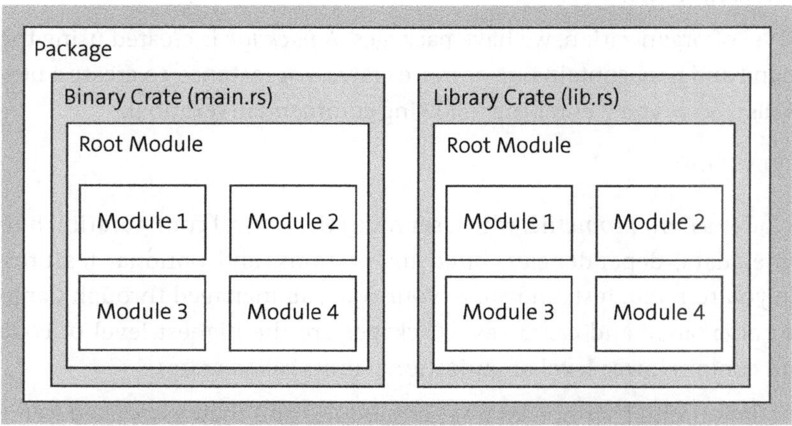

Figure 6.2 Structure of a Typical Rust Package

A binary crate will produce an executable. In contrast, a library crate will produce a library for sharing purposes.

Let's go through an example to better understand this organizational structure. We'll first create a new package using the cargo new command, as in the following example:

```
c:\> cargo new my_package
```

The package will be visible in the current directory. If you open the package, notice a manifest file, called *Cargo.toml*, similar to the code shown in Listing 6.1.

```
[package]
name = "my_package"
version = "0.1.0"
edition = "2021"

[dependencies]
```

Listing 6.1 Contents of the Cargo.toml File

The *Cargo.toml* file contains the general information such as package, version, and edition and the list of dependencies the crate depends upon. The dependencies list is currently empty. We'll learn more about this list in Section 6.5.

Running the cargo run command in the terminal executes the binary crate of *main.rs*. The executable can be found in the *target\debug* folder, which contains the *.exe* file with the name of the package.

Our package or project may contain multiple executable binary crates. By convention, executables are created inside the *bin* folder. Let's add a *bin* folder and add one binary crate to it. Listing 6.2 shows the package's structure.

```
my_package/
├── Cargo.toml
├── src/
│   ├── main.rs
├── bin/
│   └── my_binary.rs
└── target/
```

Listing 6.2 Package Structure after Adding a Binary Crate in the bin Folder

Remember the rules of packages: The package must have at least one crate. The cargo command ensures compliance by creating one default binary crate, in the *src* folder, called *main.rs*. However, you can have multiple binary crates added to the *bin* folder. When you save the package, the compiler will throw an error in the newly created binary file stating, for example, "main function not found in crate my_binary." Each binary crate must have a main function, which serves as an entry point for executing the crate. Let's add the main function to my_binary now, as in the following example:

```
fn main() {
    println!("Hello from my_binary");
}
```

If we execute my_package using the cargo run command now, you'll see an error of "cargo run could not determine which binary to run." This error makes sense because we now have two binary crates and Cargo needs further information regarding which one it should execute. A list of available binaries is provided in the error message. The first available binary is my_binary, and the second one is my_package.

To run a specific binary, you would use the bin flag and then mention the name of the binary. For instance, to execute the main.rs binary, enter the following command in the terminal:

```
c:\> cargo run --bin my_package
```

You might be surprised that we don't have any binary crate called my_package in our package directory, yet Cargo shows it as an available binary. This inconsistency is due to the conventions followed by cargo. In particular, cargo adheres to the rule that, if a *main.rs* file is present in the source directory, this file serves as the crate root for a binary crate bearing the package name.

In the same way, if a file named *lib.rs* exists in the source directory, this file is treated as the crate root for a library crate with the package's name. For instance, let's add a library crate in the source directory. The updated package structure after adding the library will look as shown in Listing 6.3.

```
my_package/
├── Cargo.toml
├── src/
│   ├── main.rs
│   └── lib.rs
├── bin/
│   └── my_binary.rs
└── target/
```

Listing 6.3 The Structure of my_package after Adding a Library Crate

In this case, since a library crate with the name of *lib.rs* exists in the source directory; therefore, this crate will be treated as the crate root for a library crate with the my_package name.

6.2 Module Basics

As covered in the previous section, modules are the finest level of code organization. We'll start this section by explaining the need for modules, followed by the basics of creating modules. Next, we'll elaborate on the difference between relative and absolute paths for accessing items contained within a module. Finally, we look at the issue of privacy and the use declaration for importing items.

6.2.1 Motivating Example for Modules

Consider the code shown in Listing 6.4, which is inside a library crate representing an online store and provides basic services such as creating products, maintaining customer's data, and processing an order.


```
struct Product {
    id: u64,
    name: String,
    price: f64,
    category: Category,
}
enum Category {
    Electronics,
    Clothing,
    Books,
}

impl Product {
    fn calculate_tax(&self) -> f64 {
        self.price * 0.1
    }
    fn product_price(&self) -> f64 {
        self.price + self.calculate_tax()
    }
}
struct Customer {
    id: u64,
    name: String,
    email: String,
}
struct Order {
    id: u64,
    product: Product,
    customer: Customer,
    quantity: u32,
}

impl Order {
    fn calculate_discount(&self) -> f64 {
```

```
        if self.quantity > 5 {
            0.1
        } else {
            0.0
        }
    }

    fn total_bill(&self) -> f64 {
        let discount = self.calculate_discount();
        let total_before_discount = self.product.product_price() *
self.quantity as f64;
        total_before_discount - (total_before_discount * discount)
    }
}
```

Listing 6.4 Library for an Online Store

Let's walk through this code. First, we define a Product struct that stores information related to a product, such as its name, price, and category. The category details are further specified using a Category enum. The id field is used for internal processing and record-keeping. Next, we have several functions defined on the Product struct, including sales_tax and price. The sales_tax function calculates a 10% tax on the product, while the price function returns the final price of the product, including taxes.

We've also defined a Customer struct to store customer-related information. Finally, an Order struct tracks orders. This struct includes a few functions, such as calculate_discount, which applies a 10% discount if the number of ordered items exceeds five. The total_bill function computes the customer's final total bill.

6.2.2 Creating Modules

This code forms the foundation of our library, but there's room for improvement. Currently, the file contains multiple levels of abstraction. We have high-level functions that we intend to expose from our library to the outside world, such as product_price and total_bill. However, the file also includes lower-level functions like calculate_tax and calculate_discount, which are internal and should not be exposed to the outside world.

Moreover, the code mixes concerns. We have logic dealing with products, customers, and orders all in one place, which can lead to confusion and maintenance difficulties. To improve the structure of this code, we should refactor it by separating product-related, customer-related, and order-related logic into distinct modules, as shown in Listing 6.5.

```
mod product {
    struct Product {
        id: u64,
        name: String,
        price: f64,
        category: Category,
    }
    enum Category {
        Electronics,
        Clothing,
        Books,
    }

    impl Product {
        fn calculate_tax(&self) -> f64 {
            self.price * 0.1
        }

        fn product_price(&self) -> f64 {
            self.price + self.calculate_tax()
        }
    }
}

mod customer {
    struct Customer {
        id: u64,
        name: String,
        email: String,
    }
}

mod Order {
    struct Order {
        id: u64,
        product: Product,
        customer: Customer,
        quantity: u32,
    }

    impl Order {
        fn calculate_discount(&self) -> f64 {
            if self.quantity > 5 {
                0.1
            } else {
```

```
                0.0
            }
        }
        fn total_bill(&self) -> f64 {
            let discount = self.calculate_discount();
            let total_before_discount = self.product.product_price() *
self.quantity as f64;
            total_before_discount - (total_before_discount * discount)
        }
    }
}
```

Listing 6.5 Code from Listing 6.4 Refined by Introducing Modules

To declare a module, use the `mod` keyword, followed by a name for the module.

The `Category` enum is currently under the `product` module with no methods, but later on, we may expect some functionality. Therefore, a best practice is to include this enum in a submodule of the `product` module, as shown in Listing 6.6.

```
mod product {
    struct Product {
        id: u64,
        name: String,
        price: f64,
        category: Category,
    }

    mod category {
        enum Category {
            Electronics,
            Clothing,
            Books,
        }
    }
    impl Product {
       ...
}
...
```

Listing 6.6 Submodule of category Defined inside the product Module

6.2.3 Relative and Absolute Paths of Items

The code is now more organized, but it contains some errors. Let's focus on fixing them, one by one.

We'll first deal with the errors inside the Order struct, as shown in Listing 6.7.

```
...
mod Order {
    struct Order {
        id: u64,
        product: Product,    // Error
        customer: Customer, // Error
        quantity: u32,
    }
...
```

Listing 6.7 Error in the Order Struct

The error is, "cannot find type Product in this scope." We have a similar error for the Customer. Certainly, this error message is logical, since the Product and Customer details are stored in a different module, distinct from the current module. For the Order module to access this information, you must provide clear instructions on how to locate these items within the module's hierarchy by specifying the fully qualified names of the items. If an item is within the same module, you can use its *relative path*. However, since the Customer and Product modules are not in the same module as the Order, you must specify their ***absolute paths***. The category: Category field in the product struct uses the relative path for the Category enum. Since it is in the same module, the complete path is not necessary.

The absolute paths for the Product and Customer structs are as follows:

```
crate::product::Product
crate::customer::Customer
```

The absolute path starts with the crate module, then navigates to the relevant module, and finally points to the struct of interest, using two colons (::) between each part. The updated code for the Order struct is shown in Listing 6.8.

```
...
mod Order {
    struct Order {
        id: u64,
        product: crate::product::Product,    // Error
        customer: crate::customer::Customer, // Error
        quantity: u32,
    }
...
```

Listing 6.8 Update Code for the Order Struct

By default, a module called `crate` serves as the root module of our module tree. To specify an absolute path, start from the root module `crate` and then navigate down the module tree to locate the specific item.

Relative Paths

Relative paths start from the current module and use `self` or no prefix to navigate within the module hierarchy. Listing 6.9 shows an example.

```
mod utilities {
    pub mod math {
        pub fn multiply(a: i32, b: i32) -> i32 {
            a * b
        }
    }
    pub fn calculate() {
        let result = math::multiply(3, 4); // relative path with no prefix
        println!("Multiplication result: {}", result);
        let result_self = self::math::multiply(5, 6); // relative path
                                                       // with self prefix
        println!("Result using `self`: {}", result_self);
    }
}
```

Listing 6.9 Example Using Relative Paths

6.2.4 Privacy in Modules

The errors shown earlier in Listing 6.7 can be fixed by specifying the paths, but now, we see additional errors in the code shown earlier in Listing 6.8. The error is "`struct Product` is private," and we have similar error for the `Customer` struct. Let's now make the two structs public, as shown in Listing 6.10.

```
mod product {
    pub struct Product { // Changed to public
        id: u64,
        name: String,
        price: f64,
        category: Category,
    }
  ...
}
```

```
mod customer {
    pub struct Customer { // Changed to public
        id: u64,
        name: String,
        email: String,
    }
}
...
```

Listing 6.10 Structs Product and Customer Changed to Public

By default, everything inside a module is private and not accessible from outside the module. To make an item public, that is, visible to code outside the module, you must use the `pub` keyword.

You might think that making the entire module public would automatically make all its internal content public as well. However, for Rust, this automatic visibility is not the case. For example, if you make the `customer` module `pub` instead of making the `Customer` struct `public`, the compiler will still return an error. Making a module public does not automatically make all of its functions, structs, enums, and other items public. Each item within the module has its own visibility, which can be controlled independently.

Unfortunately, we've got a few more errors to handle. In the `total_bill` function, the error is "`product_price` is private," as shown in Listing 6.11.

```
...
mod Order {
    ...
    impl Order {
     ...
        fn total_bill(&self) -> f64 {
            let discount = self.calculate_discount();
            let total_before_discount = self.product.product_price() *
self.quantity as f64; // Error
            total_before_discount - (total_before_discount * discount)
        }
    }
}
```

Listing 6.11 Error in the Function total_bill

Let's fix this error by making the `product_price` function public, as shown in Listing 6.12.

```
mod product {
    ...
```

```
    impl Product {
        ...
        pub fn product_price(&self) -> f64 {
            self.price + self.calculate_tax()
        }
    }
}
```

Listing 6.12 Fixing the Error from Listing 6.11

We've got one final error in the `Product` struct: "cannot find type `Category` in this scope." This error is shown in Listing 6.13.

```
mod product {
    pub struct Product {
        id: u64,
        name: String,
        price: f64,
        category: Category, // Error
    }
    ...
}
...
```

Listing 6.13 Error in the Product Struct

Let's use the fully qualified name for the item. In this case, that name is the relative path to the item since `Category` is in the submodule of the current module. The updated code is shown in Listing 6.14.

```
mod product {
    pub struct Product {
        id: u64,
        name: String,
        price: f64,
        category: category::Category, // Error
    }
    ...
}
...
```

Listing 6.14 Fixing the Code from Listing 6.13

Sadly, this fix leads to another error, "`enum Category` is private." The `Category` enum is in a submodule or child module of `category`, with the `product` module as its parent. In Rust's module system, a child module can access the contents of its parents, but parent modules cannot access the items within their child modules. To enable access, we'll make the Category enum public, as shown in Listing 6.15.

```
mod product {
    ...
    mod category {
        pub enum Category {
            Electronics,
            Clothing,
            Books,
        }
    }
    ...
}
...
```

Listing 6.15 Fixing the Error from Listing 6.14 by Making the Category Enum Public

Finally, our code is now correct!

6.2.5 The Use Declaration for Importing or Bringing Items into Scope

In large and complex code, perhaps with multiple levels of submodules, the fully qualified names can get quite long. To simplify the code, the use declaration creates local name bindings for specific paths, thus making your code more concise and more readable.

For example, let's use the use declaration and create a local name binding for the Category enum. The syntax is shown in Listing 6.16.

```
mod product {
    use category::Category;
    pub struct Product {
        id: u64,
        name: String,
        price: f64,
        category: Category, // This is now simplified
    }
   ...
}
...
```

Listing 6.16 The use Declaration for Creating Local Name Bindings

Thus, whenever the word Category is used within this module's scope, it refers to the Category enum, which is defined inside the category module.

An alternative way to express this relationship is by saying that the use declaration brings the Category enum into scope. Some people might refer to this process as *importing*. However, this term suggests bringing something in from an external source, which isn't entirely accurate in this context. The correct terminology is that the use declaration "brings items into scope" that are located somewhere else within the module tree.

Thus, we can also simplify the paths in the order module, as shown in Listing 6.17.

```
mod Order {
    use crate::product::Product;
    use crate::customer::Customer;
    struct Order {
        id: u64,
        product: Product,
        customer: Customer,
        quantity: u32,
    }
  ...
}
...
```

Listing 6.17 Simplifying the Path in the order Module

Let's focus once more on the Category enum. The Category enum is public; however, the module in which it is contained is not public. Consider our example shown in Listing 6.18.

```
mod product {
...
    mod category {
        pub enum Category {
            Electronics,
            Clothing,
            Books,
        }
    }
   ...
}
...
```

Listing 6.18 Category Enum in the Code

Notice how the Category enum is a private submodule of the parent product module. Public items of a private submodule can only be accessed by the parent module. For instance, let's try bringing the category module into the scope of the crate module via the use declaration, as in the following example:

```
use crate::product::category; // Error
mod product {
  ...
}
...
```


The compiler throws an error: "module category is private."

6.3 Visualizing and Organizing Modules

Understanding and organizing modules in Rust is crucial for structuring complex projects. Cargo provides useful tools to help you visualize a module's structure and organize code within a typical file system, which enhances clarity and maintainability. We'll explore these tools in the following sections.

6.3.1 Cargo Modules for Visualizing Module Hierarchy

Cargo includes a nice utility for visualizing the module tree, called cargo-modules. This is basically a Cargo plugin that allows you to visualize how modules within a project are organized and how they relate to each other. You can install it with the following command in the terminal:

```
c:\> cargo install cargo-modules
```

This command fetches the cargo-modules crate from crates.io and updates the Rust package registry. The following command visualizes the available module tree:

```
c:\> cargo modules structure
```

Note

This command may require you to update Rust, which can be performed using the rustup update command.

Assuming that we are running the command for the package created in the previous section (i.e., my_package), we have a library crate, a *main.rs* binary, and *my_binary.rs* binary. Thus, the command cargo modules structure will create the targets shown in Listing 6.19.

```
Targets present in package:
- my_package (--lib)
- my_binary (--bin my_binary)
- my_package (--bin my_package)
```

Listing 6.19 Result of Command in cargo modules structure

Let's specify the library since we do not have any meaningful code in the binary files with the following command:

```
c:\> cargo modules structure --lib
```

This command creates a module tree for the library, as shown in Listing 6.20.

```
crate my_package
├── mod order: pub(crate)
│   └── struct Order: pub(self)
│       ├── fn calculate_discount: pub(self)
│       └── fn total_bill: pub(self)
├── mod customer: pub(crate)
│   └── struct Customer: pub
└── mod product: pub(crate)
    ├── struct Product: pub
    │   ├── fn calculate_tax: pub(self)
    │   └── fn product_price: pub
    └── mod category: pub(self)
        └── enum Category: pub
```

Listing 6.20 Module Tree for the Library

At the top level, we have the `crate` module. The `crate` module is the root of the module tree and is automatically created for every crate. The content of the `crate` module is either a *lib.rs* file, if you're working with the library crate, or it may be a binary such as *main.rs*, if you're working with a binary crate.

Inside the `crate` module, we have individual modules. The last module, which is the `product` module, contains a submodule of `category`. Alongside the individual modules, we have information regarding the visibility of the module. The following list describes the different visibility options:

- `pub(crate)`: Items are only available within the current crate.
- `pub(self)`: Items are available within the current module in which they are defined.
- `pub`: Items are available from outside the module.

For instance, `order` is `pub(crate)` means that the `order` item is available within the current `crate`. Struct `Order` is `pub(self)`, meaning that the struct is visible within the current module. Finally, Product and Customer structs are pub, which means that they are available outside the modules in which they are defined.

6.3.2 Organizing Code Using a Typical File System

Let's look at the structure of my_package again, as shown in Listing 6.21.

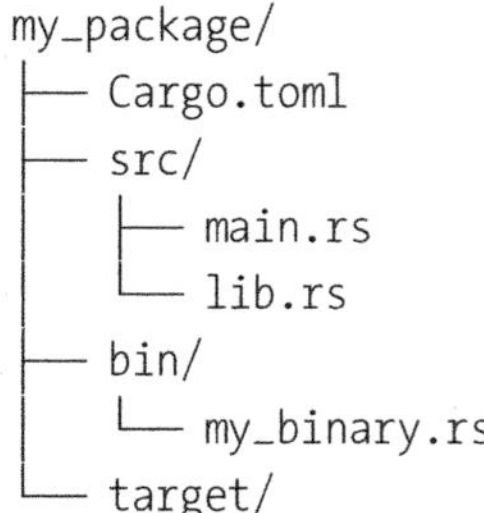

Listing 6.21 Structure of my_package

To simplify this structure, let's delete the my_binary and the bin folder.

As shown in Listing 6.21, notice how the modules are not mapped to the file system. Instead, modules are explicitly defined using the mod keyword. For instance, the product module is not in a separate file, nor are the order and customer modules. They all reside in one place within a single file, unlike some other programming languages, such as Python or JavaScript. The files themselves don't define the borders of a module in Rust.

Rust allows you to organize code using the convention of a typical file system, with two methods in which this organizing can be done. They're a bit different if you're coming from some other programming language.

Include Module in New File

The first method is to simply make a new file with the same name as that of the module and include the contents of the module inside the file. Let's apply this method to the code contained in the library of the online store created in Section 6.2.

For a simple module with no submodules, such as the customer module and order module, the process is straightforward. In the source directory, we'll create files with the name of the modules having an *.rs* extension and put the contents of module inside it. The updated file structure is shown in Listing 6.22.

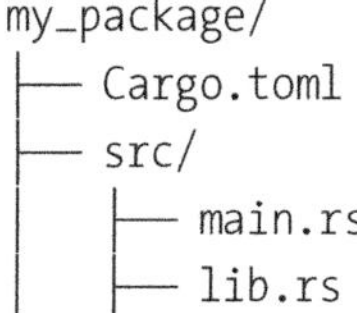

```
|   ├── customer.rs
|   └── order.rs
└── target/
```

Listing 6.22 Updated File Structure after Adding the customer and order Files

The *customer.rs* file will resemble the code shown in Listing 6.23.

```
pub struct Customer {
    id: u64,
    name: String,
    email: String,
}
```

Listing 6.23 Code inside the customer.rs File

In the same way, the contents of the *order.rs* file will resemble the code shown in Listing 6.24.

```
use crate::customer::Customer;
use crate::product::Product;
struct Order {
    id: u64,
    product: Product,
    customer: Customer,
    quantity: u32,
}
impl Order {
    fn calculate_discount(&self) -> f64 {
        if self.quantity > 5 {
            0.1
        } else {
            0.0
        }
    }
    fn total_bill(&self) -> f64 {
        let discount = self.calculate_discount();
        let total_before_discount = self.product.product_price() *
self.quantity as f64;
        total_before_discount - (total_before_discount * discount)
    }
}
```

Listing 6.24 Code inside the order.rs File

Notice that the code inside the *customer.rs* and *order.rs* files does not define modules in code; that is, we do not have `mod customer` or `mod order` statements in the two files. This omission is because these lines will be added to the *lib.rs* file so that they are tied to the code library. If we don't tie them up to the library, then they will be isolated pieces of code. In the *lib.rs* file, we'll declare these modules, as shown in Listing 6.25.

```
mod product {
    pub struct Product {
    ...
    }
    mod category {
        ...
    }
    impl Product {
        ...
}
mod customer;
mod order;
```

Listing 6.25 Updated Code in the lib.rs File

The declarations of the `customer` and `order` inside the *lib.rs* file ensure that these modules are in our module tree. These modules were submodules of the `crate` module. So this declaration ensures that it is the case. Moreover, submodules have to be defined in the parent module, regardless of whether they are defined in line or in a separate file. When Rust encounters the declaration of the submodule `customer` or `order` and it notices that the contents are not in line, it will look for a file called *customer.rs* and *order.rs* to fetch their respective contents.

Next, let's take care of the modules containing submodules, such as the `product` module in this case. The parent module can be defined in the same way as before. We'll create a file in the source directory with the name of the module and then put the contents of the module. Let's do the same with the submodule. The updated file structure will now look as depicted in Listing 6.26.

```
my_package/
├── Cargo.toml
├── src/
│   ├── main.rs
│   ├── lib.rs
│   ├── customer.rs
│   ├── product.rs
```

```
│  ├── category.rs
│  └── order.rs
└── target/
```

Listing 6.26 Updated Package File Structure after Adding product and category Files

The code inside the `category.rs` file is shown in Listing 6.27.

```
pub enum Category {
    Electronics,
    Clothing,
    Books,
}
```

Listing 6.27 Code inside the category.rs File

Similarly, the code inside the `product` file will resemble the code shown in Listing 6.28.

```
pub struct Product {
    id: u64,
    name: String,
    price: f64,
    category: category::Category,
}

mod category; // Error

impl Product {
    fn calculate_tax(&self) -> f64 {
        self.price * 0.1
    }

    pub fn product_price(&self) -> f64 {
        self.price + self.calculate_tax()
    }
}
```

Listing 6.28 Code inside the product.rs File

The compiler throws an error for the *product.rs* file: "unresolved module." The compiler further nicely suggests "to create the module `category`, create file `"src\product\category.rs"`."

To fix this error, create a folder inside the `\src` folder, with the parent module name, which is `product` in this case, and then move the *category.rs* file to inside the folder. The updated file structure is shown in Listing 6.29.

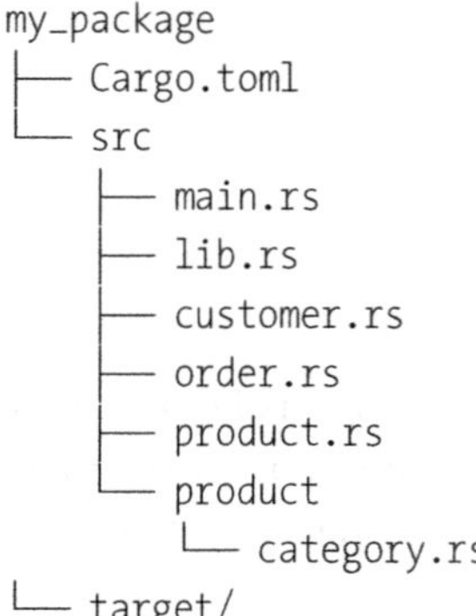

Listing 6.29 File category.rs Moved to the New Folder of product

The error goes away. Since all the code is now arranged in files, the code in the library will look like the following example:

```
mod product;
mod customer;
mod order;
```

In summary, for simple modules with no submodules, we can arrange them in separate files in the source directory. However, for the submodules that we want to arrange in separate files, we must follow the convention of typical file systems.

Organizing Modules in Separate Folders

The second approach is to make separate folders for each of the modules and then have a special file inside each folder called *mod.rs* containing the module contents. For instance, we can create folders corresponding to the `customer` and `order` modules. Each folder will contain *mod.rs* file containing the code of the respective module.

For the modules containing the submodules, like the `product` module in our case, the *mod.rs* file will contain the parent module code. Any submodules will have their own files in the same folder under the same name as that of the submodule. The file structure after making the folders is shown in Listing 6.30.

```
my_package
├── Cargo.toml
└── src
    ├── main.rs
    ├── lib.rs
    ├── customer
    │   └── mod.rs
    ├── order
    │   └── mod.rs
    │
```

```
    └── product
        ├── mod.rs
        └── category.rs
└── target/
```

Listing 6.30 Updated File Structure Based on the Second Method

The second approach, however, comes with a downside. Imagine a bunch of *mod.rs* files opened up in the code editor; telling which module you're modifying can be extremely difficult. I personally prefer the first approach. However, your personal choice dictates how you would like your code to be organized.

6.4 Re-Exporting and Privacy

In this section, we'll explore the concepts of re-exporting and privacy, which play crucial roles in controlling how items are accessed and exposed across modules. Re-exporting allows developers to reorganize and simplify module hierarchies, making external application programming interfaces (APIs) cleaner and more intuitive using the `pub use` keyword. Additionally, we'll delve into the privacy rules for structs, explaining how fields can be selectively made public or kept private to encapsulate data effectively.

6.4.1 Re-Exporting with a Pub Use Expression

Let's demonstrate how the library we've created can be used in the `main` function. To keep things simple, we'll use the basic version of the library without organizing it into separate files. We'll proceed by creating instances of the `Product` and `Customer` structs within `main`. To do this, we'll need to bring the `Product` and `Customer` structs into scope via a `use` declaration, as follows:

```
use my_package::{customer::Customer, product::Product};  // Error
fn main() {}
```

The compiler is giving us errors indicating that the modules are private. This can be fixed by making these modules public in the *lib.rs* file, as indicated in Listing 6.31.

```
pub mod product {
    ...
}
pub mod customer {
    ...
}
```

```
mod order {
    ...
}
```

Listing 6.31 The customer and product Modules Made Public

The errors have now disappeared. However, this approach is not efficient. In `main`, we only need to use the `Customer` and `Product` structs, not their entire modules. Additionally, we shouldn't expose and make everything public unnecessarily. You can bring specific items from inside a module into the scope of code, outside the crate, without making the entire module public. We can achieve this selective visibility by using the `pub` keyword alongside the `use` statement, as shown in Listing 6.32.

```
pub use customer::Customer;
pub use product::Product;
mod product { // not needed to be pub anymore
    ...
}
mod customer { // not needed to be pub anymore
    ...
}
mod order {
    ...
}
```

Listing 6.32 Re-exporting the Product and Customer Structs Using pub use

The `pub` keyword before the `use` declaration enables you to re-export an item from our top-level module, without needing to make the module public. Only the specific item will be made visible to the outside world. Note that we do not need to make the `customer` and `product` modules public anymore.

In `main`, instead of mentioning the full path, we can now use the reduced paths, as in the following examples:

```
use my_package::{Customer, Product};
```

We intend to create an instance of the `Product` in `main`. Since the `Product` struct needs the `Category` enum, let's also re-export it from library, as shown in Listing 6.33.

```
pub use customer::Customer;
pub use product::{category::Category,Product};  // Error
mod product {
    ...
}
mod customer {
    ...
```

```
}
mod order {
    ...
}
```

Listing 6.33 Trying to Re-export Category Enum

The compiler throws an error that the module category is private. The module `category` is not at the top level, but with re-exporting, we can make it available from the top-level modules. To properly re-export, we'll first re-export the `Category` enum from the `product` module to the `crate` module by using the `pub` keyword, as shown in Listing 6.34.

```
pub use customer::Customer;
pub use product::{category::Category,Product};
mod product {
    pub use category::Category;
    ...
}
mod customer {
    ...
}
mod order {
    ...
}
```

Listing 6.34 Re-exporting Category to Top-Level Module

Error: Latest Rust Versions

This syntax might throw an error in the newer versions of Rust (1.80 or later). In the latest versions, you must indicate the re-export of public items from a submodule by mentioning the `crate` module, as shown in Listing 6.35.

```
pub use customer::Customer;
pub use product::Product;
pub use crate::product::Category; // this syntax will work on newer versions
mod product {
    pub use category::Category;
    ...
}
...
```

Listing 6.35 Re-exporting Public Items from the crate Module

In summary, the pub use declaration is quite handy when you want to create a public interface for an item defined in a different module, and you want external code to access that item without needing to navigate the entire module hierarchy.

6.4.2 Privacy of Structs

Let's create an instance of the Product struct in the main function, as shown in Listing 6.36.


```
use my_package::{Category, Customer, Product};
fn main() {
    let product = Product {
        id: 1, // Error
        name: String::from("Laptop"), // Error
        price: 799.99, // Error
        category: Category::Electronics, // Error
    };
}
```

Listing 6.36 Attempting to Create an Instance of the Product

We get a bunch of errors stating that the fields are private. Let's inspect the Product struct in *lib.rs*, as shown in Listing 6.37.

```
pub use crate::product::Category;
pub use customer::Customer;
pub use product::Product;

mod product {
    pub use category::Category;
    pub struct Product {
        id: u64,
        name: String,
        price: f64,
        category: category::Category,
    }
...
}
...
```

Listing 6.37 The Struct Product in lib.rs

The Product struct is public. In Rust, making a struct public does not make its fields public. You have two options to fix this error, which we'll discuss next.

Making All the Fields Public

The first approach is to make all the fields public, as shown in Listing 6.38.

```
pub use crate::product::Category;
pub use customer::Customer;
pub use product::Product;

mod product {
    pub use category::Category;
    pub struct Product {
        pub id: u64,
        pub name: String,
        pub price: f64,
        pub category: category::Category,
    }
...
}
...
```

Listing 6.38 Fixing the Errors from Listing 6.36 by Making the Fields Public

However, this solution is not always ideal. You may want to keep certain fields private. For example, the `id` field might be for internal use and record-keeping, so we shouldn't allow external code to modify it.

Adding a New Constructor Function

Thus, we've come to our second approach, where we'll keep some fields private while still providing controlled access. To achieve this set up, create a new associated function called `new` (or create a constructor function) inside the implementation block. You can use this new constructor to create a new instance of the struct. Consider the example shown in Listing 6.39.

```
pub use crate::product::Category;
pub use customer::Customer;
pub use product::Product;
mod product {
    ...
    impl Product {
        pub fn new(id: u64, name: String, price: f64, category: Category) ->
Self {
            Self {
                id,
                name,
                price,
                category,
```

```
            }
        }
        ...
    }
}
...
```

Listing 6.39 Adding a New Constructor Function for Creating a New Instance of the Product

The inputs to the function are the field values for the struct, and the output is an instance of the `Product`. Inside the function, the field values are initialized from the values passed in. Since the variable names passed in are the same as the names of the struct fields, we can simply use the variable names.

> **Privacy in Enums versus Structs**
>
> The privacy rules for enums are slightly different from the rules that govern structs. When an enum is made public, all of its variants automatically become public as well. Unlike structs, you cannot set the individual variants of an enum to be public independently.

Let's add a similar constructor (`new`) function for creating a new `Customer` in the *lib.rs* file. Consider the example shown in Listing 6.40.

```
...
mod customer {
    ...
    impl Customer {
        fn new(id: u64, name: String, email: String) -> Self {
            Self { id, name, email }
        }
    }
}
...
```

Listing 6.40 A New Constructor Function for the Customer Struct

In `main`, you can create instances of the `Product` and `Customer` using these new constructor functions, as shown in Listing 6.41.

```
use my_package_book::{Category, Customer, Product};
fn main() {
    let product = Product::new(1, String::from("Laptop"), 799.99,
Category::Electronics);

    let customer = Customer::new(1, String::from("Alice"),
```

```
String::from("alice@example.com"));
}
```

Listing 6.41 Creating Product and Customer Using the New Constructor Functions in main

In the same way, you can use the Order struct in main by first re-exporting it, then making the Order struct public, and finally adding a new constructor function for creating a new order.

6.5 Using External Dependencies

At this point in the chapter, you've successfully created and refined a library. Now, consider a situation where the user of our library has created several sets of products and is interested in finding the common products between these sets. The Rust standard library does not provide an intersection function for vectors.

One way to address this is to have the users write the code themselves. However, this approach can be time-consuming, as it involves writing and testing the code. Another option would be to request the library developers to implement the function, but this could be even more costly.

A more efficient and cost-effective solution, especially for commonly occurring programming tasks, is to use an external dependency. Fortunately, Rust makes this easy thanks to its public crate registry, crates.io, accessible through the link *https://crates.io*. It is searchable, meaning that we can search for something similar to what we are looking for. Figure 6.3 shows the result of searching "array intersection."

Next to the search result, we have the number of downloads for the crates in the recent past and all time. The more the numbers, the more popular the crate is. We'll use the array_tool since it has relatively higher numbers. When you select a particular crate and open it, a page appears similar to the one in Figure 6.4. It contains details about the usages, documentation and how to include it in your project. In the text contained in the box in Figure 6.4, it mentions to either run the command cargo add array_tool on the terminal or add the line array_tool = "1.0.3" to your *Cargo.toml* file. Let's use the second method and add the line beneath dependencies in the *Cargo.toml* file or our package as depicted in Listing 6.42.

```
[package]
name = "my_package"
version = "0.1.0"
edition = "2021"

[dependencies]
array_tool = "1.0.3"
```

Listing 6.42 Cargo.toml File after Adding the array_tool

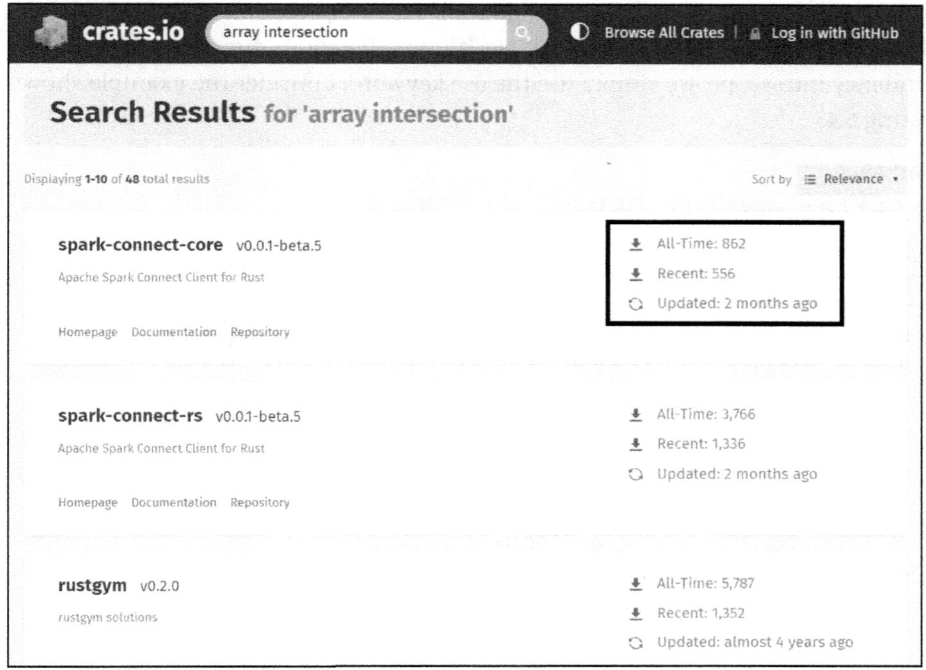

Figure 6.3 Search Results on crates.io

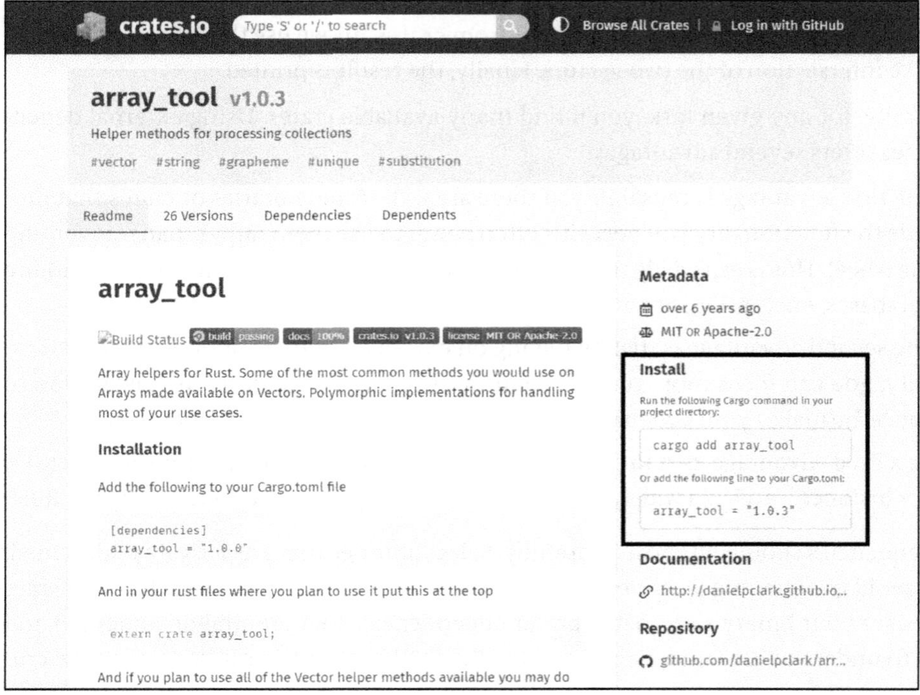

Figure 6.4 Page of a Typical Crate from crates.io

When we add a dependency to our *Cargo.toml* file, cargo automatically informs the Rust compiler about it, allowing us to use it immediately in our code. To bring the dependency into scope, we simply use the use keyword. Consider the example shown in Listing 6.43.

```
use array_tool::vec::*;
use my_package::{Category, Customer, Order, Product};
fn main() {
    let product1 = Product::new(1, String::from("Laptop"), 799.99,
Category::Electronics);
    let product2 = Product::new(2, String::from("T-Shirt"), 20.0,
Category::Clothing);
    let product3 = Product::new(3, String::from("Book"), 10.0,
Category::Books);

    let set1: Vec<&Product> = vec![&product1, &product2];
    let set2: Vec<&Product> = vec![&product2, &product3];
    let intersection = set1.intersect(set2);
    println!("The intersection is: {:?}", intersection);
}
```

Listing 6.43 Using the intersection Method from the array_tool

The code first creates three products and then defines a couple of vectors. Then following the documentation of the array_tool from crates.io, we use the intersect method to take intersection of the two vectors. Finally, the result is printed.

Typically, for any given task, you'll find many available crates. Using external dependencies offers several advantages:

- The first advantage is reusability. If there are well-tested libraries or crates that provide the functionality you need, it's often better to use them rather than reinventing the wheel. This is especially true for common tasks like working with HTTP, handling databases, encryption, or parsing.
- The second advantage is that by relying on external libraries for non-core functionality, you can focus more on the unique aspects of your project. This separation of concerns makes your codebase more maintainable and easier to understand.
- As a final advantage, certain crates become community standards for specific tasks. For instance, serde is a widely used crate for serialization and deserialization in Rust.

Dependencies should be chosen carefully. Select libraries that are actively maintained, and avoid overloading your project with excessive dependencies, as each added crate increases your binary size. Lastly, prefer dependencies that are well-documented and easy to understand, as understanding how they work internally can sometimes be crucial.

6.6 Publishing Your Crate

In the previous section, we learned about integrating dependencies from crates.io into our project. In this section, we'll explore the reverse process, i.e., publishing our package to crates.io, enabling others to utilize it.

6.6.1 Creating an Account on crates.io

We'll first login with a GitHub account on crates.io. On the main page of crates.io, in the right corner, we'll click on the menu and then go to account setting. Make sure that you have verified your email. If this is done properly, green text will appear, indicating to you that your profile is verified, as shown in Figure 6.5.

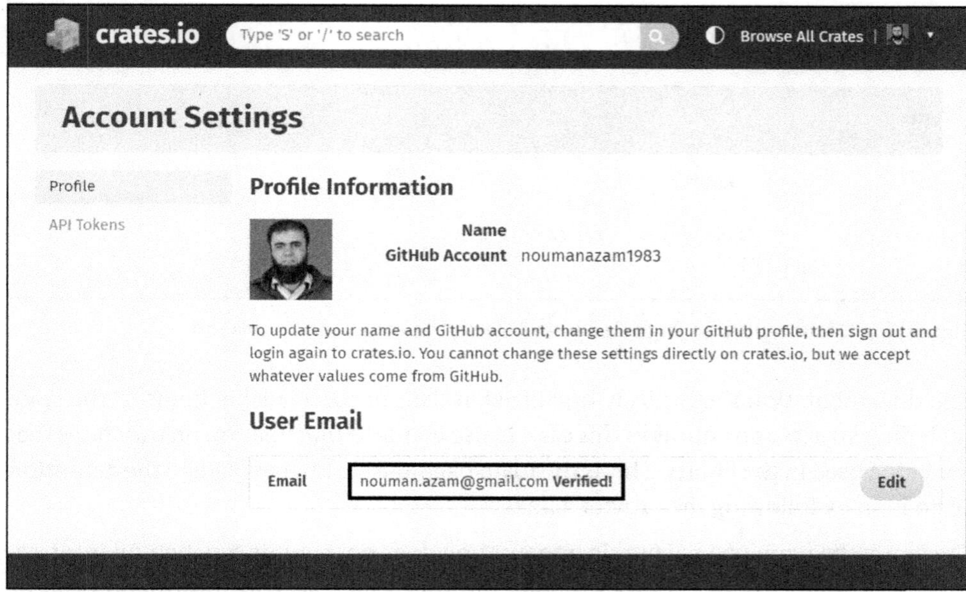

Figure 6.5 Checking the Status of Your Account on crates.io

To publish something on crates.io, we need to have an API token. New tokens are created by clicking **API Tokens** on the left side, as shown in Figure 6.5. Once created, make sure that you copy your token otherwise you may not be able to do so later on, and you'll have to create another token.

Next, we'll use the API token to login from the terminal using the following command:

```
c:\> cargo login complete_api_token
```

We can now publish our packages to crates.io.

6.6.2 Adding Documentation before Publishing

Before publishing, it's important to check how the published version will appear on crates.io. To preview the documentation, we can use the cargo doc command with the --open flag, as follows:

```
c:\> cargo doc --open
```

The command will open up the documentation page for the crate, as shown in Figure 6.6.

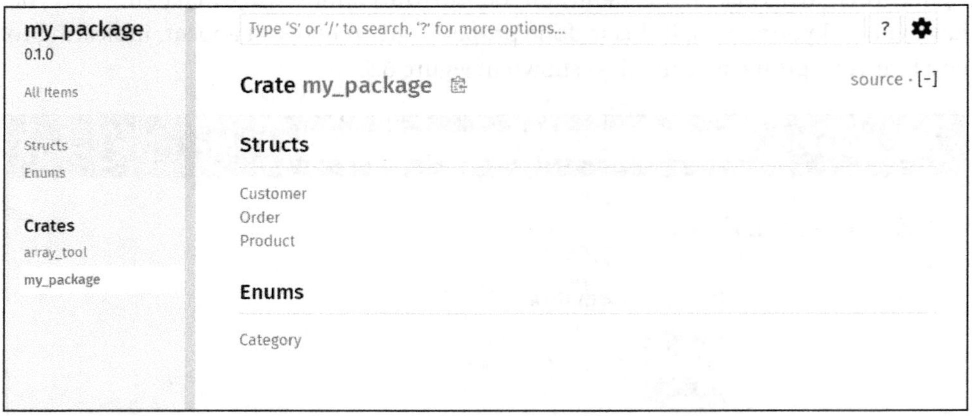

my_package
0.1.0
All Items
Structs
Enums
Crates
array_tool
my_package
Type 'S' or '/' to search, '?' for more options...
?
Crate my_package
source · [-]
Structs
Customer
Order
Product
Enums
Category

Figure 6.6 Documentation Page Generated Using the cargo doc Command

The documentation file is nicely organized. It lists all the available items in the crate including structs and enums in this case. It also contains information on the crates that are being used in the library. This is the main page and we can navigate to the definition of an item by following the provided links.

Currently the main page of our documentation does not contain any helpful information regarding what the library is about and how to use it. This information can be added using documentation comments. The documentation comments begin with three slashes, distinguishing them from standard comments that start with two slashes. To document items such as modules, structs, enums, and functions, we can add the documentation comment above the item. Consider the example shown in Listing 6.44.

```
...
mod product {
    pub use category::Category;
    /// Struct for storing product relatedion information
    pub struct Product {
    ...
    }
```

```
    mod category {
        /// Enum for representing product categories.
        pub enum Category {
       ...
        }
    }
}
...
```

Listing 6.44 Adding Documentation Comments to the Product Struct and Category Enum

If we generate the documentation again, you may notice the documentation details in front of the `Product` struct and `Category` enum on the main page.

The items that are private are not visible in the documentation and therefore any documentation comments added to them will also be not visible in the documentation. We can add sections in the documentation comments using a hash (#) followed by the name of the section. Some typical types of sections include an example section to show how to use the code, a test section for having some test cases, and a panic section to explain why the function might panic. Code blocks can also be added to the documentation comments, using the syntax of three back ticks (```). For instance, consider the code shown in Listing 6.45.

````
...
mod product {
    pub use category::Category;
    /// Struct for storing product relatedion information
    pub struct Product {
        id: u64,
        pub name: String, // make it pub so that the code in doc comments work
        price: f64,
        category: Category,
    }
    ...

    impl Product {
        /// # Example
        /// ```
        /// use my_package::Category;
        /// use my_package::Product;
        /// let some_product = Product::new(1, String::from("Laptop"), 799.9,
Category::Electronics);
        /// assert_eq!(some_product.name, String::from("Laptop"));
        ///
````

```
        /// ```
    ...
}
```

Listing 6.45 Using Documentation Comments to Add Section and Code

`assert_eq!` is a macro which tests if the two things passed in are equal. The interesting aspect of code blocks within documentation comments is that cargo test can execute these code snippets as tests. To run the test, we'll use the following command:

```
c:\> cargo test
```

In the test report section corresponding to doc-test, you may note that the test ran and passed successfully. We'll cover testing in detail in Chapter 7.

6.6.3 Publishing Your Crate

Once the documentation is added, we can proceed with publishing our crate using the following command:

```
c:\> cargo publish
```

This will throw a warning and an error. The warning is "`cargo.toml` file has no `description`, `license`, `license-file`, `documentation`, `homepage` or `repository`." Let's update the *Cargo.toml* file, as shown in Listing 6.46.

```
[package]
name = "my_package"
version = "0.1.0"
edition = "2021"
description = "Online store library"
license = "MIT"

[dependencies]
array_tool = "1.0.3"
```

Listing 6.46 Updated Cargo.toml after Adding the Description and License Information

The package name must be distinct. The names on crates.io are assigned on a first-come first-served basis. If someone else has already taken the same package name, you need to choose a different name.

Next, let's take care of the error, which states "files in the working directory contain changes that were not yet committed into git." There are two options now: You may commit the changes to Git and then try publishing again, or alternatively, you can mention the allow dirty flag and proceed with the publishing. Let's choose the second option, as follows:

```
c:\> cargo publish --allow-dirty
```

Once published, it will take some time to appear under your account on crates.io. Under the dashboard you'll be able to see it.

Once you publish a version to crates.io, you cannot unpublish it. It will remain available permanently. However, you can prevent future projects from using a specific version of your package by yanking it. To do this, simply go to crates.io and open up your dashboard and then select the appropriate crate. Next, click on the version of your crate. This will show you the crate details with the option to yank the crate.

If we want to publish newer versions of your package, we can simply change the version number in the `Cargo.toml` file and publish again. Do not reveal your API token provided by crates.io to anyone. If it gets leaked, you should revoke it by opening your account settings and then under the **API Tokens**, click the **Revoke** button.

6.7 Practice Exercises

In this section, we'll provide some practice questions to help reinforce your learning. The exercise questions cover almost all the concepts introduced in the chapter. Solutions to each exercise will follow in Section 6.8.

1. **Fixing visibility issues in nested modules**
   Fix the code so that it compiles correctly. The code consists of nested modules with a struct and an enum. The goal is to ensure that the struct `A` in module `m1` can access the enum `D` defined in module `m2`.

```
mod m1 {
    struct A {
        d: m2::D,
    }
    mod m2 {
        enum D {
            B,
            C,
        }
    }
}
fn main(){}
```

2. **Resolving module visibility and path issues**
   Fix the code so that it compiles correctly. The code involves nested modules with a structure and an enum in `m1`, and another structure in `m3` that needs to reference the enum from `m2`. Ensure that `m3` can access the enum `D` defined in `m2`.

```
mod m1 {
    struct A {
        d: m2::D,
    }
    mod m2 {
        pub enum D {
            B,
            C,
        }
    }
}
mod m3 {
    struct C {
        e: crate::m1::m2::D,
    }
}
fn main(){}
```

3. **Fixing module import and function usage**
   Fix the code so that it compiles correctly. The code involves a module seasons that contains an enum Season and a function is_holiday. Ensure that the main function can access the Season enum and the is_holiday function from the seasons module.

```
mod seasons {
    pub enum Season {
        Spring,
        Summer,
        Autumn,
        Winter,
    }

    pub fn is_holiday(season: &Season) -> bool {
        match season {
            Season::Summer => true,
            _ => false,
        }
    }
}

fn main() {
    let current_season = Season::Autumn;
    if is_holiday(&current_season) {
        println!("It's a holiday season! Time for a vacation!");
    } else {
```

```
        println!("Regular work season. Keep hustling!");
    }
}
```

4. **Fixing access to private fields in a struct**
   Fix the code so that it compiles correctly. The code defines a module University with a Student struct. Ensure that you can access the fields of Student in the main function by either making the fields public or accessing them appropriately.


```
mod University {
    pub struct Student {
        name: String,
        marks: u8,
        grade: char,
    }
}

use University::Student;

fn main() {
    let mut student_1 = Student {
        name: String::from("Alice"),
        marks: 75,
        grade: 'A',
    };
    println!("{} got {} grade", student_1.name, student_1.grade);
}
```

5. **Re-exporting functions for simplified access**
   Re-export the items from the graphics module so that the calculate_area and show_area functions are easily accessible. Ensure that the main function can use these functions correctly by fixing the import statements.

```
mod graphics {
    // Re-export the 'show_area' function for easier access
    // Re-export the 'calculate_area' function for easier access
    pub mod shapes {
        pub fn calculate_area(radius: f64) -> f64 {
            std::f64::consts::PI * radius * radius
        }
    }
    pub mod display {
        pub fn show_area(shape: &str, area: f64) {
            println!("The area of the {} is: {}", shape, area);
        }
    }
```

```
}

use ___::calculate_area; // fix this line
use ___::show_area; // fix this line
fn main() {
    let radius = 3.0;
    let area = calculate_area(radius);
    show_area("circle", area);
}
```

6.8 Solutions

This section provides the code solutions for the practice exercises in Section 6.7. The code is largely self-explanatory; however, we have included comments and additional explanations wherever necessary to enhance understanding.

1. **Fixing visibility issues in nested modules**

```
mod m1 {
    struct A {
        d: m2::D,
    }
    mod m2 {
        pub enum D {    // Child module items are not visible to the
                        // parent module

            B,
            C,
        }
    }
}
fn main(){}
```

2. **Resolving module visibility and path issues**

```
mod m1 {
    struct A {
        d: m2::D,
    }
    pub mod m2 {
    /* Public items of a private child module, are only accessible by the
       parent. We need to make the child module m2 pub, so that its public
       items can be used outside the parent module.*/

        pub enum D {
            B,
```

```
            C,
        }
    }
}

mod m3 {
    struct C {
        e: crate::m1::m2::D,
    }
}
fn main(){}
```

3. **Fixing module import and function usage**

```
mod seasons {
    pub enum Season {
        Spring,
        Summer,
        Autumn,
        Winter,
    }

    pub fn is_holiday(season: &Season) -> bool {
        match season {
            Season::Summer => true,
            _ => false,
        }
    }
}
use seasons::{is_holiday, Season};
fn main() {
    let current_season = Season::Autumn;
    if is_holiday(&current_season) {
        println!("It's a holiday season! Time for a vacation!");
    } else {
        println!("Regular work season. Keep hustling!");
    }
}
```

4. **Fixing access to private fields in a struct**

```
mod University {
    pub struct Student {
        pub name: String, // fields need to be made public
        pub marks: u8,
        pub grade: char,
```

```
    }
}

use University::Student;

fn main() {
    let mut student_1 = Student {
        name: String::from("Alice"),
        marks: 75,
        grade: 'A',
    };
    println!("{} got {} grade", student_1.name, student_1.grade);
}
```

5. **Re-exporting functions for simplified access**

```
mod graphics {
    pub use self::display::show_area;
    pub use self::shapes::calculate_area;
    pub mod shapes {
        pub fn calculate_area(radius: f64) -> f64 {
            std::f64::consts::PI * radius * radius
        }
    }
    pub mod display {
        pub fn show_area(shape: &str, area: f64) {
            println!("The area of the {} is: {}", shape, area);
        }
    }
}

use graphics::calculate_area;
use graphics::show_area;
fn main() {
    let radius = 3.0;
    let area = calculate_area(radius);

    show_area("circle", area);
}
```

6.9 Summary

In this chapter, we explored various techniques for organizing code effectively in Rust. We started by discussing the importance of code organization and introduced modules as a way to structure your code, using an online store example to illustrate their value. You learned how to create modules, how to navigate to items using relative and absolute paths, and how to handle privacy within modules. This chapter also covered the `use` declaration for bringing items into scope, ultimately improving code readability.

We then examined ways to visualize and organize modules, using Cargo to view module hierarchies and adopting typical file system structures to maintain clarity. We discussed re-exporting and privacy, with a focus on managing the visibility of struct fields and the addition of constructors to control access. We also emphasized the use of external dependencies, highlighting how they enhance functionality without reinventing the wheel. Finally, we covered the process of publishing a crate to crates.io, from creating an account to adding documentation comments and publishing the crate itself. We hope this chapter equips you with tools essential for organizing and managing your code in larger Rust projects.

Next, we'll explore unit testing, starting with how to write tests to verify the functionality of individual Rust components.

Chapter 7
Testing Code

Testing isn't just about finding bugs—it's about building confidence in your code. In this chapter, you'll learn how Rust's robust testing framework helps you ensure correctness, optimize performance, and maintain control over your software's behavior.

Get ready to discover the importance of testing and how to implement it effectively in Rust. The chapter begins with unit testing basics, teaching you how to write tests to verify individual components. It then explores testing for panic to ensure code handles unexpected conditions gracefully. We also cover controlling test execution and running integration tests, highlighting the differences between testing isolated components and entire systems. This chapter concludes with benchmark testing to measure and optimize code performance, ensuring you can deliver software that performs reliably and efficiently.

7.1 Unit Testing

This section delves into unit testing, a fundamental practice for ensuring the correctness of individual components in Rust programs. We begin by examining the structure of a typical test case, followed by step-by-step guidance on writing test functions and executing tests. Additionally, we explore how you can leverage the Result enum for test outcomes and how to handle scenarios where functions are expected to panic, thus ensuring comprehensive and robust test coverage.

7.1.1 A Typical Test Case

A test case is automatically inserted in the library crate when we create a package containing a library crate, as in the following command:

```
c:\> cargo new testing --lib
```

The `--lib` flag tells Cargo that the package will contain a library crate in the source directory. Opening up the *lib.rs* file, you'll see the code shown in Listing 7.1.

```
pub fn add(left: u64, right: u64) -> u64 {
    left + right
}

#[cfg(test)]
mod tests {
    use super::*;

    #[test]
    fn it_works() {
        let result = add(2, 2);
        assert_eq!(result, 4);
    }
}
```

Listing 7.1 Code in the lib.rs of the Newly Created Project

The code for a test is automatically generated in the *lib.rs* file. Let's walk through this code next.

We have a module called test. The module is annotated with the configuration attribute #[cfg(test)]. This attribute is for conditional compilation, specifically for tests. When applied to a module, this attribute indicates to Cargo that the contained code should only compile when running tests through cargo test. This attribute thus stops the code from being compiled during the regular build process.

Within the module, we have a test function named it_works. In Rust, test functions are marked with the test annotation. Inside this function, we define a variable called result to hold the output of the add function, which simply computes the sum of two input numbers. Finally, the assert_eq macro is used to compare the result with the value 4.

7.1.2 Writing a Test Function

Test functions are written to test the functionality of our code and are typically included in a library. Consider the code in Listing 7.2 in the library (delete the previous code).

```
mod shapes {
    pub struct Circle {
        radius: f32,
    }
    impl Circle {
        pub fn new(radius: f32) -> Circle {
            Circle { radius }
        }
```

```
        pub fn contains(&self, other: &Circle) -> bool {
            self.radius > other.radius
        }
    }
}
```

Listing 7.2 Shapes Module Containing Code Related to Circles

This code defines a set of methods for the `Circle` struct within the `shapes` module. First, we implement a public constructor function named `new` that allows us to create a new instance of `Circle` by specifying its `radius`. This function returns a `Circle` struct initialized with the provided `radius`. Additionally, we define another public method called `contains` that checks whether one circle instance can contain another. This method takes a reference to another `Circle` and compares its `radius` with the `radius` of the current instance. The method returns a Boolean value indicating whether the current circle has a larger `radius` than the `other circle`, meaning the former circle can contain the latter.

Now that our code is defined, let's add a simple test to test the functionality provided in the code. Consider the code shown in Listing 7.3.

```
...
#[cfg(test)]
mod tests {
    use super::*;
    #[test]
    fn larger_circle_should_contain_smaller() {
        let larger_circle = shapes::Circle::new(5.0);
        let smaller_circle = shapes::Circle::new(2.0);
        assert_eq!(larger_circle.contains(&smaller_circle), true);
    }
}
```

Listing 7.3 Test Function for Testing the Functionality from Listing 7.2

As explained earlier, test functions are annotated with the `#[test]` attribute and are contained within a test module, which is annotated with a test configuration attribute (`#[cfg(test)]`). The test module is the child module of the root module of the file; therefore, we need to bring all the items from the parent module into scope. The keyword `super` in the statement `use super::*;` refers to the parent module, and the star (*) tells the compiler to import everything.

The test function `larger_circle_should_contain_smaller`, as the name suggests, is checking whether the larger circle contains the smaller circle. A typical structure for a test involves three key steps:

1. Prepare the necessary inputs and set up the initial conditions.
2. Invoke the function or method being tested.
3. Verify the result by comparing it against the expected output using assertions.

Following this pattern, we create a couple of circle instances called `larger_circle` and `smaller_circle`. Next, the function `contains` is called inside an `assert` statement to assert that the `larger` circle should contain the smaller one. If the assertion passes, the macro evaluates to `true`, allowing the program to continue execution and the test is passed. If it evaluates to `false`, the macro triggers a panic causing the test to fail. A *panic* refers to a situation where a program encounters an unrecoverable error at runtime. When a panic occurs, the program's normal execution is abruptly terminated. We'll discuss testing with panics in more detail in Section 7.1.5.

We'll add one more test to the `test` module and then generate the testing report using the `cargo` command. Consider the code shown in Listing 7.4.

```
...
#[cfg(test)]
mod tests {
    use super::*;
     ...
    #[test]
    fn smaller_circle_should_not_contain_larger() {
        let larger_circle = shapes::Circle::new(5.0);
        let smaller_circle = shapes::Circle::new(2.0);
        assert_eq!(smaller_circle.contains(&larger_circle), false);
     }
```

Listing 7.4 Another Test Added to the tests Module

The test is similar to the test shown earlier in Listing 7.3; however, now, we're testing the opposite logic (i.e., the `smaller_circle` should not be contained within the `larger_circle`). We are asserting that the function should return a `false` through the following statement:

```
assert_eq!(!smaller_circle.contains(&larger_circle), true);
```

This statement has the same result as the `assert` statement shown earlier in in Listing 7.4. The exclamation mark (`!`) will alter the return `bool` value of the function.

7.1.3 Executing Tests

To execute the tests included in the library, enter the `test` command, as in the following example:

```
c:\> cargo test
```

When you execute this command, you'll see an output to the terminal similar to the output shown in Listing 7.5.

```
running 2 tests
test tests::larger_circle_should_contain_smaller ... ok
test tests::smaller_circle_should_not_contain_larger ... ok
test result: ok. 2 passed; 0 failed; 0 ignored; 0 measured; 0 filtered out;
finished in 0.00s
   Doc-tests testing1
running 0 tests
test result: ok. 0 passed; 0 failed; 0 ignored; 0 measured; 0 filtered out;
finished in 0.00s
```

Listing 7.5 Sample Output after Executing a cargo test Command

Two tests ran and their status is ok, meaning that they both passed. The other section is for documentation (doc) tests if they are included in the code files. A *doc test* in Rust is a test embedded within the documentation of a code item, allowing examples in comments to be compiled and executed to ensure they remain accurate and functional. We touched on this topic earlier in Chapter 6, Section 6.6.2. Let's see what happens if we change the condition inside the contains method (shown earlier in Listing 7.2). Consider the example shown in Listing 7.6.

```
mod shapes {
    ...
    impl Circle {
        ...
        pub fn contains(&self, other: &Circle) -> bool {
            self.radius < other.radius // Condition changed
        }
    }
}
...
```

Listing 7.6 Code from Listing 7.2 with Condition inside the contains Method Changed

If you execute the test command again, you'll see the following output:

```
running 2 tests
test tests::smaller_circle_should_not_contain_larger ... FAILED
test tests::larger_circle_should_contain_smaller ... FAILED
...
```

Now, you understand how basic testing works. You can pass custom failure messages to the assert_eq! macro by adding another parameter. Consider the code shown in Listing 7.7.

```
#[cfg(test)]
mod tests {
    ...
    fn larger_circle_should_contain_smaller() {
        ...
        assert_eq!(
            larger_circle.contains(&smaller_circle),
            true,
            "Custom failure message" // Custom message
        );
    }
    ...
}
```

Listing 7.7 Adding a Custom Failure Message to the assert_eq! Macro

Note

Macros enable code generation and metaprogramming, allowing developers to write concise, reusable code patterns that are expanded at compile time. For more information on macros, see Chapter 15.

This custom failure message will only appear in the terminal when the test fails.

Let's briefly consider a few more macros, namely, `assert_ne!` and `assert!`, through some examples:

```
assert_ne!(larger_circle.contains(&smaller_circle), false);
assert!(larger_circle.contains(&smaller_circle));
```

The first assertion will check that the `larger_circle`, containing a smaller circle, cannot be equal to `false`. The `assert!` macro checks if a Boolean expression is `true`. The second statement is therefore simply checking that the `larger_circle` contains the smaller circle, which in this case is `true`. Note that the second statement does not contain an explicit second argument.

7.1.4 Testing with Result Enum

The use of the `Result` enum (introduced in Chapter 5, Section 5.4) in connection to the `test` function is interesting. The `test` function we've created so far passes if the assertion passes and fails if the assertion fails. You can, however, make your test functions return a `Result` enum. This approach is typically used when the function being tested itself returns a `Result` enum.

Let's implement a new version of the constructor function defined earlier in Listing 7.2 with the code shown in Listing 7.8.

```
mod shapes {
...
    impl Circle {
        ...
        pub fn new_1(radius: f32) -> Result<Circle, String> {
            if radius >= 0.0 {
                Ok(Circle { radius })
            } else {
                Err(String::from("radius should be positive"))
            }
        }
    }
}
```

Listing 7.8 New Version of the Constructor Is Added to the Code from Listing 7.2

The constructor `new_1` returns a `Result` enum. This function will return an `Err` if a negative value or a zero is passed for the radius parameter.

In the `test` module, we'll add another test that will attempt to create a `Circle` with a negative value using the `new_1` constructor function, as shown in Listing 7.9.

```
#[cfg(test)]
mod tests {
    ...
        #[test]
        fn should_not_create_circle() -> Result<(), String> {
            let some_circle = shapes::Circle::new_1(-1.0)?;
            Ok(())
    }
}
```

Listing 7.9 A Test for Testing the new_1 Constructor Function

A few things to note in this scenario. First, instead of using the assert macro, we return the `Ok` variant if the test passes. If the test fails, we would return an `Err` variant, but we're not doing that explicitly. The question mark operator (`?`) handles this for us through error propagation. Specifically, if the statement evaluates to an `Ok` variant, control passes to the next statement, which returns `Ok`. However, in the case of an error, the function will return early with an `Err` variant. The `?` operator was covered earlier in Chapter 5, Section 5.4.4.

7.1.5 Testing Panics

In the previous section, we used assert macros and the Result enum in our test functions. Besides these capabilities, you can also check inside a test function to ensure that a function panics and does not complete its normal execution.

Let's create a new test function. This function will assert that the constructor should panic when a negative value is passed as the radius. To specify that a function should panic, you can use the #[should_panic] attribute, as in the code example shown in Listing 7.10.

```
#[cfg(test)]
mod tests {
    ...
#[test]
#[should_panic]
fn should_not_create_and_panic() {
    let some_circle = shapes::Circle::new(-11.0);
}
}
```

Listing 7.10 Test Function That Expects the Constructor to Panic

If we run the test using the cargo test command, the test of interest (should_not_create_and_panic) where a panic is expected failed. In the test output, we see a message explaining that the test did not panic as expected. This happened because the new constructor function currently accepts negative values.

Therefore, we must now define another version of the constructor, named new_2, one that panics when a negative or zero value is passed in. The moment of panic occurs when the program matches the radius value inside the function. Consider the code shown in Listing 7.11.

```
mod shapes {
...
    impl Circle {
        ...
        pub fn new_2(radius: f32) -> Circle {
            match radius {
                ..=0.0 => panic!("radius should be positive"),
                _ => Circle { radius },
            }
        }
    }
}
```

Listing 7.11 Another Constructor Function Added to Circle Implementation

Inside the new_2 function, we're matching on the radius. If the radius is a negative value or zero, the function will panic. Otherwise, the function will create and return a Circle. In the code, inside the match arm, we are returning a panic. This technique is not typically recommended for real-world coding. In professional programming, you should handle errors gracefully by returning a Result type, thus preventing unexpected program crashes. The use of panicking in this example is purely for demonstration purposes.

Now, inside the code shown earlier in Listing 7.10, let's call the new_2 constructor function instead of the new constructor function, as shown in Listing 7.12.

```
#[cfg(test)]
mod tests {
    ...
#[test]
#[should_panic]
fn should_not_create_and_panic() {
    let some_circle = shapes::Circle::new_2(-11.0);
}
}
```

Listing 7.12 Code from Listing 7.10 Updated by Calling new_2 Constructor Function

If we execute the test command again and inspect the test report, you may note that the test has passed. This result is correct because the test panicked as expected and, therefore, the test passed.

We can also add an expected parameter to the #[should_panic] attribute to ensure that the test panics due to an expected message (that the radius is less than –10), as shown in Listing 7.13.

```
#[cfg(test)]
mod tests {
    ...
#[test]
#[should_panic(expected = "is less than -10.0")]
fn should_not_create_and_panic() {
    let some_circle = shapes::Circle::new_2(-11.0);
}
}
```

Listing 7.13 Expected Parameter Added to #[should_panic]

When we run the test again, it fails, even though the function panicked. The reason the test failed is that the actual panic message didn't match the expected string. The actual panic message and the expected message are displayed in the test output:

```
panic message: `"should be positive"`,
expected substring: `"is less than -10.0"`
```

We'll now modify the new_2 function to ensure the test passes. We'll add a couple more arms to the new_2 function defined earlier in Listing 7.11. Now, we have the code shown in Listing 7.14.

```
mod shapes {
    ...
    impl Circle {
      ...
      pub fn new_2(radius: f32) -> Circle {
            match radius {
                ..=-10.0 => panic!("is less than -10.0"),
                -10.0..=0.0 => panic!("is between -10.0 and 0.0"),
                _ => Circle { radius },
            }
        }
        pub fn contains(&self, other: &Circle) -> bool {
            self.radius > other.radius
        }
    }
}
```

Listing 7.14 Modified new_2 with Added Arms from Listing 7.11

The first arm will panic if the radius is less than -10, and the second arm will panic if the radius is between -10 and 0, including 0. The panic message in the first arm is matching the expected string in the #[should_panic]. If we now update the call to the function with a value of -11, the test will pass.

Testing Considerations

A few important side notes regarding testing include the following:

- First, test functions can be standalone functions. In other words, they don't need to exist inside a test module to work. For example, if we move some tests outside the test module, they will still run. However, a best practice is to keep them all in a test module, as doing so keeps the code clean and organized.
- Second, functions inside the test module that lack the #[test] attribute aren't considered test functions. Instead, they act as helper functions, assisting test functions by setting up preconditions, performing common tasks, or verifying behaviors. This distinction keeps your actual test functions focused on their purposes.
- Third, Rust's testing framework allows test functions to access private functions, as long as they are within the same module. For example, as shown in Listing 7.15, a

private function without the pub keyword (i.e., some_fn) is called inside a test function.

```
fn some_fn() {}
#[cfg(test)]
mod tests {
    ...
    #[test]
    fn larger_circle_should_contain_smaller() {
        some_fn();
        ...
        }
}
```

Listing 7.15 Calling a Private some_fn Function inside a Test Function

Note that testing private functions is a hotly debated topic in the software development community, with strong arguments both for and against.

7.2 Controlling Test Execution

In this section, we'll go over some configuration options while executing tests, such as the following:

- **Viewing only library test reports**
  The first thing to note is that, if your package has multiple binary crates and a single library crate, the `cargo test` will generate separate reports for each of the binary crate and the library crate. To view only the test reports for the test in the library, we'll provide the additional flag of `--lib` to `cargo test`, as follows:

  ```
  C:\> cargo test --lib
  ```

  This approach can be helpful when working on larger projects with many binary crates.

- **Generating test output**
  The test functions, when passed, do not generate any output from the test functions they are testing. For instance, let's add a print statement to the `contains` method shown earlier in Listing 7.2, resulting in the code shown in Listing 7.16.

  ```
  mod shapes {
      ...
      impl Circle {
      ...
          pub fn contains(&self, other: &Circle) -> bool {
              println!("Checking if the circle is contained");
  ```

```
                self.radius > other.radius
            }
        }
}
```

Listing 7.16 Print Statement Added to the contains Method from Listing 7.2

Now, if we execute the tests, notice how the tests `larger_circle_should_contain_smaller` and `smaller_circle_should_not_contain_larger` did not generate any output. You can see the output by specifying the additional argument of `--show-output`:

```
C:\> cargo test --lib -- --show-output
```

- **Filtering tests**

  The test report in the previous case will contain the output corresponding to each testing function. To test a specific test, use the full name of the test function in the `cargo test` command:

  ```
  C:\> cargo test larger_circle_should_contain_smaller
  ```

  This command will only run the specified test, and the test report will indicate that remaining tests were filtered out.

  To run multiple tests, we can filter out tests by specifying part of a test name. Any tests containing the specified part will be executed. For example, if we specify `should_not`, all tests with that phrase in their name will run, as in the following command:

  ```
  C:\> cargo test --lib should_not
  ```

- **Ignoring tests**

  Sometimes, you want to ignore some tests while executing others. Perhaps, you have tests that are platform specific, that are not completed yet, or that are so huge they require lots of time and computation. In these cases, you'll want to execute tests separately. So, let's add another test function, one that represents some test that is computationally intensive. This test is shown in Listing 7.17.

```
#[cfg(test)]
mod tests {
    ...
    #[test]
    #[ignore]
    fn huge_test() {
        // code that runs for hours
    }
}
```

Listing 7.17 Computationally Expensive Test Simulated Using huge_test

The annotation of #[ignore] will ignore the test during the regular testing of functions using the cargo test command. If we run the command, you may see a line in the testing report similar to the following:

```
test tests::huge_test ... ignored
```

Moreover, in the test result section of the report, notice that we have one ignored test. To separately execute ignored tests, use the additional flag of --ignored, as in the following command:

```
C:\> cargo test -lib -- --ignored
```


This command will only execute the tests that include the annotation #[ignored].

7.3 Integration Tests

You can use integration tests to verify that different units of code work together as expected. Unlike unit tests, which focus on testing individual modules or functions in isolation, integration tests aim to identify issues that may arise when these units interact with each other.

In contrast to unit tests, integration tests exist as files separate from the code file containing the library and should only test the public interface. More specifically, integration tests are stored in a top-level directory of the package called *tests*. Cargo knows how to look for integration tests inside this directory.

Let's create a new package called integration_test, with the following command:

```
c:\> cargo new integration_test
```

Next, we'll add a *lib.rs* file to the *src* directory of the package, which will contain the code of the online store library we developed in Chapter 6, Section 6.2.1. We'll also add a *tests* folder to the top-level directory of the package. The *tests* folder will contain an *order_test.rs* file. The directory structure is shown in Listing 7.18.

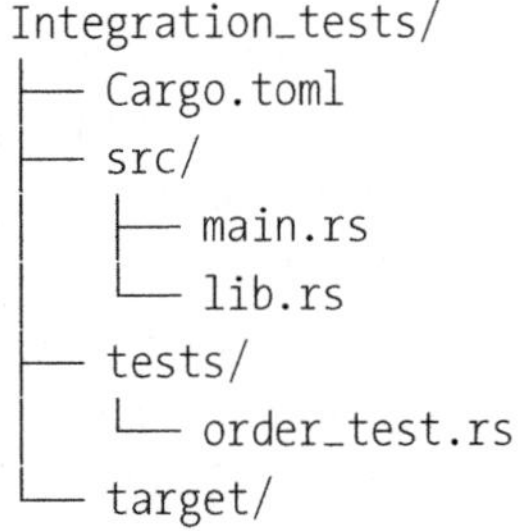

```
Integration_tests/
├── Cargo.toml
├── src/
│   ├── main.rs
│   └── lib.rs
├── tests/
│   └── order_test.rs
└── target/
```

Listing 7.18 Directory Structure of the integration_test Packages

Cargo will compile each file in the *tests* directory as a separate crate. The file *order_test* is created to test the order module. Let's add some tests to this file, as shown in Listing 7.19.

```
use integration_tests::{Category, Customer, Order, Product};
#[test]
fn test_total_bill_without_discount() {
    let product = Product::new(1, String::from("Book"), 19.9, Category::Books);
    let customer = Customer::new(1, String::from("Bob"),
                              String::from("bob@example.com"));
    let order = Order::new(2, product, customer, 3);
    assert_eq!(format!("{:.2}", order.total_bill()), "65.67");
}

#[test]
fn test_total_bill_with_discount() {
    let product = Product::new(1, String::from("Book"), 19.99,
Category::Books);
    let customer = Customer::new(1, String::from("Bob"),
String::from("bob@example.com"));
    let order = Order::new(2, product, customer, 10);
    assert_eq!(format!("{:.2}", order.total_bill()), "197.90");
}
```

Listing 7.19 Code in order_test.rs

We have a couple of tests in the file. In the first line, we are importing the relevant modules. Note that the order module uses items from the modules of product and customer; therefore, we've also brought them into scope. We can write test functions, similar to how we wrote unit tests in Section 7.1.2. We have a couple of tests in the file. Both tests are checking the function of total_bill in the order module (for details of the total_bill, see Chapter 6, Section 6.2.1, especially Listing 6.4). The first test, test_total_bill_without_discount, is checking the total bill without applying any discount. The test function first performs some setup by creating Product, Customer and Order instances and then it asserts that the total_bill should be 65.67 dollars. As per the business logic defined in *lib.rs* in Chapter 6, Section 6.2.1, Listing 6.4, since the number of ordered items is less than 5, the discount does not apply. The total price is calculated as 19.9 times 3, plus a 10% tax, resulting in a total of 65.67 dollars. The :.2 inside the parentheses is a format specifier that indicates we want a floating-point number to two decimal places. Since the format! macro returns a string, we need to compare the result against a string. The second test function checks the total_bill when a discount applies.

Integration tests are executed using the same cargo test command. After executing an integration test, you may see that the test report contains a report for the tests in the *lib.rs* file, which corresponds to unit tests, and then a separate report for the integration tests, which are in the *order_test* file. You can shrink down the testing report to only contain information about specific integration test using the `--test` flag, as in the following command:

```
c:\> cargo test --test order_test
```

The report now only contains the specific integration test result.

Every file in the *tests* directory is treated as a separate crate and considered a test file by Cargo. Sometimes, however, you want to create multiple crates to isolate and organize our tests more meaningfully. However, this approach comes with a downside. In some cases, test functions in different files within the *tests* directory may need to share code, such as test setup code. For instance, consider another file in the *tests* directory, called *helpers*, containing a public function called `common_setup`. The updated directory structure after adding this file is shown in Listing 7.20.

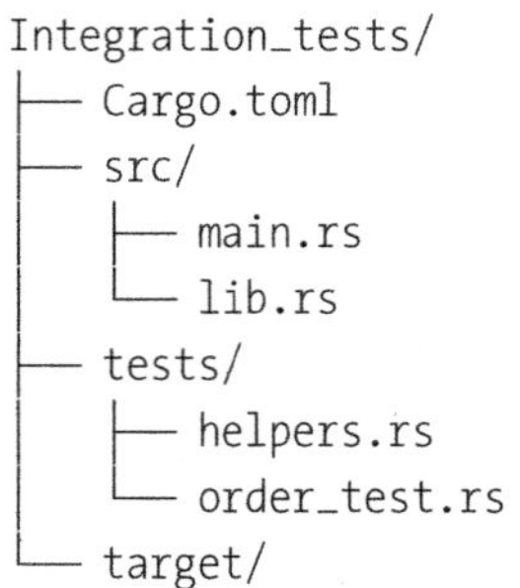

```
Integration_tests/
├── Cargo.toml
├── src/
│   ├── main.rs
│   └── lib.rs
├── tests/
│   ├── helpers.rs
│   └── order_test.rs
└── target/
```

Listing 7.20 Updated File Structure from Listing 7.18 with the Addition of the helpers.rs File

The contents of the *helpers.rs* file is created in the following code:

```
pub fn common_setup();
```

Now, when we execute the `cargo test` command, note how the *helpers.rs* file was executed as if it was a test file.

To fix this problem so that Cargo does not treat the *helpers.rs* file as a test file, we'll create `helpers` module using a *mod.rs* file. Inside the `tests` directory, add another folder named `helpers`, which will contain the *mod.rs* file. The *mod.rs* file will now contain the code in the same code previously in the *helpers.rs* file. Finally, we'll delete the *helpers.rs* file. The updated directory structure is shown in Listing 7.21.

```
Integration_tests/
├── Cargo.toml
├── src/
│   ├── main.rs
│   └── lib.rs
│
```

```
├── tests/
│   ├── helpers/
│   │   └── mod.rs
│   └── order_test.rs
└── target/
```

Listing 7.21 The Updated Directory Structure after Adding the mod.rs File

In the new directory structure, the *mod.rs* is not a top-level file in the `test` directory; therefore, Cargo will not treat it as a test file. You can now use the *helpers.rs* file as a child module in the *order_test.rs* file and can use its functions. Consider the code shown in Listing 7.22.

```
use integration_tests::{Category, Customer, Order, Product};
mod helpers;
#[test]
fn test_total_bill_without_discount() {
    helpers::common_setup();
    ...
}
...
```

Listing 7.22 Using the common_setup Function from the helpers Module

> **Note**
> Integration tests cannot directly import items from a binary crate. To address this problem, a common pattern in Rust is to keep the binary crate small and move most of the code into a library crate.

7.4 Benchmark Testing

Benchmarking is the process of comparing the performance of multiple programs that perform the same task. Benchmark testing can involve comparing different implementations of the same program or different versions of the same implementation. The primary goal is to determine if a change has improved or worsened the program's speed.

When benchmarking, various factors can be considered, such as processing time, memory usage, and disk access time. In this section, we'll primarily focus on processing time.

To illustrate benchmarking in Rust, let's consider a scenario where we want to compare two different implementations of sorting elements in an array. These implementations are included as the functions shown in Listing 7.23, which are contained in the library file.

```
pub fn sorting_algo_1<T: PartialOrd>(arr: &mut Vec<T>) {
    let mut swapped = false;
    for i in 0..(arr.len() - 1) {
        if arr[i] > arr[i + 1] {
            arr.swap(i, i + 1);
            swapped = true;
        }
    }
    if swapped {
        sort_algo_1(arr);
    }
}

pub fn sorting_algo_2<T: Ord>(arr: &mut Vec<T>) {
    let len = arr.len();
    for left in 0..len {
        let mut smallest = left;
        for right in (left + 1)..len {
            if arr[right] < arr[smallest] {
                smallest = right;
            }
        }
        arr.swap(smallest, left);
    }
}
```

Listing 7.23 Two Implementations of the Sorting Algorithm in Our lib.rs File

The first implementation, `sorting_algo_1`, and the second implementation, `sorting_algo_2`, represent two different approaches to sorting. Both functions take a mutable reference to a vector as input and sort the elements of that vector within their respective bodies.

To measure performance and ensure performance doesn't degrade, we can use benchmarking. While Rust has built-in benchmarking tests, at the time of writing, these features are unstable and only available in the nightly version of Rust. Therefore, we'll use the `criterion` library for our benchmarking.

To integrate `criterion` into our project, open the *Cargo.toml* file and add `criterion` as a development dependency, with the following lines:

```
[dev-dependencies]
criterion = "0.4.0"
```

You'll also need to configure your benchmark target, which is the name of the file that contains our benchmark test. Let's use the name `sorting_benchmark` and add the following lines to the *Cargo.toml* file:

```
[[bench]] ]
name = "sorting_benchmark"
harness = false
```

The line `harness = false` will disable the default benchmarking system, so we can use `criterion`'s provided benchmarking system.

Next, we'll create a folder named *benches* in the root directory of our project. This specific name (benches) is important because Cargo uses it to locate the benchmark files. Inside this folder, we'll create a new file named *sorting_benchmark.rs*. The updated directory structure is shown in Listing 7.24.

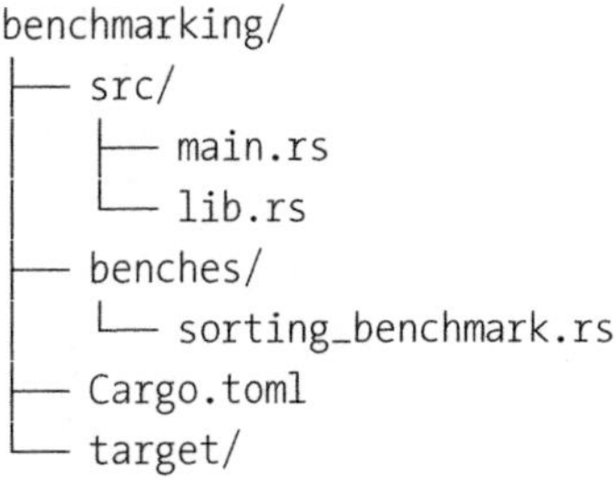

```
benchmarking/
├── src/
│   ├── main.rs
│   └── lib.rs
├── benches/
│   └── sorting_benchmark.rs
├── Cargo.toml
└── target/
```

Listing 7.24 File Structure of the Package

This name of the file inside the *benches* folder must be the same as the name mentioned earlier in the *Cargo.toml* file.

Let's now add code to the *sorting_benchmark.rs* file. First, we'll bring our sort array functions from the library into scope, as well as some items from the criterion crate into scope, with the following lines:

```
use benchmarking::{sort_algo_1, sort_algo_2};
use criterion::{criterion_group, criterion_main, Criterion};
```

Next, we'll set up our benchmarking test function, as shown in Listing 7.25.

```
fn sort_benchmark(c: &mut Criterion) {
    let mut numbers: Vec<i32> = vec![
       1, 2, 3, 6, 5, 4, 8, 52, 2, 1, 5, 4, 4, 5, 8, 54, 2, 0, 55, 5, 2, 0, 5,
5, 5, 21];

    // This creates a benchmark
    c.bench_function("Sorting Algorithm", |b| {
```

```
        b.iter(|| sort_algo_1(&mut numbers))
    });
}
```

Listing 7.25 Setting Up the Benchmarking Test Function in sorting_benchmark.rs

The `sort_benchmark` function takes a mutable reference to a `Criterion` struct, allowing us to configure and execute benchmarks. Within this function, we first define a vector of unsorted numbers that will be sorted by the algorithm we're benchmarking.

To create a new benchmark, we call the `bench_function` method on the `Criterion` instance `c`. This method takes two arguments: The first one is an identifier, which is a string that describes the benchmark, and the second one is a closure, which contains the code to be benchmarked. Inside the closure, we have access to a `Bencher` instance `b`. The `Bencher` type is essentially a struct provided by the `Criterion` crate for benchmarking purposes. It is responsible for repeatedly running a given function or closure and measuring its performance, such as execution time, to facilitate performance analysis. We use the `iter` method on this instance to repeatedly execute the code we want to benchmark. In our case, we call the `sort_algo_2` function within the `iter` closure. This function will be executed multiple times to ensure accurate performance measurements.

By executing this benchmark, we can gather data on the performance of the `sort_algo_2` algorithm and compare its performance to other sorting algorithms or implementations. Finally, we'll use a couple of macros to finalize the code for benchmarking:

```
criterion_group!(benches, sort_benchmark);
criterion_main!(benches);
```

The `criterion_group!` macro defines a collection of functions to be executed with a common criterion configuration. The first argument is the name of the group, and the subsequent arguments are the names of the functions to include. In our case, we have only one function (i.e., `sort_benchmark`). The `criterion_main!` macro expands to a `main` function that runs all the benchmarks in a specified group. In this case, we're using this `main` function to run the benchmarks in the `benches` group. We'll discuss macros in more detail in Chapter 15.

Now that our *sorting_benchmark.rs* file is complete, we can use the following `cargo bench` command to run the benchmark tests:

```
c:\> cargo bench
```

The benchmarking report will provide information about the `100` runs of the same program. This data will include the average running time in nanoseconds, giving you a clear understanding of the program's performance. A sample report might look like the following:

```
Sorting Algorithm       time:   [299.38 ns 301.41 ns 303.83 ns]
Found 10 outliers among 100 measurements (10.00%)
  4 (4.00%) high mild
  6 (6.00%) high severe
```

If you run the cargo bench command again, you'll probably notice in the report that "no change in performance is detected." This message indicates that the performance of the program is quite stable and does not change during different runs.

Now, let's change the algorithm from algo_1 to algo_2, as shown in Listing 7.26, and test performance again.

```
fn sort_benchmark(c: &mut Criterion) {
  ...
   c.bench_function("Sorting Algorithm", |b| {
        b.iter(|| sort_algo_2(&mut numbers)) // changed to algo_2
    });
}
```

Listing 7.26 Algorithm from Listing 7.25 Is Now Changed

Notice how performance has regressed; the report clearly indicates that the performance of algo_1 is superior to that of algo_2. In the report, you'll also see the percent change in performance. In this case, the difference is quite large, which is a clear indication that the implementation should prefer the use of algo_1 over the use of algo_2.

7.5 Practice Exercises

In this section, we'll provide some practice questions to help reinforce your learning. The exercise questions cover almost all the concepts introduced in the chapter. Solutions to each exercise will follow in Section 7.6.

1. **Handling struct and enum visibility in Rust modules**
   The following code provided defines two modules, m1 and m2, with a struct A that contains a field of type m2::D. However, the code fails to compile due to visibility issues. The struct A is trying to access the enum D from the inner module m2, but D is not publicly accessible outside of m2. Your task is to fix the visibility of the enum D and any other necessary components so that the code compiles successfully.

```
mod m1 {
    struct A {
        d: m2::D,
    }
    mod m2 {
        enum D {
```

```
                B,
                C,
            }
        }
    }

    fn main(){}
```

2. **Resolving module imports and visibility in Rust**
   In the following code, the module m1 contains struct A, which references enum D from the inner module m2, which is marked as pub. Similarly, module m3 contains a struct C that attempts to use the same enum D from m1::m2. However, the code does not compile due to unresolved visibility and module referencing issues. Your task is to modify the code so that the C struct in m3 can properly access the enum D from m1::m2, and the program compiles without errors.

```
mod m1 {
    struct A {
        d: m2::D,
    }
    mod m2 {
        pub enum D {
            B,
            C,
        }
    }
}
mod m3 {
    struct C {
        e: crate::m1::m2::D,
    }
}
fn main(){}
```

3. **Bringing required items into scope for enum and function access**
   In the following code, the seasons module defines the Season enum and the is_holiday function, which checks whether a given season is a holiday. However, in the main function, both Season::Autumn and is_holiday are not in scope, causing the code to fail compilation. Your task is to complete the code by correctly bringing the Season enum and the is_holiday function into the scope of main so that the program runs without errors.

```
mod seasons {
    pub enum Season {
        Spring,
        Summer,
```

```
        Autumn,
        Winter,
    }

    pub fn is_holiday(season: &Season) -> bool {
        match season {
            Season::Summer => true,
            _ => false,
        }
    }
}

fn main() {
    let current_season = Season::Autumn;
    if is_holiday(&current_season) {
        println!("It's a holiday season! Time for a vacation!");
    } else {
        println!("Regular work season. Keep hustling!");
    }
}
```

4. **Resolving field visibility in struct definitions**
In the following code, the University module defines a Student struct with private fields: name, marks, and grade. When attempting to create an instance of Student in the main function, the code fails to compile due to restricted access to these private fields. Your task is to modify the code so that the name, marks, and grade fields of the Student struct are accessible, thus allowing the code to compile and run successfully.

```
mod University {
    pub struct Student {
        name: String,
        marks: u8,
        grade: char,
    }
}
use University::Student;
fn main() {
    let mut student_1 = Student {
        name: String::from("Alice"),
        marks: 75,
        grade: 'A',
    };
    println!("{} got {} grade", student_1.name, student_1.grade);
}
```

5. **Properly re-exporting functions for simplified access**
   In the following code, the graphics module contains two submodules: shapes, which defines a calculate_area function, and display, which provides a show_area function. However, these functions are not directly accessible in main, and re-exporting them from the top level of graphics is required. Your task is to modify the code by correctly re-exporting both calculate_area and show_area and fixing the use statements, so that the code compiles and runs properly.

```
mod graphics {
    // Re-export the 'show_area' function for easier access
    // Re-export the 'calculate_area' function for easier access
    pub mod shapes {
        pub fn calculate_area(radius: f64) -> f64 {
            std::f64::consts::PI * radius * radius
        }
    }
    pub mod display {
        pub fn show_area(shape: &str, area: f64) {
            println!("The area of the {} is: {}", shape, area);
        }
    }
}

use ___::calculate_area; // fix this line
use ___::show_area; // fix this line
fn main() {
    let radius = 3.0;
    let area = calculate_area(radius);
    show_area("circle", area);
}
```

7.6 Solutions

This section provides the code solutions for the practice exercises in Section 7.5. The code is largely self-explanatory; however, we have included comments and additional explanations wherever necessary to enhance understanding.

1. **Handling struct and enum visibility in Rust modules**

```
mod m1 {
    struct A {
        d: m2::D,
    }
    mod m2 {
```

```
        pub enum D { // Child module items are not visible to parent module
                     // by default
            B,
            C,
        }
    }
}
fn main(){}
```

2. **Resolving module imports and visibility in Rust**

```
mod m1 {
    struct A {
        d: m2::D,
    }
    pub mod m2 {
    /* Public items of a private child module are only accessible by
       parent module. we need to make the child module m2 pub, so that
       its public items can be used outside the parent module. */
        pub enum D {
            B,
            C,
        }
    }
}
mod m3 {
    struct C {
        e: crate::m1::m2::D,
    }
}
fn main(){}
```

3. **Bringing required items into scope for enum and function access**

```
mod seasons {
    pub enum Season {
        Spring,
        Summer,
        Autumn,
        Winter,
    }

    pub fn is_holiday(season: &Season) -> bool {
        match season {
            Season::Summer => true,
            _ => false,
```

```
        }
    }
}
use seasons::{is_holiday, Season};
fn main() {
    let current_season = Season::Autumn;
    if is_holiday(&current_season) {
        println!("It's a holiday season! Time for a vacation!");
    } else {
        println!("Regular work season. Keep hustling!");
    }
}
```


4. **Resolving field visibility in struct definitions**

```
mod University {
    pub struct Student {
        pub name: String, // fields need to be made public
        pub marks: u8,
        pub grade: char,
    }
}
use University::Student;
fn main() {
    let mut student_1 = Student {
        name: String::from("Alice"),
        marks: 75,
        grade: 'A',
    };
    println!("{} got {} grade", student_1.name, student_1.grade);
}
```

5. **Properly re-exporting functions for simplified access**

```
mod graphics {
    pub use self::display::show_area;
    pub use self::shapes::calculate_area;
    pub mod shapes {
        pub fn calculate_area(radius: f64) -> f64 {
            std::f64::consts::PI * radius * radius
        }
    }
    pub mod display {
        pub fn show_area(shape: &str, area: f64) {
            println!("The area of the {} is: {}", shape, area);
        }
```

```
    }
}
use graphics::calculate_area;
use graphics::show_area;
fn main() {
    let radius = 3.0;
    let area = calculate_area(radius);

    show_area("circle", area);
}
```

7.7 Summary

In this chapter, we explored the fundamentals of testing in Rust, starting with unit tests, and you learned how to write and execute test functions effectively. We delved into using the `Result` enum for more robust testing scenarios that allow for better handling of success and failure cases. We introduced you to the concept of testing panics, emphasizing its importance in verifying that your code behaves as expected during erroneous conditions.

We then examined how to control test execution, exploring Rust's flexibility in running specific tests or in modifying their behaviors. Moving beyond unit testing, we highlighted integration tests as a means to test the interactions between different parts of a program to ensure all the components work together seamlessly. Finally, we covered benchmark testing to measure the performance of your code and to identify areas for optimization.

With a solid understanding of these testing techniques, you now have the tools to build and maintain reliable, efficient Rust programs. In the next chapter, we'll explore generics and traits, which provide the flexibility and abstraction needed for creating reusable code.

PART II

Intermediate Language Concepts

Chapter 8
Flexibility and Abstraction with Generics and Traits

Flexibility in programming allows for creativity. In this chapter, we'll explore how generics and traits can make your code both versatile and expressive.

This chapter introduces generics, which allow you to write flexible and reusable code by parameterizing types. We'll then cover traits, which define shared behavior across types. You'll learn about trait bounds, supertraits, and trait objects to create more abstract and powerful designs. This chapter also discusses derived traits and marker traits, as well as associated types in traits, helping you understand when to use associated types versus generic parameters. These concepts enable the creation of more versatile and maintainable code.

8.1 Generics

With generics in Rust, you can define functions, structs, enums, and traits with placeholders for data types, making the code more flexible and reusable. You can use generics to abstract over different types, essentially acting as types that will be specified later.

You'll learn how to use generics in the following sections, as we walk through scenarios involving implementation blocks, multiple implementations, duplicates, free functions, and monomorphization.

8.1.1 Basics of Generics

Let's start with an example to learn the basics. Consider the code shown in Listing 8.1.

```
struct Point {
  x: i32,
  y: i32,
}
fn main() {
  let origin = Point {x: 0, y:0};
}
```

Listing 8.1 Struct Point with an Instance Defined in main

The struct Point represents a point, which is specified by its x-axis and y-axis coordinate values. Let's try to create another instance of the Point in main, with floating values, as follows:

```
Let p1 = Point {x: 1.0, y: 4.0}; // Error
```

We'll get an error of "mismatched types." This error arose because an i32 was expected but a floating number was encountered.

Generics come into play in situations like this. Using generics allows a Point struct to support various concrete types. Let's modify the Point struct to make the x and y fields generic instead of using specific concrete types, as shown in Listing 8.2.

```
struct Point<T> {
    x: T,
    y: T,
}
```

Listing 8.2 Point Struct Redefined Using Generics

To use generics within struct fields, define the generic types in angle brackets (< >) following the struct's name. You can choose any meaningful name for the generic type, though a common practice is to use the letter T to represent a type. If you need additional generics, you can continue with the subsequent letters in the alphabet. For more descriptive names, use CamelCase to convey the purpose of the generic type.

Now, you can use the generic T instead of the concrete types for the coordinates of a point. The following code in main will now compile:

```
fn main() {
    let origin = Point {x: 0, y:0};
    let p1 = Point {x: 1.0, y: 4.0};
}
```

You can now easily create points with the concrete types of i32 and f64.

Looking at the type annotations for the points in your code editor, notice how the origin has a type Point<i32>, where the generic type T in the type Point<T> is replaced with a specific type i32. In contrast, the point p1 has a type Point<f64>, where the generic type T is substituted with the concrete type of f64.

Let's create one more point (p2) with coordinates of a different type in main:

```
let p2 = Point {x: 5, y: 5.0};
```

An error of "mismatched types" arises again. An integer was expected; however, a floating number was encountered. This is because the current struct definition with generics means that x and y can be of any type, but they must be of the same type, represented by T. The first two points meet this requirement, but the third point does not.

The problem can be solved by introducing another generic called U. Consider the redefined struct Point with an added generic, shown in Listing 8.3.

```
struct Point<T, U> {
    x: T,
    y: U,
}
```

Listing 8.3 Point Struct Redefined Using Two Generics

With the y-axis type now changed to generic U, the code in main shown in Listing 8.4 now compiles.

```
fn main() {
    let origin = Point {x: 0, y:0};
    let p1 = Point {x: 1.0, y: 4.0};
    let p2 = Point {x: 5, y: 5.0};
}
```

Listing 8.4 Code in main with Three Points Containing Different Field Types

Note that, for a specific instance of the struct, we have only a single realization of the generic. The types of the three points and the generic realization are listed in Table 8.1.

Variable Name	Type	Realization of (T, U)
origin	Point<i32,i32>	(i32,i32)
p1	Point<f64,f64>	(f64,f64)
p2	Point<i32,f64>	(i32,f64)

Table 8.1 Variable Types

8.1.2 Generics in Implementation Blocks

Let's try to add an implementation block to Point, as in the following code:

```
impl Point {} // Error
```

An error arises in this case because we are missing generics. To add generics, we'll mention the generics after the impl keyword and also after the name of the struct, as in the following code:

```
impl<T, U> Point<T, U> {}
```

Let's now add functionality to the implementation block. For instance, let's add a new constructor function and update the code in main based on the added function, as shown in Listing 8.5.

```
impl<T, U> Point<T, U> {
    fn new(x: y, y: U) -> Point<T, U> {
        Point {x, y}
    }
}
fn main() {
    let origin = Point::new(0, 0);
    let p1 = Point::new(1.0, 4.0);
    let p2 = Point::new(5, 5.0);
}
```

Listing 8.5 Addition of new Constructor to the Implementation Block and Updated Code in main

The benefit of adding the generic information to the implementation block (i.e., <T, U>) is that doing so tells Rust that this implementation is for the Point with generic types T and U. In this way, Rust can more easily differentiate it from other implementations of the Point for some other types.

8.1.3 Multiple Implementations for a Type: Generics versus Concrete

You can have multiple implementations for a type such as Point in some cases. For instance, let's add another implementation of the Point struct, as shown in Listing 8.6.

```
impl Point<i32, i32> {
    fn printing(&self) {
        println!("The values of the coordinates are {}, {}", self.x, self.y);
    }
}
```

Listing 8.6 New Implementation for a Point

This new implementation uses concrete types of i32 instead of generics and contains a simple printing function.

Specialization

Using separate implementation blocks for a type based on the different generic types it can take allows you to achieve a concept known as *specialization*, which enables you to group methods for specific types. In this way, you can optimize or customize behavior for those types, in this case, for i32.

The implementation shown in Listing 8.6 has several differences compared to the implementation shown earlier in Listing 8.5:

- The implementation `impl Point<i32, i32>` is specialized for `Point` instances (also called a *specialized implementation*) where both `x` and `y` are of type `i32`, whereas the second implementation, `impl<T, U> Point<T, U>`, is generic and applies to `Point` instances with any types `T` and `U`.
- The specialized implementation defines the printing method, which is only available for `Point<i32, i32>` instances. The generic implementation provides a new function for creating `Point` instances with any types.
- The specialized implementation is less flexible since it only works for `i32` type, while the generic implementation supports a wider range of type configurations.
- The specialized implementation focuses on functionality unique to `i32` types, while the generic implementation provides shared functionality for all type configurations.

The `printing` function defined in the new implementation block shown in Listing 8.6 is only available for the `origin`. In the code editor, if you write `p1` and then try to see the methods after the dot, `printing` is not listed, meaning that the function is not available. However, you we write `origin` and then try to see the method after the dot, we may note that printing is available. Let's try to access the `printing` function after `p1`, as in the following example:

```
p1.printing(); // Error
```

The compiler will throw an error, "no method of printing for point with float, float types."

8.1.4 Duplicate Definitions in Implementation Blocks

You are not allowed to define duplicate functions in multiple implementation blocks of the same type. For instance, let's redefine the `new` constructor function inside the implementation block, as shown in Listing 8.7.

```
impl Point<i32, i32> {
    ...
    fn new(x: i32, y: i32) -> Point<i32, i32> { // Error
        Point {x, y}
    }
}
```

Listing 8.7 The new Constructor Function Added to the Code from Listing 8.6

The compiler will throw an error, "duplicate definitions with name `new`." This error arises because, when the function is called for an instance of `Point` with both x-axis and y-axis being of i32 type, such as the origin point in this case, the compiler has no way to decide which specific function should be invoked. More concretely, the compiler is confused as to whether it should invoke the function in the generic implementation block of the function inside the concrete implementation block. If you really want to create the points for i32 differently, then you should some other name, such as new_1 in our example shown in Listing 8.8.

```
impl Point<i32, i32> {
    ...
    fn new_1(x: i32, y: i32) -> Point<i32, i32> {
        Point {x, y}
    }
}
```

Listing 8.8 Fixing the Code from Listing 8.7 by Renaming the Constructor Function

Functions with same name are, however, allowed in different concrete implementation blocks, corresponding to different types. For instance, consider the code shown in Listing 8.9.

```
impl Point<i32, i32> {
    fn printing(&self) {
        println!("The values of the coordinates are {}, {}", self.x, self.y);
    }
    ...
}
impl Point<f64, f64> {
    fn printing(&self) {
        println!("The values of the coordinates are {}, {}", self.x, self.y);
    }
}
...
```

Listing 8.9 Functions with the Same Name in Different Concrete Implementations Are Allowed

The printing method is common in the implementation of impl Point<i32,i32> and impl Point<f64,f64>. However, the code compiles with no issues.

8.1.5 Generics and Free Functions

Generics are also quite frequently used in functions that are not tied to any structs, enums, or traits. Such functions are also known as *free functions*. Consider the code shown in Listing 8.10.

```
fn add_points<T, U>(p1: &Point<T, U>, p2: &Point<T, U>) -> Point<T, U> {
    unimplemented!();
}
```

Listing 8.10 Generic Function Not Tied to Any Struct, Enum, or Trait

This generic function uses the `Point` definition defined in the code shown earlier in Listing 8.3. The generics with free functions follow a syntax similar to `impl` blocks. After the name of the function, you mention the generics inside angled brackets. Next, you can use the generic to define the types for the parameters to the function as well as the return type. Inside the function, the `unimplemented!` macro indicates that we are currently not interested in the implementation. This macro will not create any compile time errors; however, it will panic at runtime. Note that, if you plan to implement it later, a better approach is to use the `todo!` macro. For more information on macros, see Chapter 15.

8

8.1.6 Monomorphization

Generics do not incur any runtime performance cost due to a feature called *monomorphization*. During compile time, specialized concrete implementations of the generic function are created for each set of generic type parameters used in the code. For instance, consider the function shown in Listing 8.11.

```
fn main() {
    let origin = Point::new(0, 0);
    let p1 = Point::new(1.0, 4.0);

    add_points(&origin, &origin);
    add_points(&p1, &p1);
}
```

Listing 8.11 Code in main for Calling the Function add_points

In the code, the `origin` has a type `Point<i32, i32>` while the `p1` point has a type `Point<f64, f64>`. In other words, the call to the function `add_points(&origin, &origin)` the generics of `<T, U>` will be `<i32, i32>`, and in the call to function `add_points(&p1, &p1)`, the generics will be `<f64, f64>`. This combination will lead to the generation of two implementations under the hood for the function, as shown in Listing 8.12.

```
fn add_points_i32(p1: &Point<i32, i32>, p2: &Point<i32, i32>) -> Point<i32,
i32> {
    unimplemented!();
}
fn add_points_f64(p1: &Point<f64, f64>, p2: &Point<f64, f64>) -> Point<f64,
```

```
f64> {
    unimplemented!();
}
```

Listing 8.12 Specific Implementations Generated for the Generic Function Based on Generic Types Used in main

The specific implementations will only be created for the types that are used in the code. The creation of these specific functions is also referred to as *static dispatch*. Static dispatch refers to the fact that the methods or function calls are resolved at compile time, thus ensuring predictable and efficient code execution without runtime overhead. More details about this topic will be covered in Section 8.2.6. The call sites are updated with the concrete implementation. As a result, in `main`, the calls will be made for invoking the specific implementations, as shown in Listing 8.13.

```
fn main() {
    ...

    add_points(&origin, &origin);  // add_points_i32(&origin, &origin);
    add_points(&p1, &p1); // add_points_f64(&p1, &p1);
}
```

Listing 8.13 Call Sites Updated with the Specific Implementations of the Function

Monomorphization occurs at compile time, and therefore, there is no runtime cost. However, one concern with monomorphization is potential code bloat. Using a generic function with many different types can generate multiple copies of a function, leading to larger binary sizes. We'll discuss a solution to this issue later in Section 8.2.6, called *dynamic dispatch*.

8.2 Traits

In this section, we'll explore the concept of traits in Rust and how they enable *polymorphism*. We'll start by explaining why and how to use traits. Then, we'll dive into the details, covering topics like default implementations, trait bounds, supertraits, trait objects, derived traits, marker traits, and associated types.

8.2.1 Motivating Example for Traits

Let's consider two structures, `Square` and `Rectangle`, as shown in Listing 8.14.

```
struct Square {
    side: f32,
    line_width: u8,
```

```
    color: String,
}
struct Rectangle {
    length: f32,
    width: f32,
    line_width: u8,
    color: String,
}
```

Listing 8.14 Defining the Struct Square and Rectangle

The Square struct has a field called side, which defines the square's dimensions. Additionally, this struct has the fields line_width and color, which can be useful for rendering the square. Similarly, the Rectangle struct contains the fields length and width, which describe the dimensions of the rectangle. This struct also includes line_width and color as additional fields.

Now, suppose you want to add functionality to compute the area for both shapes. A straightforward approach would be to implement separate blocks for both Square and Rectangle and add an area method to each. Consider the code shown in Listing 8.15.

```
impl Square {
    fn calculate_area(&self) {
        println!("The area is: {}", self.side * self.side);
    }
}
impl Rectangle {
    fn area(&self) -> f32 {
        self.length * self.width
    }
}
```

Listing 8.15 Adding Functionality for Computing the area to the Square and Rectangle Structs

While this approach works, certain drawbacks arise. First, no standardization exists in how the area method is structured for both shapes. For example, the Square implementation returns no value, while the Rectangle implementation does. Moreover, the method names differ between the two implementations.

This lack of consistency is problematic. Not only must you remember the differences in method signatures, but you must also ensure that the correct methods are called for the correct shapes. What we desire instead is a consistent interface that can be used for both shapes. This leads us to the concept of polymorphism.

Polymorphism allows us to define a unified interface, such as the area method, without worrying about the specific types implementing it. In object-oriented languages, you might solve this problem using inheritance. For instance, you could create a common

Shape class with shared fields like line_width and color, then have Rectangle and Square inherit from this class. However, Rust does not support traditional inheritance. Instead, Rust achieves polymorphism through traits.

8.2.2 Traits Basics

Traits allow you to define shared behaviors across multiple types, similar to interfaces in languages like Java. Let's define a trait called Shape to serve as our common interface for the area method, as shown in Listing 8.16.

```
trait Shape {
    fn area(&self) -> f32;
}
```

Listing 8.16 Defining a Trait Shape with a Common Method Area

Our trait is created using the trait keyword followed by a name for the trait, in this case, Shape. In this trait, we declare the area method without a body, indicating that any type implementing this trait must provide its own definition of area. In other words, we are simply declaring an interface that must be implemented by the types that have this trait.

Next, let's implement this trait for both Rectangle and Square. Traits are implemented for types using an implementation block with the syntax impl trait_name for type. Following this syntax, try to implement the trait for the Square with the following code:

```
Impl Shape for Rectangle {} // Error
```

You'll see an error saying that "not all trait items implemented, missing the method of area." When a trait is implemented for a type, all its methods must be defined for that type. Let's add the implementation for the method of area, as shown in Listing 8.17.

```
impl Shape for Square {
    fn area(&self) -> f32 {
        let area_of_square = self.side * self.side;
        println!("Square area: {}", area_of_square);
        area_of_square
    }
}
```

Listing 8.17 Implementing the Shape Trait for Square

In the same way, you can add implementation of Shape for Rectangle, as shown in Listing 8.18.

```
impl Shape for Rectangle {
    fn area(&self) -> f32 {
        let area_of_rect = self.length * self.width;
        println!("Rectangle area: {}", area_of_rect);
        area_of_rect
    }
}
```

Listing 8.18 Implementing the Shape Trait for Rectangle

Unlike our previous approach shown earlier in Listing 8.15, you now have a consistent interface for the area method. If you attempt to deviate from the trait's signature, Rust will throw an error, thus ensuring that the method implementations remain consistent across types. For example, changing the area method for Square to return no value instead of an f32 (shown earlier in Listing 8.17) will produce the following error:

```
method `area` has an incompatible type for trait
expected signature `fn(&Rectangle) -> f32`
found signature `fn(&Rectangle)
```

You can now create instances of both Rectangle and Square in the main function and invoke their area methods, as shown in Listing 8.19.

```
fn main() {
    let r1 = Rectangle {
        width: 5.0,
        length: 4.0,
        line_width: 1,
        color: String::from("Red"),
    };
    let s1 = Square {
        side: 3.2,
        line_width: 1,
        color: String::from("Red"),
    };
    r1.area();
    s1.area();
}
```

Listing 8.19 Invoking the area Method on Both Rectangle and Square

As expected, the method calls follow a consistent syntax, making your code easier to read and maintain.

8.2.3 Default Implementations

In addition to method declarations, traits can also provide default method implementations for some or all of their methods. These default options allow types that implement a trait to inherit the default behavior without explicitly defining those methods. However, the type can override the default implementation if customized behavior is needed.

Default implementations for methods are specified by adding the body of the function inside a trait. Let's add a `perimeter` method with a default implementation to the `Shape` trait, as shown in Listing 8.20.

```
trait Shape {
    fn area(&self) -> f32;
    fn perimeter(&self) -> f32 {
        println!("Perimeter not implemented, returning dummy value");
        0.0
}
```

Listing 8.20 Default Implementation of the Perimeter Method in the Shape Trait

In this example, the method is not only declared but also defined with its body inside the trait definition. The inclusion of a method body inside a trait definition is what is referred to as the default implementation. With a default implementation, you can define a default behavior for a method in a trait, which can be optionally overridden by types implementing the trait. By default, this method returns a placeholder value and prints a message indicating that the perimeter is not implemented. If you wish, you can override this default implementation for specific types. For example, let's override the `perimeter` method for `Rectangle`, while leaving `Square` to use the default implementation, as shown in Listing 8.21.

```
impl Shape for Rectangle {
    fn perimeter(&self) -> f32 {
        let perimeter_of_rect = 2.0 * (self.length + self.width);
        println!("Rectangle Perimeter: {}", perimeter_of_rect);
        perimeter_of_rect
    }
}
```

Listing 8.21 Overriding the perimeter Method for Rectangle

Let's now call the `perimeter` methods in the `main` function by adding the following two lines to the code shown earlier in Listing 8.19:

```
r1.perimeter();
s1.perimeter();
```

The Rectangle will use its custom perimeter method, while Square will rely on the default implementation. This distinction thus provides flexibility in the method's behavior, while still ensuring that all types have access to a consistent interface.

One key point to note is that, unlike classic inheritance, where both data and functionality can be shared, with traits only functionality can be shared. However, workarounds exist for sharing data among types. For instance, consider our Square and Rectangle structs, shown earlier in Listing 8.14. We can extract the common fields of color and line_width into a separate struct and then make that struct a field for the two types. The updated definitions of the structs are shown in Listing 8.22.


```
struct drawing_info {
    line_width: u8,
    color: String,
}
struct Square {
    side: f32,    info: drawing_info,
}
struct Rectangle {
    length: f32,
    width: f32,
    info: drawing_info,
}
```

Listing 8.22 Extracting the Common Fields in a New Struct

The common fields of the Square and Rectangle are extracted in a separate struct of drawing_info. The new struct of drawing_info is next shared among the two types by it as a field for the two structs.

Inheritance versus Composition

Rust emphasizes composition over inheritance. *Composition* allows you to create new types by combining existing ones and their functionalities, unlike inheritance, which relies on a hierarchical relationship where new types derive behavior from parent types, often leading to tightly coupled and less flexible designs. An approach emphasizing composition leads to more compact, reusable modules, ultimately making the code easier to maintain and understand.

8.2.4 Trait Bounds

Listing 8.23 shows the code that we used in the previous section.

```
struct Square {
    side: f32,
    line_width: u8,
    color: String,
}
struct Rectangle {
    length: f32,
    width: f32,
    line_width: u8,
    color: String,
}
trait Shape {
    fn area(&self) -> f32;
    fn perimeter(&self) -> f32 {
        println!("Perimeter not implemented, returning dummy value");
        0.0
    }
}
impl Shape for Rectangle {
    fn area(&self) -> f32 {
        ...
    }
    fn perimeter(&self) -> f32 {
        ...
    }
}
impl Shape for Square {
    fn area(&self) -> f32 {
        ...
    }
}
```

Listing 8.23 Code Considered in the Previous Section

Now, we want to add a general-purpose function called `shape_properties` to this code. The function will take an instance of `Square` or `Rectangle` and return its respective `area` and `perimeter`. Consider the definition of the function shown in Listing 8.24.

```
fn shape_properties<T>(object: T) {
    object.area(); // Error
    object.perimeter(); // Error
}
```

Listing 8.24 Generic Function to Display the area and perimeter of Any Object

This function throws an error, "no method named area, found for type parameter T," further clarifying that the "method area not found for this type parameter." A similar error is highlighted for the method perimeter. These errors make sense because T can represent any concrete type, and currently, we don't know whether that type implements the area method or the perimeter method. What we actually want is for T to represent any type, as long as it implements the Shape trait.

To achieve this goal, you can introduce a *trait bound*, specified after the generic type by using a colon followed by the trait's name, as shown in Listing 8.25.

```
fn shape_properties<T: Shape>(object: T) {
    object.area();
    object.perimeter();
}
```


Listing 8.25 Trait Bound Added to the Generic T

This trait bound means that T is now restricted to types that implement the Shape trait, which is exactly what we need.

Binding a generic type with a trait has two effects:

- First, you can limit the generic type to those that conform to the specified trait, in this case, Shape.
- Second, you can allow instances of the generic type to access the trait's methods, like area and perimeter, as defined by the Shape trait.

Multiple trait bounds can be mentioned by adding a plus sign (+), followed by the name of another trait, as in the following example:

```
fn shape_properties<T: Shape + SomeOtherTrait>(object: T) {}
```

Sometimes, trait bounds are created using a slightly different syntax, as in the following example:

```
fn shape_properties<T>(object: impl Shape) {
    ...
}
```

This syntax, called the impl trait syntax, is a shorthand for specifying that the object parameter implements the Shape trait. This syntax essentially has the same meaning as the syntax T: Shape shown in Listing 8.25, but this latter option is easier to read. For instance, in this case, we can read this statement as, for "any object that implements the Shape trait."

The last way to define the trait bound is to use a where clause, as in the following example:

```
fn shape_properties<T>(object: T) where T: Shape,
{
    ...
}
```

The `where` clause syntax explicitly specifies that the generic type `T` must implement the `Shape` trait, offering flexibility to define trait bounds in a clean and organized manner. This approach is particularly helpful in cases where multiple trait bounds are involved, and you want to make your functions easier to read.

To summarize, you have three options for creating a trait bound in a function signature, as shown in Table 8.2.

Direct Trait Bound	`fn function_name<T: Trait_name>(param: T)`
Impl Trait Syntax	`fn function_name(param: impl Trait_name)`
Using the Where Clause	`fn function_name<T>(param: T)` `where T: Trait_name {}`

Table 8.2 Different Syntaxes for Trait Bounds

Type Doesn't Satisfy the Trait Bounds

Types that don't satisfy the trait bounds of generic function cannot be passed into a generic function. Let's walk through an example to understand this limitation by adding one more type to the code shown earlier in Listing 8.23. Add another shape through the following lines:

```
struct Circle {
    radius: f32,
}
```

This new type (`Circle`) does not implement the `Shape` trait and therefore cannot access the shared method defined in the `Shape` trait. Finally, let's call the `shape_properties` function in `main` with some types, as shown in Listing 8.26.

```
fn main() {
    let r1 = Rectangle {
        width: 5.0,
        length: 4.0,
        line_width: 1,
        color: String::from("Red"),
    };
    let s1 = Square {
        side: 3.2,
        line_width: 1,
        color: String::from("Red"),
```

```
    };
    let c1 = Circle { radius: 5.0 };
    shape_properties(r1);
    shape_properties(s1);
    shape_properties(c1); // Error
}
```

Listing 8.26 Using the Function shape_properties in main

The first two calls were successful; however, the call for an instance of Circle creates an error. The error says that the trait bound Circle: Shape is not satisfied. This error makes sense because the Circle does not have any implementation for Shape, and therefore, the trait bound cannot be satisfied. For the other types, the bound is satisfied.

8

You can also use the impl trait syntax to mention a return value from a function. For instance, consider the code shown in Listing 8.27.

```
fn returns_shape() -> impl Shape {
    let sq = Square {
        side: 5.0,
        line_width: 5,
        color: String::from("Red"),
    };
    Sq
}
```

Listing 8.27 Function Returning Something That Implements the Shape Trait

The output of the function returns_shape indicates that this function can return anything that implements the Shape trait. The function returns an instance of the Square. This code works because the returning type is a Square, which implements the Shape trait.

A couple things to note include the following:

- First, only the impl trait syntax can be used for function return values, not the other syntaxes we listed in Table 8.2 for specifying trait bounds.
- Second, the impl trait syntax for return values only works with a single concrete type. In other words, if you have a condition based on which you return different types, the compiler will not allow this and will throw an error. For example, consider a revised version of the function returns_shape, as shown in Listing 8.28.

```
fn returns_shape() -> impl Shape {
    let sq = Square {
        side: 5.0,
        line_width: 5,
```

```
            color: String::from("Red"),
        };
        let rect = Rectangle {
            length: 5.0,
            width: 10.0,
            line_width: 5,
            color: String::from("Red"),
        };
        let x = false;
        if x {
            sq
        } else {
            Rect // Error
        }
    }
```

Listing 8.28 Function from Listing 8.27 Revised to Return Different Types Based on Condition

In the function body, we are creating an instance of `Square` and `Rectangle` and then, based on a condition, we'll either return a `Square` or `Rectangle`. The compiler throws an error, "`if` and `else` have incompatible types." This error occurs because the `impl trait` syntax, when used for a return value, can only specify a single concrete type. To allow the function to return different types, you must use trait objects, which we'll discuss in Section 8.2.6.

8.2.5 Supertraits

In Rust, a *supertrait* refers to a more generalized trait that encompasses the functionalities of other traits. While using supertraits might resemble the concept of inheritance, it specifically refers to the idea of extending the capabilities of one trait by requiring the implementation of other traits. The traits that a certain trait depends on are known as its supertraits.

In the following sections, we first cover the basic syntax and use of supertraits. Then, we'll explore the advantages of using supertraits for reducing the list of trait bounds.

Using Supertraits

Let's add a more generalized trait called `Draw` to the code shown earlier in Listing 8.23. The updated code, after additions from the previous section, is shown in Listing 8.29.

```
struct Square { ... }
struct Rectangle { ... }
struct Circle { ... }
```

```
trait Draw {
    fn draw_object(&self);
}
trait Shape { ... }
impl Shape for Rectangle { ... }
impl Shape for Square { ... }
fn shape_properties<T: Shape>(object: T){ ... }
fn returns_shape() -> impl Shape { ... }
fn main() {  ... }
```

Listing 8.29 Updated Code from Listing 8.23 after Additions in the Previous Section and an Added Draw Trait

The `Draw` trait contains a single function of `draw_object`. You can now make the `Draw` trait a supertrait of `Shape` by using the following syntax:

```
trait Shape: Draw { … }
```

The code throws some errors, specifically in the following lines:

```
impl Shape for Rectangle { ... } // Error
impl Shape for Square { ... }    // Error
```

An error is thrown that "the trait bound `Rectangle: Draw` is not satisfied," further elaborating that "the trait `Draw` is not implemented for `Rectangle`." Similar errors can also be seen for the `Square`.

A key requirement for supertraits is that all types implementing the base trait, like the `Shape` trait in this case, must also implement the supertrait, such as the `Draw` trait in our example. Thus, since the since `Rectangle` and `Square` implement the `Shape` trait, they must also implement the `Draw` trait. The code shown in Listing 8.30 implements the `Draw` trait for the two types.

```
impl Draw for Square {
    fn draw_object(&self) {
        println!("Drawing Square");
    }
}
impl Draw for Rectangle {
    fn draw_object(&self) {
        println!("Drawing Rectangle");
    }
}
```

Listing 8.30 Implementation of the Draw Trait for Square and Rectangle

You can mention multiple traits as supertraits. The syntax is similar to mentioning multiple generic bounds. For instance, if you want to add another supertrait to the Shape trait, you can use the following syntax:

```
impl Shape: Draw + SomeOtherTrait {}
```

Reducing the List of Trait Bounds

One benefit of supertraits is that they help reduce the list of trait bounds in functions using generics. We currently have one such function called shape_properties (shown earlier in Listing 8.29). In this example, we have is a single trait bound of Shape on the generic T in the function signature.

Let's add a few more trait bounds by first defining some additional traits, as shown in Listing 8.31.

```
trait OtherTrait {}
impl OtherTrait for Rectangle {}
impl OtherTrait for Square {}

trait SomeOtherTrait {}
impl SomeOtherTrait for Rectangle {}
impl SomeOtherTrait for Square {}
```

Listing 8.31 Some Arbitrary Traits Added to the Code and Implemented for Rectangle and Square

Note that these traits are empty, containing no items. Such traits are called *marker traits*, which we'll cover in Section 8.2.8. Since the traits do not contain any functions, the implementation is empty.

Next, let's assume that the generic T in the function shape_properties requires the added traits of OtherTrait and SomeOtherTrait. This can be done by mentioning them as trait bounds, as in the following example:

```
fn shape_properties<T: Shape + OtherTrait + SomeOtherTrait>(object: T) { ... }
```

We currently have three bounds on the generic, but in many cases, this list could become longer. Supertraits allow us to simplify this list by specifying additional traits as supertraits of the Shape trait, as in the following example:

```
trait Shape: Draw + OtherTrait + SomeOtherTrait { ... }
```

As a result, now, any type implementing the Shape trait must also provide implementations for the supertraits of Draw, OtherTrait, and SomeOtherTrait. In the shape_properties function, we no longer need additional trait bounds:

```
fn shape_properties<T: Shape>(object: T) { ... }
```

We only need to mention the single bound of Shape because these are already guaranteed as supertraits of Shape. This results in cleaner, more concise code.

In summary, supertraits are particularly useful for organizing functionality in a hierarchical structure by providing higher-level, generalized functionality.

8.2.6 Trait Objects

Trait objects provide a way to achieve dynamic dispatch in Rust, enabling polymorphism by allowing different types to be handled through a common interface. This section will explore the concept of trait objects, including their syntax and their behavior, followed by a discussion of the flexibility they offer in designing versatile and reusable Rust programs.

Using Trait Objects

Let's continue to modify our code from earlier in Listing 8.29, as shown in Listing 8.32.

```
struct Square { ... }
struct Rectangle { ... }
struct Circle { ... }

trait Draw { ... }
trait Shape: Draw { ... }
impl Shape for Rectangle { ... }
impl Shape for Square { ... }
impl Draw for Rectangle { ... } // Added
impl Draw for Square { ... } // Added
fn shape_properties<T: Shape>(object: T){ ... }
fn returns_shape() -> impl Shape { ... }
fn main() {
    let r1 = Rectangle {
        ...
    };
    let s1 = Square {
        ...
    };
    shape_properties(r1);
    shape_properties(s1);
}
```

Listing 8.32 Updated Code from Listing 8.29

In this example, the shape_properties function is called twice in main. Behind the scenes, the compiler generates specific code for each concrete type used with generic functions or methods. In this case, two versions of the function are being constructed:

one for Rectangle and another one for Square. The generated code will look something like Listing 8.33.

```
fn shape_properties_rect(object: Rectangle) {
    object.area();
    object.perimeter();
}
fn shape_properties_sq(object: Square) {
    object.area();
    object.perimeter();
}
```

Listing 8.33 Compiler-Generated Code Based on the Function Calls to shape_properties in main

The call sites in main are next replaced with the following calls to these specialized versions:

```
shape_properties_rect(r1);
shape_properties_sq(s1);
```

This process of generating specific, specialized versions of functions for each type is called *monomorphization* or *static dispatch*, which we introduced in Section 8.1.6. The key advantage of static dispatch is that this approach enhances performance by removing runtime overhead since there's no need to determine which function version to call. For example, in the specialized version of shape_properties_rect, the compiler knows exactly that it should call the area and perimeter functions defined for the Rectangle.

In contrast to static dispatch, with *dynamic dispatch*, the specific implementations are not generated at compile time. Dynamic dispatch is based on trait objects. Table 8.3 summarizes the key differences between static dispatch and dynamic dispatch.

Static Dispatch	Dynamic Dispatch
Compile-time resolution of function calls	Runtime resolution of function calls
Generally faster due to no runtime overhead	Slower due to overhead of determining the method to call at runtime
Generates specific code for each concrete type	Generates a single function call that resolves to the appropriate method at runtime
Used with generics and trait bounds	Used with trait objects (e.g., Box<dyn Trait>)

Table 8.3 Differences between Static and Dynamic Dispatch

Static Dispatch	Dynamic Dispatch
Less flexible because all types must be known at compile time	More flexible, allows for varying types at runtime
Typically requires more code size due to specialization	May use less code but involves indirection, i.e., accessing data or functionality through references or pointers

Table 8.3 Differences between Static and Dynamic Dispatch (Cont.)

To illustrate the basics of dynamic dispatch, let's consider another version of the `shape_properties` function, called `shape_properties_dynamic`, as shown in Listing 8.34.

```
fn shape_properties_dynamic(object: Box<dyn Shape>) {
    object.area();
    object.perimeter();
}
```

Listing 8.34 Dynamic Version of the shape_properties Function

The body of the function is the same as the `shape_properties`. The function signature, however, is different. The first thing to note is that we are no longer using generics. Therefore, we do not have any generic bounds. The type of the input object is `Box<dyn shape>`. Note that `Box` is a smart pointer, which we'll cover in Chapter 10, Section 10.2. For now, just understand that it points to heap-allocated data. The `dyn` keyword defines a trait object, where `dyn` stands for dynamic dispatch. A trait object must be stored behind a pointer, and in this case, we used the `Box` smart pointer, which is a common choice.

With trait objects, you can define a type that implements a trait without knowing the exact type at compile time because references are resolved at execution time. As a result, specialized versions of the function won't be generated, and function resolution will occur at runtime. The `dyn shape` means that the function expects something that implements the shape, and the `Box<dyn shape>` ensures that it must be behind a pointer.

In `main`, you can now call the function `shape_properties_dynamic` by passing in instances of `Rectangle` and `Square`, wrapped inside the box pointers. The code shown in Listing 8.35 illustrates this step.

```
fn main() {
    let r1 = Rectangle {
        ...
    };
    let s1 = Square {
        ...
    };
```

```
    shape_properties_dynamic(Box::new(r1));
    shape_properties_dynamic(Box::new(s1));
}
```

Listing 8.35 Calling the shape_properties_dynamic in main

Box::new creates a new box smart pointer, which involves allocating memory on the heap to store the value and then returns a pointer to that value. Since references are resolved at execution time and not at compile time, the exact resolution of function call will be made at execution time. At compile time, the function only expects something that implements the Shape trait; it does not need to know the exact type since the type is behind a reference.

As mentioned earlier, a key requirement for a trait object is that it must be behind a pointer. In this example, we used a Box smart pointer, but any type of pointer can be utilized, including a simple reference. Later in this book, we'll explore more types of pointers, such as Rc and Ref, which can also be employed in this context.

Flexibility with Trait Objects

An essential advantage of trait objects is flexibility. To illustrate this flexibility, let's reconsider the function returns_shape shown earlier in Listing 8.28, repeated in Listing 8.36 for easy reference.

```
fn returns_shape() -> impl Shape {
    let sq = Square {
        side: 5.0,
        line_width: 5,
        color: String::from("Red"),
    };
    let rect = Rectangle {
        length: 5.0,
        width: 10.0,
        line_width: 5,
        color: String::from("Red"),
    };
    let x = false;
    if x {
        sq
    } else {
        Rect // Error
    }
}
```

Listing 8.36 Function returns_shape Considered in Listing 8.28

The compiler throws an error, "incompatible types." Recall from Section 8.2.4 that the syntax of impl Shape represents a generic with a trait bound. When using a generic as a return type, the compiler requires that the generic must be substituted with one concrete type at compile time. However, in this case, we have two concrete types, which causes the error.

To fix this problem, in other words, to add the ability to return multiple concrete types and therefore enjoy more flexibility in returning more types, we'll use trait objects. Simply change the return type to a Box trait object and return either a square or rectangle wrapped inside a box, as shown in Listing 8.37.


```
fn returns_shape() -> Box<dyn Shape> {
    ...
    let x = false;
    if x {
        Box::new(sq)
    } else {
        Box::new(Rect)
    }
}
```

Listing 8.37 Function returns_shape Updated to Return Multiple Types Using Trait Objects

The function returns_shape is now more flexible because it can return multiple types implementing the shape trait. The exact type of the return value will be determined at execution time, not at compile time.

8.2.7 Derived Traits

In Rust, derived traits refer to traits that can be automatically implemented for a type using the derive attribute. These traits are applied to structs and enums to provide default implementations for certain common behaviors.

Consider the program shown in Listing 8.38.

```
struct Student {
    name: String,
    age: u8,
    sex: char,
}
fn main() {
    let s_1 = Student {
        name: String::from("ABC"),
        age: 35,
        sex: 'M',
    };
```

```
    let s_2 = Student {
        name: String::from("XYZ"),
        age: 35,
        sex: 'M',
    };
}
```

Listing 8.38 Student Struct and Its Instances Defined in main

The code defines a `Student` struct and in `main` we have a couple of instances of the struct. Let's try to print the `s_1` inside a print statement in debug mode, as follows:

```
println!("Student: {:?}", s_1); // Error
```

The colon question mark syntax (`:?`), which you've seen quite often throughout this book, prints out a type with debug formatting. However, this syntax only works if the type that you are printing implements the `Debug` trait. Currently, we have no implementation of `Debug` for the `Student`; therefore, we get an error that "`Student` doesn't implement `Debug`." This error message also tells us that we can either "add the `#[derive(Debug)]` to `Student` or manually `impl Debug for Student`." Note that the `Debug` trait is defined in the Rust standard library.

Let's use the first solution and derive the `Debug` trait for the student by adding the following line before the definition of the `Student` struct in Listing 8.38:

```
#[derive(Debug)]
```

This addition fixes the error. What happens behind the scenes is that the `derive` attribute automatically implements the specified trait for the given struct. In this case, it implements the `Debug` trait for the `Student` struct, thus providing a basic default implementation so that we don't have to manually write the code ourselves.

The derived traits are available for common behaviors such as comparisons, cloning, and initializing instances with default values. Let's look at the traits used for comparisons.

If you add the following print statement to code in the listing, you'll get an error:

```
println!("s_1 and s_2 are equal: {}", s_1 == s_2);  // Error
```

The error states that "an implementation of `PartialEq` might be missing for `Student`," and it further gives a suggestion to "consider annotating `Student` with `PartialEq`." Let's derive the `PartialEq` for student, which we can achieve in the same way as we derived the `Debug` trait earlier, by adding the following line before the `Student` struct definition:

```
#[derive(Debug, PartialEq)]
```

This change fixes the error in the print statement. According to the default implementation, two struct instances will be considered to be the same if they have the same field values. The print statement will therefore print the following output:

```
s_1 and s_2 are equal: false
```

To overwrite that default implementation, you must provide your own implementation for the trait. Consider the code shown in Listing 8.39.

```
impl PartialEq for Student {
    fn eq(&self, other: &Self) -> bool {
        self.age == other.age
    }
}
```

Listing 8.39 Implementing the PartialEq for Student

In this implementation, we are redefining equality based on the age field. Two Students are equal if they will have the same age values. Based on the new implementation, the two instances defined in main in earlier in Listing 8.38 are now considered to be equal. In contrast to the eq method, the ne method in the PartialEq trait checks whether two values are not equal, as the counterpart to the eq method, which checks for equality. The ne method returns the Boolean value true if the two values are not equal, and false if they are equal. By implementing this method, types can define how inequality comparisons should work between instances of that type.

8.2.8 Marker Traits

A marker trait is a trait that doesn't require any methods to be implemented. More accurately, its body is empty. Let's add a marker trait called Properties to the code in Listing 8.38:

```
trait Properties {}
```

A marker trait is typically used to add metadata or impose constraints on a type. It helps communicate additional information about a type to the compiler without needing any actual functionality to be implemented. This can be enhanced by adding supertraits (Section 8.2.5) to marker traits. Let's add a few supertraits from the standard library to the Properties, as follows:

```
trait Properties: PartialEq + Default + Clone {}
```

Now, any type that implements the Properties marker trait must also provide implementations for PartialEq, Default, and Clone due to the requirements of the supertraits. This ensures that the type not only carries the metadata of the marker trait but also satisfies these additional traits. For instance, if we want Student to have the properties indicated by the marker trait of Properties, we'll implement it for the Student by adding the following line:

```
impl Properties for Student {} // Error
```

This statement throws an error because the Student does not have all the traits implemented; that is, the Default and Clone are not implemented. To fix this error, we'll derive these traits for the Student by adding the following line before the definition of the Student struct:

```
#[derive(Debug, PartialEq, Default, Clone)]
```

In this way, we can ensure that the Student has all the traits indicated by the Properties trait.

This approach can be quite handy in situations where you have many different types but you want certain types to exhibit some essential behaviors. By using marker traits with supertraits, specific traits are implemented by a type, thus allowing for consistent behavior across those types.

8.2.9 Associated Types in Traits

Associated types in Rust traits define an abstract or placeholder type that is later determined by the specific implementing type. This mechanism provides flexibility, allowing each implementing type of the trait to specify the concrete type to be used in place of the associated type. Let's look at a motivating example to understand the need and use of the associated types.

Consider a scenario where we want to determine how far our car will travel in three hours at a specific speed. We'll define a few structs to represent key units of measurement. The first struct will be named Km (short for kilometers), with a single field representing the distance, and the second struct is named Kmh (kilometers per hour), also containing a single field representing speed. The two structs are shown in Listing 8.40.

```
#[derive(Debug)]
struct Km {
    value: u32,
}
#[derive(Debug)]
struct Kmh {
    value: u32,
}
```

Listing 8.40 Structs Km and Kmh Definitions

Defining structs with single fields in this manner adds type safety, thus preventing accidental comparisons between kilometers and kilometers per hour, which could lead to errors.

Since not all regions use the metric system, we'll also account for those familiar with miles and miles per hour by introducing two additional structs for Miles and Mph, as shown in Listing 8.41.

```
#[derive(Debug)]
struct Miles {
    value: u32,
}
#[derive(Debug)]
struct Mph {
    value: u32,
}
```

Listing 8.41 Structs Miles and Mph Definitions

Next, we'll implement methods to calculate how far the car travels in three hours. First, we'll implement a method for Kmh. Listing 8.42 shows the implementation.

```
impl Kmh {
    fn distance_in_three_hours(&self) -> Km {
        Km {
            value: self.value * 3,
        }
    }
}
```

Listing 8.42 Implementation of Distance Traveled in Three Hours for Kmh

The method distance_in_three_hours takes self as an input, where self refers to the instance of Kmh. The method returns the total distance, calculated as speed multiplied by three. In other words, if a car is traveling at a certain speed in kilometers per hour, the output will be the distance in kilometers after three hours. In the same way, you can implement a method for calculating distance in three hours for the Mph. Listing 8.43 shows the implementation.

```
impl Mph {
    fn distance_in_three_hours(&self) -> Miles {
        Miles {
            value: self.value * 3,
        }
```

```
    }
}
```

Listing 8.43 Implementation of Distance Traveled in Three Hours for Mph

While this code works and compiles successfully, something feels redundant. We've implemented the same concept twice, using two nearly identical methods. This redundancy suggests an opportunity for improvement: We can unify these methods using a trait.

Both implementations share the functionality of calculating the distance traveled in three hours, differing only in their return types. To address this similarity, we can define a trait that encapsulates the shared behavior. This trait will be called `DistanceThreeHours`. Let's define the trait with a common method in both implementations (i.e., `distance_in_three_hours`), as in the following example:

```
trait DistanceThreeHours {
    fn distance_in_three_hours(&self) -> ?; // what should be the output type
}
```

The trait contains the function common in the two implementations. If we look at the two implementations in Listing 8.42 and Listing 8.43, the input is the same, i.e., `&self`. However, what about the output? In one implementation, it is `Km` and in the second it is `Miles`. Using a concrete type like `Km` would prevent the trait from working for `Miles`, and vice versa. Associated types can come into play to work around this problem.

With associated types in traits, you can define placeholder types within a trait, where the concrete type is determined by the implementing type. This approach provides flexibility for trait implementers to choose the specific types that make sense for their implementation. Associated types within a trait are declared using the following syntax:

```
type name_of_associated_type
```

Let's add the associated type to the definition of the trait, as in the following example:

```
trait DistanceThreeHours {
    type Distance;
    fn distance_in_three_hours(&self) -> Self::Distance;
}
```

The trait now uses the associated type of `Distance`, which means that `Distance` is now a new trait item. Note that the output of the method `distance_in_three_hours` is now an associated type of `Distance`. The self-colon syntax (`Self::Distance`) is used to refer to the associated type within a trait. It is now up to the specific implementation of the trait to provide a concrete type for the associated type of `Distance`.

Let's implement the trait for Kmh. In this implementation, the associated type will be set to Km. The code shown in Listing 8.44 illustrates the implementation.

```
impl DistanceThreeHours for Kmh {
    type Distance = Km;
    fn distance_in_three_hours(&self) -> Self::Distance {
        Self::Distance {
            value: self.value * 3,
        }
    }
}
```

Listing 8.44 Implementation of the DistanceThreeHours for Kmh

Next, we'll implement the trait for Mph, with the associated type set to Miles. The code shown in Listing 8.45 illustrates the implementation.

```
impl DistanceThreeHours for Mph {
    type Distance = Miles;
    fn distance_in_three_hours(&self) -> Self::Distance {
        Self::Distance {
            value: self.value * 3,
        }
    }
}
```

Listing 8.45 Implementation of the DistanceThreeHours for Mph

Now more organized and unified, the code allows you to reuse functionality across different types while preserving their distinct characteristics. The code shown earlier in Listing 8.42 and Listing 8.43 is no longer required. The code can now be easily used in main, as shown in Listing 8.46.

```
fn main() {
    let speed_Kmh = Kmh { value: 90 };
    let distance_Km = speed_Kmh.distance_in_three_hours();
    println!(
        "At {:?}, you'll travel {:?} in 3 hours",
        speed_Kmh, distance_Km
    );

    let speed_Mph = Mph { value: 90 };
    let distance_Miles = speed_Mph.distance_in_three_hours();
    println!(
        "At {:?}, you'll travel {:?}, in 3 hours",
```

```
        speed_Mph, distance_Miles
    );
}
```

Listing 8.46 Using the Methods in main Defined in Listing 8.44 and Listing 8.45

> **Key Points Regarding Associated Types**
>
> There are a couple of important points to remember with regards to associated types:
>
> - Associated types should not be confused with generics. You use generics to write functions, structs, and enums that can operate on multiple types, while associated types in traits define abstract types that implementations must specify. In simpler terms, associated types act as placeholders for types that will be defined by the implementers of the trait. We'll explore the differences between associated types and generics in more detail in the next section.
> - The type keyword can also create type aliases, as discussed in Chapter 2, Section 2.2. However, within the context of traits, when a specific type is not defined in the trait itself, we refer to that type as an associated type.

8.3 Choosing between Associated Types and Generic Types

Both generic types and associated types allow the implementer to decide which concrete types should be used in a trait's functions and methods. In this section, we explore the circumstances under which each should be used, aiming to clarify when generics are more suitable and when associated types offer a better solution.

Consider an Addition trait with a single add method, as shown in Listing 8.47.

```
trait Addition {
    type Rhs;
    type Output;
    fn add(self, rhs: Self::Rhs) -> Self::Output;
}
```

Listing 8.47 Defining the Addition Trait with Associated Types of Rhs and Output

This trait will let's add instances of types self and Rhs using the add method.

Next, define a simple Point struct representing coordinates on a 2D plane. The struct has two fields, x and y, both of type i32, as follows:

```
struct Point {
    x: i32,
    y: i32,
}
```

Now, let's consider the implementation of the trait Addition for Point. Since the addition of a point with another point should result in another point, we'll define the associated types of Rhs and Output as Points. The implementation is shown in Listing 8.48.

```
impl Addition for Point {
    type Rhs = Point;
    type Output = Point;
    fn add(self, rhs: Self::Rhs) -> Point {
        Point {
            x: self.x + rhs.x,
            y: self.y + rhs.y,
        }
    }
}
```

Listing 8.48 Implementing the Addition Trait for Point with Rhs and Output, Both Set to Point

The method returns the addition of respective coordinates of self with the respective coordinates of rhs.

Let's say we need the ability to add an integer to Points. The integer value will be added to both coordinate values of the Point. We can accomplish this goal by adding another implementation to the same code for Point. The rhs in this case will be an integer, and the Output will be a Point. The implementation is shown in Listing 8.49.

```
impl Addition for Point { // Error
    type Rhs = i32;
    type Output = Point;
    fn add(self, rhs: Self::Rhs) -> Self::Output {
        Point {
            x: self.x + rhs,
            y: self.y + rhs,
        }
    }
}
```

Listing 8.49 Implementing the Addition Trait for Adding a Point with an Integer

The compiler throws an error, "conflicting implementations of trait Addition, for type Point." This error arises because the new implementation conflicts with the implementation already defined in the code shown earlier in Listing 8.48.

Since the Add trait is not parameterized by generic types, you can only implement it once per type, meaning you can only choose the types for both Rhs and Output once. To overcome this limitation, you must refactor Rhs from an associated type to a generic type. Doing so will allow you to implement the trait multiple times for Point, each time with a different type for Rhs. Let's modify Rhs from an associated type, as shown earlier

Listing 8.47, to a generic type, to provide the needed flexibility. The new definition of the trait is shown in Listing 8.50.

```
trait Addition<Rhs> {
   type Output;
   fn add(self, rhs: Rhs) -> Self::Output;
}
```

Listing 8.50 Redefining the Addition Trait from Listing 8.47 with a Generic Rhs

Note that the Rhs type is no longer needed because this type is now treated as a generic type, not as an associated type. The Self from rhs is also removed in the method signature since it is now considered generic, meaning it is not tied to the specific type for which we are implementing the trait.

This new implementation will now require updates to the implementation of Addition for Point, shown earlier in Listing 8.48. The updated code is shown in Listing 8.51.

```
impl Addition<Point> for Point {
    type Output = Point;
    fn add(self, rhs: Point) -> Self::Output {
        Point {
            x: self.x + rhs.x,
            y: self.y + rhs.y,
        }
    }
}
```

Listing 8.51 Updated Implementation from Listing 8.48 of Addition to Add a Point to a Point

Notice that, in the implementation line impl Addition<Point> for Point, we added a generic parameter of Point since this implementation considers rhs as Point. This parameter indicates to the compiler that, if the generic of rhs is considered a Point, then this implementation should be considered. Moreover, the type rhs is removed from the implementation since it is no longer needed, and in the method signature, the type of rhs is changed to a Point.

In the same way, let's now fix the implementation shown earlier in Listing 8.49 by setting the generic type to that of i32 and using a concrete type of i32 in the method signature. This updated implementation is shown in Listing 8.52.

```
impl Addition<i32> for Point {
    type Output = Point;
    fn add(self, rhs: i32) -> Self::Output {
        Point {
            x: self.x + rhs,
            y: self.y + rhs,
```

```
        }
    }
}
```

Listing 8.52 Updated Implementation from Listing 8.49 of Addition to Add a Point with an Integer

Comparing Listing 8.51 and Listing 8.52, note how generics allow you to have multiple implementations of the same trait for a single type, which is the `Point` type in this case. With associated types, it was not possible.

In general, deciding whether to use associated types or generics is quite straightforward. Associated types should be used when only a single implementation of the trait is needed for each type. If you need multiple implementations of the same trait for a type, generics should be used instead.

Let's further extend this implementation by considering another case. We'll add a new type called `Line`, which contains two points as its fields:

```
struct Line {
  start: Point,
    end: Point,
}
```

Now, there are situations in our program where adding two `Point` instances should yield a `Line` instead of another `Point`. This implies that the `output` type must not be limited to just a `Point` but could also be a `Line`. Given the current design of the `Addition` trait, where `Output` is an associated type, this requirement cannot be fulfilled. However, we can meet these new needs by refactoring `Output` from an associated type into a generic type. The new definition of the `Addition` trait shown earlier in Listing 8.50 uses the following lines of code:

```
trait Addition<Rhs, Output> {
    fn add(self, rhs: Rhs) -> Output;
}
```

Next, we'll update the existing two implementations to make them consistent with the new definition of the trait. First, we'll update the implementation shown earlier in Listing 8.51 to now consider both `Rhs` and `Output` as `Point`. The new implementation is shown in Listing 8.53.

```
impl Addition<Point, Point> for Point {
    fn add(self, rhs: Point) -> Point {
        Point {
            x: self.x + rhs.x,
            y: self.y + rhs.y,
```

```
        }
    }
}
```

Listing 8.53 Updated Implementation from Listing 8.51 of Addition to Add a Point with a Point

The associated type of `Output` is no longer needed. In the same way, you can add an implementation of `Addition` for adding an integer to a `Point`, as shown in Listing 8.54.

```
impl Addition<i32, Point> for Point {
    fn add(self, rhs: i32) -> Point {
        Point {
            x: self.x + rhs,
            y: self.y + rhs,
        }
    }
}
```

Listing 8.54 Updated Implementation from Listing 8.52 of Addition for Adding a Point with an Integer

Let's now add the new implementation of `Addition` for `Point`, where the `Rhs` will be `Point`, and the `Output` will be a `Line`. The new implementation is shown in Listing 8.55.

```
impl Addition<Point, Line> for Point {
    fn add(self, rhs: Point) -> Line {
        Line {
            start: self,
            end: rhs,
        }
    }
}
```

Listing 8.55 Implementation of Addition for Adding a Point with a Line

This method sets the `self` (which is a `Point`) as the starting `Point` of the `Line` and the `rhs` as the ending `Point` for the `Line`.

Let's now use the implementations in `main` and write some code for adding a `Point` to another `Point`, as shown in Listing 8.56.

```
fn main() {
    let p1 = Point { x: 1, y: 1 };
    let p2 = Point { x: 2, y: 2 };
    let p3 = p1.add(p2); // Error
    assert_eq!(p3.x, 3);
```

```
    assert_eq!(p3.y, 3);
}
```

Listing 8.56 Using the Implementation from Listing 8.55 to Add a Point with Another Point

The compiler throws an error, "consider giving p3, an explicit type," and further elaborating "type annotations needed." This error arises because the call to the add method matches multiple implementations. We are passing the rhs as a Point in the call to add method and therefore the implementations in Listing 8.53 and Listing 8.55 both match, since Rhs is realized as Point in both of these implementations. However, both the implementations return different types. To explicitly tell the compiler that we need to call the implementation where the Rhs and Output are both Points, we'll annotate the type to p3, as in the following example:

```
let p3: Point = p1.add(p2); // Error
```

This statement explicitly tells the compiler that you need to invoke the method where Rhs is a Point and the Output is a Point. This only matches a single implementation which is defined for adding a Point to another Point given earlier in Listing 8.53.

The main takeaway from this section is that you should use associated types when a trait only requires one implementation for a given type. However, when multiple implementations for the same type are necessary, generics should be preferred.

8.4 Practice Exercises

In this section, we'll provide some practice questions to help reinforce your learning. The exercise questions cover almost all the concepts introduced in the chapter. Solutions to each exercise will follow in Section 8.5.

1. **Generic enum for basic mathematical operations**
   Consider the following code in main. Write a generic enum named Operation that makes the code in main compile. The generic enum should represent the four basic mathematical operations: addition, subtraction, multiplication, and division. Each variant of the enum should store two values of the same type.

   ```
   fn main() {
       let op_1 = Operation::Addition(5, 10);
       let op_2 = Operation::Multiplication(3.5, 2.0);
       let op_3 = Operation::Subtraction(3.5, 2.0);
       let op_4 = Operation::Division(2, 3);
   }
   ```

2. **Fix the generic create method in the Container struct**
   The following code contains a compilation issue in the create function within the generic Container<T> implementation. Your task is to fix the error in the create

method of the generic Container<T> implementation so that the code compiles correctly. Additionally, ensure that the specific implementation for i32 type works as expected.

```
struct Container<T> {
    value: T,
}
impl<T> Container<T> {
    fn create(value: T) -> Container<T> {      // something wrong here
        Container { value }
    }
}
impl Container<i32> {
    fn display(&self) {
        println!("The value inside the container is: {}", self.value);
    }
    fn create(value: i32) -> Container<i32> {
        Container { value }
    }
}
fn main(){}
```

3. **Generalize the take_and_return() function to work with any type**
   The current implementation of the take_and_return() function only works for the User struct. Modify the function so that it can accept and return any type, enabling the code to compile correctly when used with both User and String types.

```
struct User {
    name: String,
    id: u32,
}
fn take_and_return(user: User) -> User { // this line needs updating
    user
}

fn main() {
    let user1 = User {
        name: "Alice".to_string(),
        id: 199,
    };
    let _user2 = take_and_return(user1);

    let str1 = String::from("Hello folks");
    let _str2 = take_and_return(str1); // we want this to compile
}
```

4. **Fix the Sound trait for Fish and ensure the program compiles**

   The Fish struct does not have a meaningful sound, so the animal_sound() method is not implemented for it. However, a compiler error arises. Modify the following code so that the Sound trait is properly implemented or excluded for Fish without causing the program to fail.

```
trait Sound {
    fn animal_sound(&self) -> String; // Consider adding some code here
}
struct Dog;
struct Cat;
struct Fish;

impl Sound for Dog {
    fn animal_sound(&self) -> String {
        "woof".to_string()
    }
}
impl Sound for Cat {
    fn animal_sound(&self) -> String {
        "meow".to_string()
    }
}
impl Sound for Fish {} /* Fish do not make any sound so we should
                           not implement the fn animal_sound(). This will
                           make compiler unhappy. */
fn main() {
    let dog = Dog;
    let cat = Cat;
    let fish = Fish;
    println!("Dog Sound: {}", dog.animal_sound());
    println!("Cat Sound: {}", cat.animal_sound());
    println!("Fish Sound: {}", fish.animal_sound());
}
```

5. **Fix the code to use proper trait bound syntax**

   The following code defines a trait Greeting and three functions that are supposed to print a greeting message. However, the trait bounds are not specified correctly, which causes the code to fail to compile. Update the code so that it compiles. Use the specific syntax as indicated to fix each line.

```
trait Greeting {
    fn greet(&self) -> String {
        "Hello from Rust!".to_string()
    }
}
```

```
fn print_greeting1<T: >(input: &T) {// Fix using trait bound
    println!("{}", input.greet());
}
fn print_greeting2(input: &impl ) {// Fix using impl trait syntax
    println!("{}", input.greet());
}
fn print_greeting3<T>(input: &T)
// Fix by using the where clause
{
    println!("{}", input.greet());
}
struct Greeter;

impl Greeting for Greeter {}
fn main() {
    let greeter_instance = Greeter;
    print_greeting1(&greeter_instance);
    print_greeting2(&greeter_instance);
    print_greeting3(&greeter_instance);
}
```

6. **Complete the function to compare horn sounds**
   The following code defines a trait VehicleHorn with a default horn sound and provides implementations for Car and Truck. However, the function compare_horn_sound() is incomplete. Update the function signature so that it accepts two arguments of any type that implements the VehicleHorn trait and compares their horn sounds.

```
pub trait VehicleHorn {
    fn horn_sound(&self) -> String {
        "beep beep".to_string()
    }
}
struct Car {}
struct Truck {}
impl VehicleHorn for Car {}
impl VehicleHorn for Truck {}
fn compare_horn_sound(vehicle_1: ??, vehicle_2: ??) -> bool { /* complete
                                                                 the function
                                                                 definition*/
    vehicle_1.horn_sound() == vehicle_2.horn_sound()
}

fn main() {
    let car = Car {};
```

```
    let truck = Truck {};
    assert_eq!(compare_horn_sound(car, truck), true);
}
```

7. **Implement missing trait for Circle**

   The Drawable and AnimatedDrawable traits are defined, but the Circle struct is missing the required implementation for the Drawable trait. Complete the following code so that Circle implements both Drawable and AnimatedDrawable, allowing it to be drawn and animated.

```
trait Drawable {
    fn draw(&self);
}
trait AnimatedDrawable: Drawable {
    fn animate(&self);
}
struct Circle;
/* some code missing here */
impl AnimatedDrawable for Circle {
    fn animate(&self) {
        println!("Animating a circle");
    }
}

fn main() {
    let circle = Circle;
    circle.draw();
    circle.animate();
}
```


8.5 Solutions

This section provides the code solutions for the practice exercises in Section 8.4. The code is largely self-explanatory; however, we have included comments and additional explanations wherever necessary to enhance understanding.

1. **Generic enum for basic mathematical operations**

```
enum Operation<T> {
    Addition(T, T),
    Subtraction(T, T),
    Multiplication(T, T),
    Division(T, T),
}
fn main() {
```

```
    let op_1 = Operation::Addition(5, 10);
    let op_2 = Operation::Multiplication(3.5, 2.0);
    let op_3 = Operation::Subtraction(3.5, 2.0);
    let op_4 = Operation::Division(2, 3);
}
```

2. **Fix the generic create method in the Container struct**

```
struct Container<T> {
    value: T,
}
impl<T> Container<T> {
    fn new(value: T) -> Container<T> {    /* we need to remove duplicate
                                             definitions for create fn */
        Container { value }
    }
}
impl Container<i32> {
    fn display(&self) {
        println!("The value inside the container is: {}", self.value);
    }
    fn create(value: i32) -> Container<i32> {
        Container { value }
    }
}
fn main() {}
```

3. **Generalize the take_and_return() function to work with any type**

```
struct User {
    name: String,
    id: u32,
}
fn take_and_return<T>(input: T) -> T {
    input
}
fn main() {
    let user1 = User {
        name: "Alice".to_string(),
        id: 199,
    };
    let _user2 = take_and_return(user1);

    let str1 = String::from("Hello folks");
    let _str2 = take_and_return(str1); // This now compiles
}
```

4. **Fix the Sound trait for Fish and ensure the program compiles**

```
trait Sound {
    fn animal_sound(&self) -> String {
        "I dont have sound!".to_string()
    }
}
struct Dog;
struct Cat;
struct Fish;
impl Sound for Dog {
    fn animal_sound(&self) -> String {
        "woof".to_string()
    }
}
impl Sound for Cat {
    fn animal_sound(&self) -> String {
        "meow".to_string()
    }
}
impl Sound for Fish {}
fn main() {
    let dog = Dog;
    let cat = Cat;
    let fish = Fish;
    println!("Dog Sound: {}", dog.animal_sound());
    println!("Cat Sound: {}", cat.animal_sound());
    println!("Fish Sound: {}", fish.animal_sound());
}
```

5. **Fix the code to use proper trait bound syntax**

```
trait Greeting {
    fn greet(&self) -> String {
        "Hello from Rust!".to_string()
    }
}
fn print_greeting1<T: Greeting>(input: &T) {
    println!("{}", input.greet());
}
fn print_greeting2(input: &impl Greeting) {
    println!("{}", input.greet());
}
fn print_greeting3<T>(input: &T)
where T: Greeting,
{
```

```
    println!("{}", input.greet());
}
struct Greeter;
impl Greeting for Greeter {}
fn main() {
    let greeter_instance = Greeter;
    print_greeting1(&greeter_instance);
    print_greeting2(&greeter_instance);
    print_greeting3(&greeter_instance);
}
```

6. **Complete the function to compare horn sounds**

```
pub trait VehicleHorn {
    fn horn_sound(&self) -> String {
        "beep beep".to_string()
    }
}
struct Car {}
struct Truck {}
impl VehicleHorn for Car {}
impl VehicleHorn for Truck {}
fn compare_horn_sound(vehicle_1: impl VehicleHorn, vehicle_2: impl
VehicleHorn) -> bool {
    vehicle_1.horn_sound() == vehicle_2.horn_sound()
}
fn main() {
    let car = Car {};
    let truck = Truck {};
    assert_eq!(compare_horn_sound(car, truck), true);
}
```

7. **Implement missing trait for `Circle`**

```
trait Drawable {
    fn draw(&self);
}
trait AnimatedDrawable: Drawable {
    fn animate(&self);
}
struct Circle;
impl Drawable for Circle {
    fn draw(&self) {
        println!("Drawing a circle");
    }
}
```

```
impl AnimatedDrawable for Circle {
    fn animate(&self) {
        println!("Animating a circle");
    }
}
fn main() {
    let circle = Circle;
    circle.draw();
    circle.animate();
}
```


8.6 Summary

This chapter delved into the powerful concepts of generics and traits, which form the backbone of Rust's type system. The first section introduced generics, covering their application in implementation blocks and exploring how multiple implementations can coexist alongside generics. We also covered some common pitfalls like duplication in implementation blocks and learned how generics simplify free functions. The section on monomorphization explained how Rust optimizes generic code at compile time.

Next, we explored traits, starting with trait bounds and moving through supertraits, trait objects, and the roles of derived traits and marker traits. We looked at the intricate relationships between traits and associated types to provide a deeper understanding of how to express complex behaviors.

A key section is the comparison between associated types and generic types, helping you make informed decisions on which option to use in different scenarios.

Next up, we'll cover Rust's functional programming features, such as closures, functional pointers, and iterators.

Chapter 9
Functional Programming Aspects

Functional programming introduces a new perspective on coding. In this chapter, we'll embrace functions as first-class citizens to enhance our toolkit for problem-solving.

9

We'll explore Rust's functional programming features in this chapter, starting with closures that capture their environment and functional pointers for more flexible function handling. The chapter delves into iterators, explaining how to use them to traverse collections efficiently, with a focus on `IntoIter` and iterating over various collection types. Combinators are introduced to show how to chain and transform iterator operations. The chapter also covers iterating through `Option` types, providing practical tools for concise and expressive functional programming in Rust.

9.1 Closures

Closures are anonymous functions that we can store in variables or pass as arguments to other functions. In the following sections, you'll learn the motivation behind using closures, then we'll discuss its basic syntax and how to pass closures to a function. Finally, we'll explain how closures capture a variable from its environment.

9.1.1 Motivating Example for Closures

Consider a business that stores information regarding its users in a struct. The definition of this struct is shown in Listing 9.1.

```
struct User {
    name: String,
    age: u8,
    salary: u32,
}
```

Listing 9.1 Struct Containing Information of Users

Our business is interested in a function that can validate users. For simplicity, assume that a user is considered valid if their `name` field is not empty. The definition of the function involves the following code:

```
fn validate_user(name: &str) -> bool {
    name.len() != 0
}
```

The validate_user function checks whether a given string name is valid by ensuring its length is not zero, returning true if the name is non-empty and false otherwise. In the main function, you can create instances of the User and call the validate_user function to check their validity, as shown in Listing 9.2.

```
fn main() {
    let person_1 = User {
        name: String::from("someone"),
        age: 35,
        salary: 40_000,
    };
    println!("User validity {}", validate_user(&person_1.name));
}
```

Listing 9.2 Using the Function validate_user in main

This code is perfectly fine; however, an alternative approach exists. Instead of creating a separate function for validating users, you can store the validation logic in a closure.

9.1.2 Basic Syntax

The function validate_user can be written in main in closure form, as shown in Listing 9.3.

```
fn main() {
    ...
    let validate_user = |name: &str| name.len() != 0;
    ...
}
```

Listing 9.3 Function validate_user in Closure Form

The input to the closure is enclosed in vertical pipes (| |). The closure body contains the same code as that of the function validate_user. The variable validate_user stores the entire closure. The function validate_user is no longer needed.

If you look at the type of the variable validate_user in your editor, which is storing the closures, notice a strange type indicated by impl Fn(&str) -> bool. In Rust, each closure has a concrete type. In this case, the type impl Fn() is followed by the signature of the closure, which means that this closure has a type that implements the Fn trait. The Fn trait has a special syntax, which provides information regarding the signature of the closure. (The syntax impl Fn() for mentioning the types was discussed in Chapter 8,

Section 8.2.4, in the lesson on trait bounds.) Besides Fn, the other two important traits a closure may implement are FnMut and FnOnce. We'll discuss these topics later in this section.

In most cases, the Rust compiler will infer the argument types and the return type. Closures are thus unlike functions, where the types of inputs and return type must always be explicitly provided. Note that we have not provided an output type in the definition of the closure shown in Listing 9.3. If the body only contains a single expression, then the curly braces around the body of the closure are optional.

Closures can be called just like functions, that is, by writing the variable name storing the closures and passing in the required inputs. For instance, the definition of the closure shown in Listing 9.3 can be called in a print statement, such as the following example:

```
println!("User validity {}", validate_user(&person_1.name));
```

Notice how this syntax is quite similar to calling a function.

9.1.3 Passing Closures to Functions

A key advantage of closures is that you can store them and then pass them as arguments or inputs to the function. This capability is possible with the help of generics and trait bounds, covered in Chapter 8.

Let's add one more closure to the code shown earlier in Listing 9.3. The updated code is shown in Listing 9.4.

```
fn main() {
    ...
    let validate_user_simple = |name: &str| name.len() != 0;     // renamed
    let validate_user_advance = |age: u8| age >= 30;
    ...
}
```

Listing 9.4 Another Closure validate_user_advance Added to the Code from Listing 9.3

The closure validate_user_advance checks whether the age is greater than or equal to 30. The closure will return a Boolean value depending on the condition.

We'll next define a function of is_valid. This function will take the two closures (defined in Listing 9.4) as inputs and will check both of them inside the function. To pass in closures as inputs to the function, we'll use generics. The definition of the function is shown in Listing 9.5.

```
fn is_valid_user<V1, V2>(name: &str, age: u8, simple_validator: V1, advance_
validator: V2) -> bool
where
```

```
    V1: Fn(&str) -> bool,
    V2: Fn(u8) -> bool,
{
    simple_validator(name) && advance_validator(age)
}
```

Listing 9.5 Passing in the Closures to a Function

The input to the function includes the variables of name and age and the two validators. The output is a bool value. The input validators (i.e., simple_validator and advance_validator) are generics V1 and V2 with some trait bounds. The simple_validator has a trait bound of Fn, followed by the signature of the closure &str -> bool. Thus, V1 could be of any type that implements the Fn trait and has the particular signature indicated by Fn(&str) -> bool. The second validator also has a trait bound of Fn followed by the signature of the closure. Inside the function, we return true if both the validators return true. The function can now be called from main, as shown in Listing 9.6.

```
fn main() {
    ...
    println!(
        "User validity {}",
        is_valid_user(
            &person_1.name,
            person_1.age,
            validate_user_simple,
            validate_user_advance
        )
    );
    ...
}
```

Listing 9.6 Calling the Function is_valid_user in main

9.1.4 Capturing Variables from the Environment

One special feature of closures is their ability to capture variables from the scope in which they are defined. Variables can be captured through an immutable borrow, a mutable borrow, or by transferring ownership. (For more information on borrowing, see Chapter 4, Section 4.3.) When environment variables are captured via an immutable borrow, the closure is considered to have implemented the Fn trait. If variables are captured through a mutable borrow, the closure implementation is considered to have implemented the FnMut trait. Finally, when a transfer of ownership occurs, the closure is said to have implemented the FnOnce trait.

Let's walk through some examples of these three traits. Consider a variable in the main function called banned_user and an updated definition of validate_user_simple, as in the code shown in Listing 9.7.

```
fn main() {
    ...
    let mut banned_user = String::from("banned user");
    let validate_user_simple = |name: &str| {
        let banned_user_name = banned_user;
        name.len() != 0 && name != banned_user_name
    };
    ...
}
```


Listing 9.7 Updated Definition of valid_user_simple

Inside the closure, the ownership of banned_user is moved to the variable banned_user_name. The closure is now checking for two conditions of name not being empty and also not being equal to the banned_user_name. In your editor, you may note that the type of the variable changes from Fn to FnOnce. This change occurs because the closure is assuming ownership of a variable from its scope or environment. Inside the closure, let's change the line to use an immutable reference instead, as in the following example:

```
let banned_user_name = &banned_user;
```

Now, the type of the closure changes back to Fn. Finally, we can use a mutable reference to the banned_user, as in the following example:

```
let banned_user_name = &mut banned_user;
```

At this point, the type of the closure will change to FnMut.

Types of Traits Implemented for Closures

In summary, depending on how the variables are captured, three types of traits might be implemented for a closure:

- For an immutable borrow, the closure implements the Fn trait.
- For a mutable borrow the closure implements the FnMut trait.
- For an ownership transfer, the closure implements the FnOnce trait.

Rust automatically infers which trait is implemented for a given closure based on how the variables from the environment are utilized.

You must pay special attention to the definitions of your functions while using variables from the environment. For instance, consider the closure definition in main, as shown in Listing 9.8.

```
let mut banned_user = String::from("banned user");
let validate_user_simple = |name: &str| { // Error
    let banned_user_name = banned_user;
    name.len() != 0 && name != banned_user_name
};
```

Listing 9.8 The Closure validate_user_simple Taking Ownership of the Variable from Environment

The compiler will throw an error, "expected a closure that implements `Fn` trait, but this closure only implements `FnOnce`." This error arises because, in the trait bounds of the function `is_valid_user` shown earlier in Listing 9.5, for `V1`, we mentioned the trait bound of `Fn`, while currently the closure is `FnOnce`. We must either use a reference to `banned_user` in the assignment inside the closure (which will change its type to that of `Fn`) or change the trait bound for `V1`. We'll change the trait bound on `V1`. The updated definition of the `is_valid_user` function is shown in Listing 9.9.

```
fn is_valid_user<V1, V2>(name: &str, age: u8, simple_validator: V1, advance_
validator: V2) -> bool
where
    V1: FnOnce(&str) -> bool,
    V2: Fn(u8) -> bool,
{
    simple_validator(name) && advance_validator(age)
}
```

Listing 9.9 Updated Definition of Function is_valid_user

Note that every closure implements the `FnOnce` trait because every closure can be called at least once. Sometimes, a closure may be using more than one variable from its environment. In such cases, to force the closure to take ownership of all the variables it is using, the `move` keyword is used before the start of the closure. Let's look at the syntax for this feature next:

```
let closure_name = move |parameters| { // closure body };
```

The `move` keyword converts any variables captured by reference or mutable reference to variables captured by value. For instance, as shown earlier in Listing 9.8, if we use the `move` keyword with a reference to the `banned_user` and try to the access the `banned_user` in the closure body, we'll see an error, "borrow of moved value," as shown in Listing 9.10.

```
let mut banned_user = String::from("banned user");
let validate_user_simple = move |name: &str| { // Error
    let banned_user_name = &banned_user;
```

```
    name.len() != 0 && name != banned_user_name
};
println!("{banned_user}"); // Error
```

Listing 9.10 Updated Closure validate_user_simple by Using the move Keyword and Using a Reference to banned_user

This error occurred because the `move` keyword enforces the transfer of ownership to inside the closure, even though the closure is using the variable through an immutable reference.

9.2 Function Pointers

Function pointers and closures in Rust are both mechanisms for referencing callable entities, but they have distinct differences. A *function pointer* is a reference to a standalone function defined in a program, and it does not capture or depend on any external variables or states. In contrast, closures are more flexible because they can capture variables from their surrounding environment, either by borrowing or owning them.

Function pointers are ideal for scenarios requiring stateless operations, such as callbacks or function dispatching. Closures, however, shine when you need to encapsulate and work with data from their defining context. Unlike closures, function pointers do not implement the `Fn`, `FnMut`, or `FnOnce` traits, though they can be converted into closures when necessary.

To understand the basics, let's return to our earlier example from the previous section, with some minor modifications. For the sake of clarity, our example code is again shown in Listing 9.11.

```
struct User {
    name: String,
    age: u8,
    salary: u32,
}
fn is_valid_user<V1, V2>(name: &str, age: u8, simple_validator: V1, advance_
validator: V2) -> bool
where
    V1: Fn(&str) -> bool,
    V2: Fn(u8) -> bool,
{
    simple_validator(name) && advance_validator(age)
}
fn main() {
    let person_1 = User {
        name: String::from("someone"),
```

```
        age: 35,
        salary: 40_000,
    };
    let validate_user_simple = move |name: &str| name.len() != 0;
    let validate_user_advance = |age: u8| age >= 30;
    println!(
        "User validity {}",
        is_valid_user(
            &person_1.name,
            person_1.age,
            validate_user_simple,
            validate_user_advance
        )
    );
}
```

Listing 9.11 Code from the Previous Section with Slight Modification

The modification in this code is that the closure `validate_user_simple` is simplified, and it no longer captures variables from its environment. In other words, this code doesn't use the variable `banned_user`, which was previously defined in `main`. Next, we'll convert `validate_user_simple` from a closure to a function and delete the definition of the closure from `main`, as shown in Listing 9.12.

```
fn validate_user_simple(name: &str) -> bool {
    name.len() != 0
}
struct User {
        ...
}
fn is_valid_user<V1, V2>(name: &str, age: u8, simple_validator: V1, advance_
validator: V2) -> bool
where
    V1: Fn(&str) -> bool,
    V2: Fn(u8) -> bool,
..

fn main() {
    let person_1 = User {
        ...
    };
    //    let validate_user_simple = move |name: &str| name.len() != 0;
    //    removed
    let validate_user_advance = |age: u8| age >= 30;
    println!(
```

```
        "User validity {}",
        is_valid_user(
            &person_1.name,
            person_1.age,
            validate_user_simple, // this is now functional pointer
            validate_user_advance
        )
    );
}
```

Listing 9.12 Updated Code from Listing 9.11 with Added Function validate_user_simple and Removed Closure Definition in main

Notice that, even though we remove the closure, the code still compiles. Instead of passing in the closure, in the print statement shown in Listing 9.12, we are now passing in a function name, which essentially serves as a functional pointer. The function `is_valid_user` was expecting a closure with a type `Fn(&str) -> bool` for the third argument; however, we passed in a functional pointer while calling it in `main` by mentioning the function name of `validate_use_simple`. Thus, we can provide a functional pointer anywhere a closure is expected. This approach works because the function pointers implement all the three closure traits, that is, `Fn`, `FnMut`, and `FnOnce`. In summary, you can pass in regular functions anywhere closures are expected.

The `simple_validator` in the function signature of `is_valid_user` is a function pointer, which is essentially is pointing to the function `validate_user_simple`. As a result, we can call the function `validate_user_simple`, with the help of a different name in the function signature (i.e., `simple_validator`).

In the same way, we can convert the closure `validate_user_advance` to a function and remove the closure body from `main`, as shown in Listing 9.13.

```
fn validate_user_advance(age: u8) -> bool {
    age >= 30
}
fn validate_user_simple(name: &str) -> bool {
    name.len() != 0
}
struct User {
        ...
}
fn is_valid_user<V1, V2>(name: &str, age: u8, simple_validator: V1, advance_
validator: V2) -> bool
...
fn main() {
    let person_1 = User {
        ...
```

```
    };
    // let validate_user_simple = move |name: &str| name.len() != 0;
    // removed
    // let validate_user_advance = |age: u8| age >= 30;
    // removed
    ...
}
```

Listing 9.13 Code from Listing 9.12 Updated by Converting validate_user_advance to a Function

Now that our closures are converted to functions, let's update the definition of the function `is_valid_user` to accept function pointers instead of generics. We'll remove the generics in the function signature and use simple functional pointer types, as shown in Listing 9.14.

```
fn is_valid_user(name: &str, age: u8, simple_validator: fn(&str) -> bool,
    advance_validator: fn(u8) -> bool) -> bool
{
    simple_validator(name) && advance_validator(age)
}
fn validate_user_advance(age: u8) -> bool {
    age >= 30
}
fn validate_user_simple(name: &str) -> bool {
    name.len() != 0
}
struct User {
        ...
}
fn main() {
    let person_1 = User {
        ...
    };
   ...
}
```

Listing 9.14 Updated Definition of is_valid_user Using Functional Pointers

Notice the types of `simple_validator` and `advance_validator`. Their types are `fn(&str) -> bool` and `fn(u8) -> bool`. Function pointers are concrete types represented with lowercase `fn`, which should not be confused with capital `Fn`, which is a closure trait. Since the function pointers are concrete types, we don't need to use generics anymore, and trait bounds are therefore no longer required.

The key requirement for a closure to be converted to a function pointer is that it must not use the variables from its environment. However, one way to overcome this requirement is to convert a closure to a function pointer by explicitly setting all the variables that are used from the environment as arguments to the function.

For instance, consider a variable banned_user defined in main, as in the following example:

```
let banned_user = "banned user";
```

Previously, the closure corresponding to validate_user_simple made use of this variable to check whether the user is not a banned user. To make sure that the function also uses this, we'll pass this local variable to the function and check if the user is not equal to the banned user. The updated function definition is shown in the following example:

```
fn validate_user_simple(name: &str, banned_user_name: &str) -> bool {
    name.len() != 0 && name != banned_user_name
}
```

In contrast to the previous definition of the function, this updated function is now checking whether, in addition to the name passed in not being empty, the name must also be different from banned_user_name. The updated definition of validate_user_simple will require an update to the function is_valid_user. The updated code for is_valid_user is shown in Listing 9.15.

```
fn is_valid_user(
    name: &str,
    age: u8,
    banned_user_name: &str, //updated
    simple_validator: fn(&str, &str) -> bool, //updated
    advance_validator: fn(u8) -> bool,
) -> bool {
    simple_validator(name, banned_user_name) && advance_validator(age)
}
```

Listing 9.15 Updated Definition of is_valid_user Based on New Definition of validate_user_simple

The function list of arguments is updated. The function now has one additional input argument of banned_user_name and an updated type for the simple_validator. This change occurs because the simple_validator function is now using two &str types of inputs.

Finally, we'll ensure that we call the function with a valid argument in the main function, as shown in Listing 9.16.

```
println!(
    "User validity {}",
    is_valid_user(
    &person_1.name,
    person_1.age,
    banned_user, // updated
    validate_user_simple,
    validate_user_advance
    )
);
```

Listing 9.16 Update to Function Call is_valid_user in main

The function call now includes the additional input of `banned_user`.

9.3 Iterators

With the *iterator design pattern*, different types can have a common interface for accessing the elements of these types sequentially. An iterator abstracts how iterations are implemented internally and how types are defined internally. Iterators are heavily used in Rust programs, and therefore, an important task is to understand how it works.

In the following sections, we'll explain the basics of the `Iterator` and the related `IntoIterator` traits. We'll also demonstrate how to iterate over collections.

9.3.1 The Iterator Trait

The Rust standard library has a trait called `Iterator`, which any type can implement. The definition of this trait resembles the code shown in Listing 9.17.

```
trait Iterator {
    type Item;
    fn next(&mut self) -> Option<Self::Item>;
}
```

Listing 9.17 Definition of Iterator Trait in Standard Library

The `Iterator` trait has an associated type called `Item` and requires an explicit implementation of the method called `next`. Associated types were discussed in Chapter 8, Section 8.2.9. The key property of associated types is that they are restricted to one concrete type per trait implementation. The `next` method provides a mechanism for obtaining the next item in the iteration. This method takes a mutable reference to `self` and returns an optional instance of the associated type item. Calling the `next` method return repeatedly will you give you the next items in the iteration wrapped around by the Some variant, until no more items are left, in which case the method will return a None variant.

Note that the Iterator trait has many methods with default implementations. However, the next method has no default implementation and must be explicitly implemented. Let's walk through an example to understand the Iterator trait.

Consider the Employee and Employee_records structs defined in the code shown in Listing 9.18.

```
struct Employee {
    name: String,
    salary: u16,
}
struct Employee_Records {
    employee_db: Vec<Employee>,
}
```

Listing 9.18 Definitions of Employee and Employee_Records Structs

Let's now implement the Iterator trait for the Employee_Records. This implementation will enable us to go over the employee names in sequential order. Listing 9.19 shows the implementation of Iterator for Employee_Records.

```
impl Iterator for Employee_Records {
    type Item = String;
    fn next(&mut self) -> Option<Self::Item> {
        if self.employee_db.len() != 0 {
            let result = self.employee_db[0].name.clone();
            self.employee_db.remove(0);
            Some(result)
        } else {
            None
        }
    }
}
```

Listing 9.19 Implementation of Iterator Trait for the Employee_Records

Let's break down the code line by line. First, we set the Item type to that of String. Since we are iterating through the names, which is of type String, we'll set the associated type of Item to String. The signature of the next method should match the signature defined in the Iterator trait (shown earlier in Listing 9.17). The next method is assumed to return the first available entry in the collection if available. Therefore, our first check determines whether the employee database contains an entry using the following statement:

```
if self.employee_db.len() != 0
```

If the database entry exists, you can grab the name of the employee at the first index and then remove the entry from the vector. Finally, return the name of the employee wrapped inside the Some variant. In any other case, the vector will be empty, so the method will return a None variant.

Now, use this implementation in main, as shown in Listing 9.20.

```
fn main() {
    let mut emp_1 = Employee {
        name: String::from("John"),
        salary: 40_000,
    };
    let mut emp_2 = Employee {
        name: String::from("Joseph"),
        salary: 30_000,
    };
    let mut emp_db = Employee_Records {
        employee_db: vec![emp_1, emp_2],
    };
    println!("{:?}", emp_db.next());
    println!("{:?}", emp_db.next());
    println!("{:?}", emp_db.next());
}
```

Listing 9.20 Using the Iterator Implementation for Employee_Records in main

This function creates a few instances of Employee, namely, emp_1 and emp_2. Next, it creates an instance of Employee_Records simulating the employee database or emp_db containing the two employee's records. Finally, calls are made to the next method on the emp_db. The first call to the next method will return the name of the first employee inside the Some variant, for instance, Some(John). The second call will return Some(Joseph), and the final call will return None.

Instead of iterating through the employee databases manually by repeatedly calling the next method, a more elegant approach is to iterate through the employee database automatically using for loops, as in the following example:

```
for employee in emp_db {
        println!("{employee}");
}
```

If we comment out the calls to the next methods shown in Listing 9.20 and use the for loop in the preceding code instead, you'll see the following output:

```
John
Joseph
```

The for loop automatically handles types that implement the Iterator trait. In this case, while emp_db is not an iterator itself, the loop smartly converts it into one, producing items of type String, which matches the type of the employee variable. The loop terminates when the None variant is encountered, and therefore, None is not printed. Additionally, all values returned by the next method are automatically unwrapped.

You can also return your own custom defined types from the call to the next method. For instance, in Listing 9.19, you can return Employee instead:

```
impl Iterator for Employee_Records {
    type Item = Employee;
    fn next(&mut self) -> Option<Self::Item> {
       ... (needs updating to work with the changes)
       }
}
```

You can make necessary changes in the code where indicated accordingly. You're completely free to implement iterators as you like, as long as you stick to the basic philosophy of the trait: The Iterator trait should have one mandatory function called next, which return the next items. Moreover, the returning items must be of the same type.

9.3.2 IntoIterator

As we've explored the Iterator trait and its pivotal role in iterating over collections, an essential feature to recognize is Rust's flexibility when dealing with data ownership during iteration. Regarding ownership as well, the IntoIterator trait allows collections to be converted into iterators, thus offering a fresh perspective on how to traverse and consume values efficiently. Let's dive into how IntoIterator works and its significance in transforming collections into iterators.

The key difference between the Iterator trait and IntoIterator trait is that the Iterator is implemented on a type, over which you can iterate. On the other hand, the IntoIterator is implemented on a type, which you can turn into an Iterator (or more specifically which implements the Iterator). This key difference may take some time to digest.

Listing 9.21 shows how the IntoIterator trait is defined in the standard library.

```
trait IntoIterator {
    type Item;
    type IntoIter: Iterator;
    fn into_iter(self) -> Self::IntoIter;
}
```

Listing 9.21 Definition of IntoIterator in the Standard Library

The `IntoIterator` trait has a single method, namely, `into_iter`, which consumes `self` and returns an `Iterator`. In contrast, the `next` method of `Iterator` trait returns an `Item`. Let's walk through an example to make things more clear.

Consider a `Book` struct with three fields: `tile`, `author`, and `genre`. The definition of the struct is shown in Listing 9.22.

```
struct Book {
    title: String,
    author: String,
    genre: String,
}
```

Listing 9.22 Definition of the Book Struct

Next, define another struct of `BookIterator` with a single field of `properties`. The definition of the struct is shown in Listing 9.23.

```
struct BookIterator {
    properties: Vec<String>,
}
```

Listing 9.23 Definition of the BookIterator Struct

An instance of the struct `BookIterator` will correspond to a book, with the details of the `book` being captured in the `properties` vector. This struct has been created with the specific intention of adding the ability to iterate over the fields of the `book`. The `properties` vector captures all the fields of the `Book` struct, as they are all of the same type, in this case, `String`. Storing them in the `properties` vector allows you to easily iterate over all the fields of a `Book`. First, you need a struct to hold the iterator state (by state, we mean the current item to be returned by the `next` method), and then you must implement the `Iterator` trait on that struct. You could include the iterator state inside the `Book` struct, but doing so would make the `Book` struct look a bit messy. Therefore, let's create the `BookIterator` struct to store the state of the iterator separately from the data in the `Book` struct.

Listing 9.24 shows the implementation of the `Iterator` for the `BookIterator`.

```
impl Iterator for BookIterator {
    type Item = String;
    fn next(&mut self) -> Option<Self::Item> {
        if !self.properties.is_empty() {
            Some(self.properties.remove(0))
        } else {
            None
        }
    }
}
```

Listing 9.24 Implementation of Iterator for BookIterator

This simple implementation of the `BookIterator` will iterate over the details of each book. The `properties` vector contains the details of a particular book, and the `next` method is returning the first entry from the `properties` vector, if the vector is not empty. Note that the statement `Some(self.properties.remove(0))` first removes the item. This removal is necessary to remain consistent with the definition of the `next` method in the `Iterator` trait, which is defined in the standard library. Next, to return the type mentioned in the method signature, we wrapped the removed item with the `Some` variant.

Recall that the `IntoIterator` trait is implemented for a type that can be converted into an `Iterator`. As a result, given a `Book` instance, you should be able to convert it into some type that implements the `Iterator`, in this case, the `BookIterator`. Listing 9.25 shows the implementation of the `IntoIterator` for the `Book`.

```
impl IntoIterator for Book {
    type Item = String;
    type IntoIter = BookIterator;

    fn into_iter(self) -> Self::IntoIter {
        BookIterator {
            properties: vec![self.title, self.author, self.genre],
        }
    }
}
```

Listing 9.25 Implementation of IntoIterator for Book

Note that the type `IntoIter` must be some type that implements the `Iterator` trait, in this case, the `BookIterator`. The method `into_iter` simply returns an instance of the `BookIterator`, with the value set to the fields of the `Book` struct. Now, when the `into_iter` method is called for an instance of the `Book`, it will consume that instance and return another type (i.e., BookIterator), which essentially converts the Book instance into an Iterator. Thus, the into_iter will consume the instance since it returns an owned type.

You can now use this implementation in main, as shown in Listing 9.26.

```
fn main() {
    let book = Book {
        title: "Digital Image Processing".to_string(),
        author: "Gonzales".to_string(),
        genre: "Science Book".to_string(),
    };
    let mut book_iterator = book.into_iter();
    println!("{:?}", book_iterator.next());
    println!("{:?}", book_iterator.next());
    println!("{:?}", book_iterator.next());
    println!("{:?}", book_iterator.next());
}
```

Listing 9.26 Using the Iterator and IntoIterator Implementation in main

The into_iter method on the book returns a BookIterator instance. Since next is implemented for the BookIterator, we can call the next method on the Book_Iterator to iterate over the individual fields of a Book.

A key advantage of IntoIterator is that a type can now be used in for loops or other iterator-consuming contexts. IntoIterator allows you to seamlessly integrate a type into a for loop. The following code shows how you can iterate over Book_Iterator using a for loop:

```
for book_info in book_iterator {
    println!("{book_info}");
}
```

One final point to note is that, in the IntoIterator implementation for the Book, the BookIterator type is not strictly necessary. We included it for illustration purposes. Instead, we could have directly used the vector, which can also be converted into an iterator. In other words, rather than returning a BookIterator, we could simply return a vector iterator. Listing 9.27 shows how this simplification can be achieved.

```
impl IntoIterator for Book {
    type Item = String;
    type IntoIter = std::vec::IntoIter<Self::Item>;
    fn into_iter(self) -> Self::IntoIter {
        vec![self.title, self.author, self.genre].into_iter()
```

```
    }
}
```

Listing 9.27 Updated Implementation of the IntoIterator for the Book Using Vectors

The into_iter method on a vector converts a vector into an Iterator over the vector. Since the into_iter method is now returning an iterator over the vector, we changed the type of the intoIter to that of the vector iterator given by std::vec::IntoIter <Self::Item>. Using the code shown in Listing 9.27, the BookIterator struct and the Iterator implementation for the BookIterator are not needed. The code in the main shown earlier in Listing 9.26 will compile with only the code shown in Listing 9.27.

9.3.3 Iterating over Collections

A common use case for iterators is to facilitate operations on elements stored in a collection. In Rust's standard library, collections can be converted into types such as Iter and IntoIter, which thus allow access to the methods provided by the Iterator trait since those types implement that trait. Two commonly used collection types are vectors and hash maps (refer to Chapter 2, Section 2.2.2, and Chapter 5, Section 5.5, respectively). Let's look at some examples of how these collection types can be used.

Iterating over Vectors

Consider the following vector:

```
let mut vec_1 = vec![45, 30, 85, 90, 41, 39];
```

Three primary methods are used to create a type over which we can iterate from a collection, depending on how you want to reference the values within the collection. The first method is called iter. The iter method gives you an Iter type that you can then use to iterate over immutable references to items in the collection. This capability can be demonstrated by calling the next method on the vector after the iter method, as follows:

```
let mut vec_1_iter = vec_1.iter();
let value_1 = vec_1_iter.next();
```

The type of value_1 is an Option<&i32> having an immutable reference to i32 values. The other two methods are iter_mut and into_iter, which enables the IterMut and IntoIter types. The iter_mut method iterates over mutable references to the items in the collection, and the into_iter method iterates over owned items.

When using the collections in a for loop, Rust will automatically infer the type of iterator based on how the values of the collection are used. For instance, let's iterate over vec_1 using immutable reference to values inside the vector:

```
for values in &vec_1 {
    println!("{values}");
}
```

The vector is borrowed immutably, allowing the for loop to create an iterator over immutable references to the values within the vector. If we borrow the vector mutably, then we would get an iterator with mutable references to the values inside the vector, as in the following example:

```
for values in &mut vec_1 {
    println!("{values}");
}
```

Finally, let's try the owned form of the vector, as in the following example:

```
for values in vec_1 {
    println!("{values}");
}
```

In this case, the loop will take over ownership of the values of the vector vec_1.

Iterating over HashMaps

Let's go over HashMaps now. Consider the following HashMap:

```
let mut person: HashMap<String, i32> = HashMap::new();
person.insert("Hannash".to_string(), 40);
person.insert("Joseph".to_string(), 44);
person.insert("Sara".to_string(), 55);
```

We can use tuples to iterate over the HashMap, as follows:

```
for (name, age) in &person {
    println!("The person {} has an age of {}", name, age);
}
```

The loop uses a tuple pattern to extract the name and age from the HashMap. The HashMap is currently borrowed immutably, which means that we get an iterator over immutable values. This reality can be confirmed by inspecting the type of the variables names and age in your code editor, which are &String and &i32 values, respectively. As a result, we have access to the immutable references to the two values.

However, let's say we change and borrow mutably instead, as in the following example:

```
for (name, age) in &mut person {
    println!("The person {} has an age of {}", name, age);
}
```

Then, we should be expecting to iterate over mutable values. However, surprisingly, if we inspect the types of the variables name and age, we obtain &String and &mut i32. We've only obtained a mutable reference to the age, and the name is still borrowed immutably. The name variable represents the keys of our HashMap, and the keys are immutable by default because changing the keys would mess up the internal state of the HashMap (see Chapter 5, Section 5.5, for more details).

Finally, you could also take ownership, in which case the owned values are assigned to the variables of names and age, as in the following example:

```
for (name, age) in person {
    println!("The person {} has an age of {}", name, age);
}
```


This code takes ownership of the values in the HashMap, and therefore, the values are no longer accessible due to this transfer of ownership.

9.4 Combinators

Now that you understand how iterators allow us to traverse collections efficiently, it's time to look at how we can further enhance iteration using combinators. Combinators are methods provided by the Iterator trait that allow us to transform, filter, or combine elements in a chainable manner. They offer a powerful way to build more complex behaviors from simple operations, making iteration not just efficient but also expressive. Let's dive into how combinators work and how they can be applied in real-world scenarios through an example.

Consider the following vector containing the names of some fruits:

```
let words = vec!["apple", "banana", "grape", "orange", "pear"];
```

Our task is to capture all the names of the fruits that start with either "a" or "b" into another vector. Additionally, all selected words should be converted to uppercase.

One way of coding this program is to use for loops, as shown in Listing 9.28.

```
let mut result: Vec<String> = vec![];
for word in words {
    if word.starts_with("a") || word.starts_with("b") {
        let uppercase_word = word.to_uppercase();
        result.push(uppercase_word);
    }
}
println!("Result: {:?}", result);
```

Listing 9.28 Using for Loop to Return All Fruit Names That Start with Letters a or b

In this code, we first created a vector of strings for storing the final result computed inside the loop. Next, we iterated through all the words to check whether the word starts with the letter "a'" or the letter "b" via the `starts_with` method. This method returns `true` if the string slice starts with the pattern indicated. Next, we convert the word to uppercase and finally add it to the vector.

Although this code works, it can be much more precise and cleaner using iterators and combinators. Combinators are compact, pure functions created for specific tasks, and they can be linked together to execute complex operations. Iterators, meanwhile, come with a variety of handy methods with default implementations.

Let's start implementing our program using combinators. First we'll create a variable of type `IntoIter` by defining a variable `result` and setting it equal to `words.into_iter()` with the following code:

```
let result: Vec<String> = words
    .into_iter()
```

We discussed the use of `into_iter` in Section 9.3.2. This method returns an iterator over the string slices in the vector of `words`. We'll next use the `filter` combinator. This combinator filters out items from an iterator. In this case, we would like to filter out all the words that start with the letter "a" or the letter "b." The updated code after applying the `filter` resembles the following code:

```
let result: Vec<String> = words
    .into_iter()
    .filter(|&word| word.starts_with("a") || word.starts_with("b"))
```

The `filter` will create a new iterator, with items being filtered out based on some condition. The filter takes a closure and executes the closure for every item in the iterator. The closure accepts one argument, which is the item, and checks it against a condition to determine if it should stay or if it needs to be filtered out. The type written in front of the filter method (in your code editor), which is `impl Iterator<item = &str>`, tells us that the output from this combinator will be an iterator, with the item being equal to a string slice.

Next, we would like to convert the strings slices in the iterator to uppercase. The `map` combinator can be used for this purpose. The updated code after applying the `map` is shown in Listing 9.29.

```
let result: Vec<String> = words
    .into_iter()
    .filter(|&word| word.starts_with("a") || word.starts_with("b"))
    .map(|word| word.to_uppercase())
```

Listing 9.29 The Code after Applying the map Combinator

The map converts items in an iterator from one type to another or from one form to another. This combinator takes an Iterator and returns a new Iterator. A closure is executed for each item in the Iterator. The closure takes one argument (the item) and returns a new type. In this case, we want to map each word to its corresponding uppercase form. The type next to the map function in your code editor is impl Iterator<item = String>, which indicates the map returns another Iterator, with the items being equal to String.

Finally, we would like to convert the iterator into a collection. We can do this conversion with the help of the collect combinator. The updated code after applying the collect is shown in Listing 9.30.

```
let result = words //Error
    .into_iter()
    .filter(|&word| word.starts_with("a") || word.starts_with("b"))
    .map(|word| word.to_uppercase())
    .collect();
```

Listing 9.30 The Code after Applying the collect Combinator

This code throws an error, "type annotations needed." The collect combinator needs further information regarding the type in which we wish to collect the items. This additional information can be provided using the *turbofish syntax*, as shown in Listing 9.31.

```
let result = words
    .into_iter()
    .filter(|&word| word.starts_with("a") || word.starts_with("b"))
    .map(|word| word.to_uppercase())
    .collect::<Vec<String>>();
```

Listing 9.31 The Code after Correcting the Error Using Turbo Fish Syntax

The syntax ::<Vec<String>> is called the turbo fish syntax. The general form of this syntax is ::<T>.

Turbo Fish

Turbofish is a playful nickname given by the Rust community due to the visual appearance of the syntax ::<T>. The :: resembles the body of a fish, and <T> looks like the fish's tail. The "turbo" part likely refers to its powerful role in disambiguating types, making it feel like an enhanced or "turbocharged" feature.

You can use the turbo fish syntax when a function defines a generic but it's unclear what concrete type to substitute for that generic. Look more closely at the definition of the collection by hovering your mouse cursor on the collection. Notice that it has the following signature:

```
pub fn collect<B>(self) -> B
```

In this case, the collect defines a generic B; however, the Rust compiler does not know what concrete type should substitute B. Therefore, you should use the turbo fish syntax to indicate that the concrete type should be a vector of strings.

An alternative to this approach would be to explicitly set the type of the result. Listing 9.32 shows how this explicit mentioning of type can be performed.

```
let result: Vec<String> = words
    .into_iter()
    .filter(|&word| word.starts_with("a") || word.starts_with("b"))
    .map(|word| word.to_uppercase())
    .collect();  // turbo fish syntax is not needed
```

Listing 9.32 Alternative to Turbo Fish Syntax

The type of result has been explicitly set to Vec<String>. In this case, additional information using the turbo fish syntax is not required. The compiler can infer the concrete type for the generic, associated with the collect combinator.

Turbo Fish versus Explicit Type Annotation

The turbo fish syntax keeps type declarations close to their logic, making it ideal for short-lived, single-use variables in method chains. The syntax is concise and reduces redundancy but may appear intimidating to beginners or may seem to clutter long chains. On the other hand, explicit type annotation provides clarity at the point of variable declaration, making it more intuitive for newcomers and suitable for variables reused over a larger scope. However, explicit type annotation can be verbose and feel disconnected from the transformation logic when used in complex chains.

In the code shown in Listing 9.32, we've implemented the same functionality as the code shown earlier in Listing 9.28, but now using combinators. The true benefits of combinators may not be clear yet, however, because it was fairly a simple example. In general, combinators will result in shorter, cleaner, clear code.

Considerations for Combinators

A few points should be noted with regard to combinators:

- First, Iterators and methods on iterators that we also refer to as combinators, such as map, filter, and collect, are lazy. In other words, they won't actually do any work until explicitly asked to iterate over items by calling the next method or by calling any other method that ultimately calls the next.

  For instance, as shown in Listing 9.32, no work is actually performed until the line in which a call is made to the collect method. When the collect method is called, it ultimately calls the next method, and at that particular point, all these method calls execute their respective transformations for each item in the Iterator.

- Second, a whole bunch of different combinators are available, and we've only covered a few important ones. We recommend browsing through the complete list in the standard Rust documentation.

9.5 Iterating through Option

After exploring how iterators and combinators allow you to efficiently traverse and manipulate collections, let's now extend these concepts to handle more specialized types like Option, which we introduced in Chapter 5, Section 5.3. Iterating through an Option type provides a way to process values that may or may not be present so you can work seamlessly with optional data. This section will show you how the Option type integrates with iterators and how you can use familiar methods to handle optional values in a more expressive manner.

Consider the following code:

```
let some_product = Some("laptop");
let mut products = vec!["cellphone", "battery", "charger"];
```

This code initializes an optional variable some_product containing the string laptop and a mutable vector named products that contains the strings cellphone, battery, and charger. The some_product variable is wrapped in an option to represent a value that might or might not exist, while the products vector is a simple collection of strings. Now, let's say we want to add the logic that, if some_product is not None, then the product should be added into the products vector. This task can be coded using a match, as shown in Listing 9.33.

```
match some_product {
    Some(product) => products.push(product),
    _ => {}
};
```

Listing 9.33 Adding the Logic That If some_product Is Not None, Then We'll Add It to products Vector

This code works but can also be simplified using the if let syntax. Recall that a match can be simplified into if let when we only care about one specific arm and can ignore the others. In this case, we don't need to consider the second arm. Let's now examine the simplified code:

```
if let Some(product) = some_product {
    products.push(product);
}
```

The if let syntax simplifies the code, but an even simpler way is available: using the extend method. The extend method extends a collection with the contents of an iterator in the following way:

```
products.extend(some_product);
```

This code will perform the same job as the code shown in Listing 9.33, but now using the if let syntax.

The reason passing an Option works is that the Option type implements the IntoIterator trait. Think of an Option as an iterator containing either zero or one element. In this case, if product is the None variant, nothing happens. However, if product is the Some variant, the value inside Some will be appended to the vector.

Since the Option enum implements the IntoIterator, this enum could be used where an Iterator is expected. For instance, you can chain it with some other iterator to form a larger iterator via the chain method. For instance, consider the following code:

```
let products_iter = products.iter().chain(some_product.iter());
```

First, the iter method is called over the products vector. Then, using the chain, we combined it with the iterator of some_product to return a longer combined chain, containing the two iterators. In this case, the option, that is, some_product, is converted into an iterator using the iter function, and the value is chained with the other iterator. You can now use a for loop to go over the values:

```
for prod in products_iter {
    println!("{:?} ", prod);
}
```

Let's turn to another example where iterating through Option can be handy. Consider the following products vector again:

```
let products = vec![Some("charger"), Some("battery"), None,
Some("cellphone")];
```

Now, we want to remove the None variants and retrieve only the Some variants from the vector. One way to accomplish this task is by using a for loop to iterate through all the values, checking each one individually. If we encounter the Some variant, we'll store the item in another vector. The code for this task should resemble the code shown in Listing 9.34.

```
let mut prod_without_none = Vec::new();
for p in products {
    if p.is_some() {
        prod_without_none.push(p.unwrap());
    }
}
```

Listing 9.34 Removing the None Variants and Keeping the Some Variants

The code initializes an empty vector prod_without_none and iterates over products, checking if each element is Some. If so, the code unwraps the value and adds it to prod_without_none, filtering out any None values. Although the code works, it could be improved by using iterators instead. With iterators, you can call the into_iter on products, which converts it into an iterator. Then, you can use the filter method to keep only the Some variants, ignoring the None variants. Next, applying the map method unwraps each Some variant to extract the actual value inside. Finally, calling collect gathers the unwrapped values into a new Vec<&str>, resulting in a vector that contains only the unwrapped product names. This updated code is shown in Listing 9.35.


```
let prod_without_none = products
        .into_iter()
        .filter(|x| x.is_some())
        .map(|x| x.unwrap())
        .collect::<Vec<&str>>();
```

Listing 9.35 Removing the None Variants and Keeping the Some Variants Using Iterators and Combinators

Both solutions shown in Listing 9.34 and Listing 9.35 are correct; however, some coding is required. Luckily, an easy way to code the same functionality is by using the flatten method, as in the following example:

```
let prod_without_none: Vec<&str> = products.into_iter().flatten().collect();
```

The flatten method in Rust converts an iterator of iterators into a single iterator. It essentially "flattens" a nested structure by taking each item from the inner iterators and combining them into a single stream. This method is particularly useful when working with collections of collections, such as a Vec<Option<Vec<T>>>, where you want to remove the nested structure and work with the individual elements directly. You can also use this method to return only the Some variants from a collection. The call to the flatten method in the previous code extracts the items stored inside the Some variants and throws away the None variants. Finally, the collect method transforms the iterator back into a collection, in this case, a vector of string slices.

9.6 Practice Exercises

In this section, we'll provide some practice questions to help reinforce your learning. The exercise questions cover almost all the concepts introduced in the chapter. Solutions to each exercise will follow in Section 9.7.

1. **Completing the closure definition**
   You're provided with an incomplete Rust program that uses a closure. Your task is to complete the code by defining the closure add_to_x so that it adds the value of y to the variable x.

```
fn main() {
    let x = 10;
    let add_to_x = |y| /* Add closure definition here */
    let result = add_to_x(5);
    println!("Result: {}", result);
}
```

2. **Implementing a mutable closure for incrementing a counter**
   You're given an incomplete Rust program that involves a mutable closure. Your task is to complete the closure increment_counter so that it increments the counter variable each time it is called.

```
fn main() {
    let mut counter = 0;
    let mut increment_counter = || /* Complete the Closure definition */
    increment_counter();
    increment_counter();
    println!("Final Counter: {}", counter);
}
```

3. **Fixing struct definitions to handle closures capturing environment values**
   The given code defines an EventHandler struct, which is designed to handle events using closures. However, the struct does not properly support closures that capture values from their environment. Your task is to fix the struct definition so that closures can modify the environment, as shown in the main function.
   Hint: The problem lies in the trait bound, as closures capturing environment values need to be mutable. Adjust the struct definition accordingly to allow this behavior.

```
struct EventHandler<T>
where
    T: Fn(), // Something wrong here.
             // Hint: Check the code in main and see how the closure is using
             // the values from its environment
{
    on_event: T,
}
```

```
impl<T> EventHandler<T>
where
    T: Fn(), // Something wrong here
{
    fn handle_event(&mut self) {
        (self.on_event)()
    }
}
fn main() {
    let mut lights_on = false;
    let mut temperature = 25;
    let mut lights_handler = EventHandler {
        on_event: || {
            lights_on = !lights_on;
            println!("Lights are now {}", if lights_on { "on" } else { "off"
});
        },
    };
    let mut temperature_handler = EventHandler {
        on_event: || {
            temperature += 5;
            println!("Temperature increased to {}°C", temperature);
        },
    };
    lights_handler.handle_event();
    temperature_handler.handle_event();
    temperature_handler.handle_event();
    lights_handler.handle_event();
    assert_eq!(temperature, 35);
    assert_eq!(lights_on, true);
}
```

4. **Completing the function signature for `sum_of_squares`**

   You're provided with a partial function signature for `sum_of_squares`, and your task is to complete it without using generics. The function will take a number, a squaring function, and an addition function to calculate the sum of squares from 1 to the given number. Complete the function signature so that `sum_of_squares` can properly accept a `u32` value for the number, a function for squaring, and a function for addition.

```
fn add(x: u32, y: u32) -> u32 {
    x + y
}
fn square(x: u32) -> u32 {
    x * x
```

```
}
fn sum_of_squares(num: ?, sq: ?, add: ?) -> u32 {
    let mut result = 0;
    for i in 1..=num {
        result = add(result, sq(i));
    }
    result
}
fn main() {
    let num = 4;
    let sum = sum_of_squares(num, square, add);
    println!("Sum of squares from 1 to {} = {}", num, sum);
}
```

5. **Updating function signature to use function pointers**
You're given a Rust program that defines an `invoker` function using a closure. Your task is to update the function signature to accept function pointers instead of closures. Then, add a `square` function that matches the required function pointer type. Update the function signature for `invoker` to utilize function pointers and ensure that the `square` function can be passed correctly to it.

```
fn invoker<O>(operation: O, operand: i32) -> i32 // This needs to be updated
where
    O: Fn(i32) -> i32,
{
    operation(operand)
}
/* A square function needs to be added here */
fn main() {
    let square = |x: i32| x * x;
    let result = invoker(square, 4);
    println!("Result is: {}", result);
}
```

6. **Implementing the next method for a custom iterator**
You're tasked with completing the implementation of the `next` method for a custom iterator called `Counter`. The `Counter` struct has a `current` value that starts at `0` and a `max` value that determines the stopping condition for the iteration. Your goal is to return the next value of `current` each time `next` is called, until `current` reaches `max`, at which point it should return `None`. Complete the `next` method so that it correctly implements the iterator behavior for the `Counter` struct.

```
struct Counter {
    current: u32,
    max: u32,
}
```

```
impl Counter {
    fn new(max: u32) -> Counter {
        Counter { current: 0, max }
    }
}
impl Iterator for Counter {
    type Item = u32;
    fn next(&mut self) -> Option<Self::Item> {
        /* Add code here */
    }
}
fn main() {
    let mut counter = Counter::new(3);
    assert!(matches!(counter.next(), Some(0)));
    assert!(matches!(counter.next(), Some(1)));
    assert!(matches!(counter.next(), Some(2)));
    assert!(matches!(counter.next(), None));
}
```

7. **Completing the `into_iter` function for struct iteration**
   Implement the `into_iter` function for a `Person` struct, which will allow you to iterate over its fields as a vector of strings. The `Person` struct contains three fields: `name`, `age`, and `occupation`. Your task is to define how the `into_iter` method will return a vector containing these fields. Complete the `into_iter` function to return a vector of `Strings` representing the fields of the `Person` struct.

```
struct Person {
    name: String,
    age: u32,
    occupation: String,
}
impl IntoIterator for Person {
    type Item = String;
    type IntoIter = std::vec::IntoIter<Self::Item>;
    fn into_iter(self) -> Self::IntoIter {
        /* Your code here
        Hint: Should return a vector of Strings,
        representing the fields of the struct  */
    }
}
fn main() {
    let person = Person {
        name: "Alice".to_string(),
        age: 30,
        occupation: "Software Engineer".to_string(),
```

```
    };
    let mut person_iterator = person.into_iter();
    while let Some(property) = person_iterator.next() {
        println!("{}", property);
    }
}
```

8. **Implementing into_iter for a struct with color components**

   You must implement the IntoIterator trait for a Pixel struct, which contains three components: r, g, and b. Your tasks include completing the associated type for Item in the trait definition and implementing the into_iter function to return an iterator over the color components of the pixel. Complete the associated Item type and implement the into_iter function to return a vector of the pixel's RGB components.

```
struct Pixel {
    r: i8,
    g: i8,
    b: i8,
}
impl IntoIterator for Pixel {
    type Item = ?; // this needs to be fixed
    type IntoIter = std::vec::IntoIter<Self::Item>;
    fn into_iter(self) -> Self::IntoIter {
        /* The function needs to be completed */
    }
}
fn main() {
    let p = Pixel {
        r: 54,
        g: 23,
        b: 74,
    };
    let p = p.into_iter();
    for component in p {
        println!("{}", component);
    }
}
```

9. **Fixing mutable iteration over a vector**

   Fix the following code so that it compiles and successfully iterates over the vector vec_1, allowing modifications to its elements. You must call the correct method that provides mutable references to the elements of the vector.

```
fn main() {
    let mut vec_1 = vec![4, 5, 6, 9, 8];
    for i in vec_1. ?? { // fix this line by making a call to relevant
```

```
function
        *i = *i * 2;
    }
    println!("{:?}", vec_1);
}
```

10. **Refactoring code using combinators**

    Refactor the given code, which calculates the sum of the squares of odd numbers in a vector, using combinators instead of a traditional for loop.

```
fn main() {
    let numbers = vec![1, 2, 3, 4, 5, 6, 7, 8, 9, 10];
    let mut result = 0;
    /* The code in the for loop needs to be replaced */
    for &num in &numbers {
        if num % 2 != 0 {
            let squared_num = num * num;
            result += squared_num;
        }
    }
    println!("Result without combinators: {}", result);
}
```


9.7 Solutions

This section provides the code solutions for the practice exercises in Section 9.6. The code is largely self-explanatory; however, we have included comments and additional explanations wherever necessary to enhance understanding.

1. **Completing the closure definition**

```
fn main() {
    let x = 10;
    let add_to_x = |y| x+y;
    let result = add_to_x(5);
    println!("Result: {}", result);
}
```

2. **Implementing a mutable closure for incrementing a counter**

```
fn main() {
    let mut counter = 0;
    let mut increment_counter = || counter +=1;
    increment_counter();
```

```
    increment_counter();
    println!("Final Counter: {}", counter);
}
```

3. **Fixing struct definitions to handle closures capturing environment values**

```
struct EventHandler<T>
where
    T: FnMut(),
{
    on_event: T,
}
impl<T> EventHandler<T>
where
    T: FnMut(),
{
    fn handle_event(&mut self) {
        (self.on_event)()
    }
}
fn main() {
    let mut lights_on = false;
    let mut temperature = 25;
    let mut lights_handler = EventHandler {
        on_event: || {
            lights_on = !lights_on;
            println!("Lights are now {}", if lights_on { "on" } else { "off"
});
        },
    };

    let mut temperature_handler = EventHandler {
        on_event: || {
            temperature += 5;
            println!("Temperature increased to {}°C", temperature);
        },
    };
    lights_handler.handle_event();
    temperature_handler.handle_event();
    temperature_handler.handle_event();
    lights_handler.handle_event();
    assert_eq!(temperature, 35);
    assert_eq!(lights_on, true);
}
```

4. **Completing the function signature for `sum_of_squares`**

```
fn add(x: u32, y: u32) -> u32 {
    x + y
}
fn square(x: u32) -> u32 {
    x * x
}
fn sum_of_squares(num: u32, sq: fn(u32) -> u32, add: fn(u32, u32) -> u32) ->
u32 {
    let mut result = 0;
    for i in 1..=num {
        result = add(result, sq(i));
    }
    result
}
fn main() {
    let num = 4;
    let sum = sum_of_squares(num, square, add);
    println!("Sum of squares from 1 to {} = {}", num, sum);
}
```

5. **Updating a function signature to use function pointers**

```
fn invoker(operation: fn(i32) -> i32, operand: i32) -> i32 {
    operation(operand)
}
fn square(x: i32) -> i32 {
    x * x
}
fn main() {
    let square = |x: i32| x * x;
    let result = invoker(square, 4);
    println!("Result is: {}", result);
}
```

6. **Implementing the `next` method for a custom iterator**

```
struct Counter {
    current: u32,
    max: u32,
}
impl Counter {
    fn new(max: u32) -> Counter {
        Counter { current: 0, max }
    }
}
```

```
impl Iterator for Counter {
    type Item = u32;
    fn next(&mut self) -> Option<Self::Item> {
        if self.current < self.max {
            let result = Some(self.current);
            self.current += 1;
            result
        } else {
            None
        }
    }
}
fn main() {
    let mut counter = Counter::new(3);
    assert!(matches!(counter.next(), Some(0)));
    assert!(matches!(counter.next(), Some(1)));
    assert!(matches!(counter.next(), Some(2)));
    assert!(matches!(counter.next(), None));
}
```

7. **Completing the `into_iter` function for struct iteration**

```
struct Person {
    name: String,
    age: u32,
    occupation: String,
}
impl IntoIterator for Person {
    type Item = String;
    type IntoIter = std::vec::IntoIter<Self::Item>;
    fn into_iter(self) -> Self::IntoIter {
        vec![self.name, self.age.to_string(), self.occupation].into_iter()
    }
}
fn main() {
    let person = Person {
        name: "Alice".to_string(),
        age: 30,
        occupation: "Software Engineer".to_string(),
    };
    let mut person_iterator = person.into_iter();
    while let Some(property) = person_iterator.next() {
        println!("{}", property);
    }
}
```

8. **Implementing `into_iter` for a struct with color components**

```
struct Pixel {
    r: i8,
    g: i8,
    b: i8,
}
impl IntoIterator for Pixel {
    type Item = i8;
    type IntoIter = std::vec::IntoIter<Self::Item>;
    fn into_iter(self) -> Self::IntoIter {
        vec![self.r, self.g, self.b].into_iter()
    }
}
fn main() {
    let p = Pixel {
        r: 54,
        g: 23,
        b: 74,
    };
    let p = p.into_iter();
    for component in p {
        println!("{}", component);
    }
}
```

9. **Fixing mutable iteration over a vector**

```
fn main() {
    let mut vec_1 = vec![4, 5, 6, 9, 8];
    for i in vec_1.iter_mut()  {
        *i = *i * 2;
    }
    println!("{:?}", vec_1);
}
```

10. **Refactoring code using combinators**

```
fn main() {
    let numbers = vec![1, 2, 3, 4, 5, 6, 7, 8, 9, 10];
    let mut result = 0;
    let result: i32 = numbers
        .iter()
        .filter(|&&num| num % 2 != 0)
        .map(|&num| num * num)
```

```
        .sum();
    println!("Result without combinators: {}", result);
}
```

9.8 Summary

This chapter introduced Rust's key functional programming aspects, beginning with closures, which allow a function to capture its surrounding environment to enable more dynamic behavior. We also explored functional pointers, which enhance flexibility in how functions are passed and utilized. We spent significant time on `Iterators`, covering the `IntoIter` trait and demonstrating how to efficiently traverse collections in Rust. We also discussed combinators, which enable the chaining and transformation of iterator operations for more concise code. Additionally, by iterating through `Option` types, we illustrated how Rust's powerful functional tools can simplify handling optional values.

Next, we'll dive into Rust's distinct memory management features, focusing on lifetimes and smart pointers for managing memory safely and efficiently.

Chapter 10
Memory Management Features

Effective memory management is crucial for efficient programming. This chapter will guide you through the concepts of lifetimes and smart pointers, both key capabilities to keeping your code safe and performant.

This chapter covers critical memory management concepts, starting with lifetimes, which ensure references are valid as long as needed. Lifetime elision is explained to simplify code by inferring lifetimes automatically. The chapter also discusses lifetimes in structs, showing how to manage complex data relationships. Smart pointers like `Box`, `Rc`, and `RefCell` are introduced, providing powerful tools for heap allocation, reference counting, and interior mutability. This chapter also explores deref coercion, a mechanism that simplifies working with smart pointers by automatically converting them to reference types, thus enabling seamless interaction with values stored on the heap. These features are essential for writing safe and efficient Rust code that manages memory correctly.

10.1 Lifetimes

Lifetimes in Rust ensure that references are valid for only as long as they are needed, preventing common issues like dangling pointers or memory leaks. This section will guide you through the concept of lifetimes, their syntax, and how they work under the hood to guarantee memory safety. By the end, you'll understand how to annotate lifetimes in functions and structs, allowing for more flexible and error-free code.

Lifetimes may be better understood by breaking them into concrete lifetimes and generic lifetimes. We'll cover both in the following sections and then explore lifetime elisions and how lifetimes work with structs.

10.1.1 Concrete Lifetimes

A *concrete lifetime* refers to the duration during which a value exists in memory. The lifetime of a value begins when it is created and ends when the value is dropped or moved from its memory location, often due to a change in ownership. You'll learn how to use concrete lifetimes in various scenarios in the following sections.

Concrete Lifetimes with Owned Values

Consider the code shown in Listing 10.1.

```
fn main() {
    let i = 5;
    let j = i;
    println!("{i}");
}
```

Listing 10.1 A Simple Program for Illustrating the Concept of Lifetimes

The lifetime of variable i starts at the line in which it is defined, that is, the line `let i = 5;`. The variable's lifetime ends when its value is dropped or cleaned up, which happens when the `main` function ends. Similarly, the lifetime of the variable j starts on the line on which it is created and ends when the `main` function ends.

Now, let's introduce an inner scope and move the variable i into this inner scope, as shown in Listing 10.2.

```
fn main() {
    {
        let i = 5;                          // i lifetime starts
    }                                       // i lifetime ends
    let j = i;
    println!("{i}"); // Error
}
```

Listing 10.2 Code from Listing 10.1 Updated by Adding an Inner Scope Containing Variable i

In this case, the lifetime of variable i starts on the line on which it is created. However, the variable's lifetime does not end at the end of the `main` function but rather ends at the end of the inner scope. At the end of the inner scope, variable i will be dropped and cleared from memory. Since variable i does not exist after the inner scope, an error arises on the print line.

The variables used in this example are all stack allocated. Let's look at an example containing heap-allocated data, as shown in Listing 10.3. (For a refresher on the difference between the stack and the heap, refer to Chapter 4, Section 4.1.1.)

```
fn main() {
    let str_1 = String::from("abc");            // str_1 lifetime starts
    let str_2 = str_1;                          // str_1 lifetime ends
    println!("str_1: {str_1}"); // Error
}
```

Listing 10.3 An Example Similar to Listing 10.1 but Using Heap-Allocated Data

Although the same as the example shown earlier in Listing 10.1, this example uses heap-allocated data instead of stack-allocated data and throws an error. This error arises because of the move of the value from str_1 to str_2. While this problem may seem obvious, but an important task is to understand it from the perspective of lifetimes.

The lifetime of the variable str_1 starts from the line in which it is defined, that is, the line let str_1 = String::from("abc");. The variable's lifetime ends on the next line when its value is moved into str_2. At this point, variable str_2 contains the String, and str_1 is invalid, which is why an error occurs when trying to print out str_1 on the next line.

The same issue arises when we send a value to a function by ownership. For instance, consider the code shown in Listing 10.4.


```
fn main() {
    let str_1 = String::from("abc");            // str_1 lifetime starts
    str_fn(str_1);                              // str_1 lifetime ends
    let str_2 = str_1; // Error
}

fn str_fn(s: String) {
    println!("s: {s}");
}
```

Listing 10.4 A Similar Issue as in Listing 10.3 but with Functions

In main, the lifetime of str_1 starts from the line in which it is defined and ends on the line when we call function str_fn. This variable's lifetime is limited to the two lines because the function takes ownership of the str_1, and therefore, the variable is not available afterwards.

Concrete Lifetimes with References

Up to this point, we've only worked with examples of owned values. Let's review some examples of references. Consider the code shown in Listing 10.5.

```
fn main() {
    let i;
    {
        let j = 5;                      // j lifetime starts from this line
        i = &j; // Error
    }                                   // j lifetime ends on this line
    println!("i: {i}");
}
```

Listing 10.5 An Example Containing References

The compiler through an error, "j does not live long enough, borrowed value does not live long enough." This situation results in what's called a *dangling reference*, which you may recall from Chapter 4, Section 4.3.2. To recap, a dangling reference occurs when a reference points to a value in memory that no longer exists. To prevent this problem, at compile time, Rust's *borrow checker* ensures that a reference's lifetime is valid within the lifespan of the borrowed value. However, in this case, we are violating that rule. To see why this violation occurred, let's review the code shown in Listing 10.5 line by line.

On the first line, variable `i` is a reference. The lifetime of the variable `i` starts from the line on which it is declared and ends at the end of the `main` function. However, the value it is borrowing (i.e., the variable `j`) has a shorter lifetime. The lifetime of `j` starts from the line in the inner scope in which the variable is defined and ends at the end of the scope. At the end of the inner scope, variable `j` will be dropped and cleaned up. As a result, in subsequent lines of code, variable `i` will be referencing invalid memory, precisely what the error message is saying: "j does not live long enough."

To fix this problem, move the print statement to inside the inner scope, as shown in Listing 10.6.

```
fn main() {
    let i;
    {
        let j = 5;
        i = &j;
        println!("i: {i}");
    }
}
```

Listing 10.6 Error from Listing 10.5 Fixed

This code does not throw an error because the variable `j` is valid until the end of the inner scope.

Let's look at an example involving a mutable reference. Consider the code shown in Listing 10.7.

```
fn main() {
    let mut vec_1 = vec![6, 5, 8, 9];
    let ref_1 = &vec_1;                          // ref_1 lifetime starts
    println!("ref 1: {:?}", ref_1);              // ref_1 lifetime ends
    let ref_2 = &mut vec_1;                      // ref_2 lifetime starts
    ref_2.push(3);
    println!("ref 2: {:?}", ref_2);              // ref_2 lifetime ends
}
```

Listing 10.7 An Example Involving Mutable References

In this example, after defining the vector vec_1, we have an immutable reference to it (i.e., ref_1), which is printed. Next, we create a mutable reference of ref_2 to the vector and add a value using this mutable reference. Finally, the program prints ref_2.

If you recall the borrowing rules from Chapter 4, Section 4.3.2, they state that we can either have one mutable reference or multiple immutable references at any given time. In other words, ultimately, immutable and mutable references should not coexist. However, in our example, we have both an immutable reference *and* a mutable reference, seemingly at the same time. The question is, why does our code still compile?

The answer to this mystery lies in the Rust concept called *non-lexical lifetimes*, which aim to relax some of the strictness imposed by traditional lifetime rules. Under this concept, lifetimes are determined by analyzing the actual usage of references in the code rather than relying strictly on scopes. In simple terms, non-lexical lifetimes are lifetimes that are not strictly bound to the scope of a variable.

As shown in Listing 10.7, the compiler notices that last usage of ref_1 is on the print line println!("ref 1: {:?}", ref_1). Therefore, its lifetime is limited to two lines only. In the same way, when the compiler looks into ref_2, it notices that its last usage is on the last print line and therefore its lifetimes is limited to the three lines shown in Listing 10.7. The lifetimes of the two references do not overlap, and therefore, they do not co-exist. As a result, Rust's borrowing rules are not violated.

However, let's see what happens when we push the print line printing ref_1 down by one line, as shown in Listing 10.8.

```
fn main() {
    let mut vec_1 = vec![6, 5, 8, 9];
    let ref_1 = &vec_1;                          // ref_1 lifetime starts
    let ref_2 = &mut vec_1; // Error             // ref_2 lifetime starts
    println!("ref 1: {:?}", ref_1);              // ref_1 lifetime ends
    ref_2.push(3);
    println!("ref 2: {:?}", ref_2);              // ref_1 lifetime ends
}
```

Listing 10.8 Code from Listing 10.7 Modified by Moving the Print Line Printing ref_1 Down by One Line

The compiler throws an error, "cannot borrow vec_1 as mutable, because it is also borrowed as immutable." The two lifetimes of ref_1 and ref_2 are overlapping in this case, and therefore, the borrowing rule that immutable and immutable references should not co-exist is violated. To fix this problem, make sure that the lifetimes of your immutable and mutable references do not overlap. The code shown in Listing 10.7 is therefore the correct form of the code.

10.1.2 Generic Lifetimes

In the previous section, we explored concrete lifetimes, which involve specifying exact scopes for references. Now, we'll shift our focus to generic lifetimes, an essential tool that allows you to write more flexible and reusable functions and structs by abstracting over the lifetime of references.

Lifetimes with One Relationship

Let's begin with the code shown in Listing 10.9.

```
fn main() {
    let int1 = 5;
    let int2 = 10;
    let picked_value = picking_int(&int1, &int2);
    println!("{picked_value}");
}
fn picking_int(i: &i32, j: &i32) -> i32 {
    if rand::random() {
        *i
    } else {
        *j
    }
}
```

Listing 10.9 A Simple Function That Randomly Returns One of the Two References Passed In

This code includes a function called picking_integer, which randomly selects one of the integers passed to it and returns the chosen integer value. This function uses the random function from the rand crate to generate a random Boolean value, which is either true or false. This function is called in the main function, and its result is stored and printed. The program compiles successfully. However, let's explore what happens if we modify the function to return a reference instead. The updated function definition is shown in Listing 10.10.

```
fn picking_int(i: &i32, j: &i32) -> &i32 { // Error
    if rand::random() {
        i
    } else {
        j
    }
}
```

Listing 10.10 Updated Function picking_int That Now Returns a Reference

We removed the dereference operator * from the returning variables of `i` and `j` since the function is expected to return a reference instead of an actual value.

The updated function definition throws an error, "missing lifetime specifier." Let's break down what went wrong when we changed the return value to a reference by considering the situation from the borrow checker's perspective. How will the borrow checker ensure that the `picked_value` in the `main` function is not a dangling reference when we print it after the function call? The borrow checker would look at the lifetime of the `picked_value`.

The `picked_value` is the return value from the function, which is a reference to an integer. So what's the lifetime of the returned reference? The target of the returning reference from the function is simply not clear in this code. It could be either variable `i` or variable `j`, and each could have a different lifetime. Because of this ambiguity, the borrow checker cannot analyze our code, thereby resulting in an error, which states in full "missing lifetime specifier, this function's return type contains a borrowed value, but the signature does not say, whether it is borrowed from `i` or `j`."

Generic Lifetimes versus Concrete Lifetimes

Lifetime specifiers, also known as *generic lifetime annotations*, provide a way to describe relationships between the lifetimes of references. They shouldn't be confused with concrete lifetimes, which are specific, but rather are used to generalize those relationships.

Let's fix this error by introducing generic lifetime annotations. Like generics (covered in Chapter 8, Section 8.1), generic lifetime annotations are defined inside angled brackets (< >) after the function name. The code shown in Listing 10.11 illustrates how you can fix the error shown earlier in Listing 10.10.

```
fn picking_int<'a>(i: &'a i32, j: &'a i32) -> &'a i32 {
    if rand::random() {
        i
    } else {
        j
    }
}
```

Listing 10.11 Code from Listing 10.10 Fixed Using Generic Lifetime Specifiers

Unlike typical generics, a generic lifetime specifier starts with a backtick (`) followed by the name of the generic. Common convention is to use a lowercase letter for the generic, starting it from the letter `a` and then subsequently down the alphabet. The syntax for adding generic lifetime specifiers is to add it after the reference, for instance, `i: &'a i32`, in our example.

Let's refocus on the code shown in Listing 10.11. What is the meaning or semantics of adding the generic lifetime specifier? In simpler terms, they basically establish a relationship between the lifetimes of the variables i, j, and the returned value. But what exactly is this relationship? Essentially, it means that the lifetime of the returned value will be equal to the shorter of the two lifetimes of the parameters i and j. For example, if i has a shorter lifetime than j, then the returned reference will have a lifetime tied to i. Conversely, if j has the shorter lifetime, the returned reference will be valid for as long as j exists. This ensures that the reference never outlives the data it refers to, preventing potential dangling references.

Let's see what happens in main. The code in main from Listing 10.9 is given again in Listing 10.12.

```
fn main() {
    let int1 = 5;                                  // int1 lifetime starts
    let int2 = 10;                                 // int2 lifetime starts
    let picked_value = picking_int(&int1, &int2);
    println!("{picked_value}");
}                                                  // int1 and int2 lifetimes end
```

Listing 10.12 Same Code in main from Listing 10.9

Considering the line on which the picking_int is called, the borrow checker now knows that the lifetime of picked_value will be the smallest lifetime of the parameters passed into the function. The first parameter is a reference to int1, which will be valid for the lifetime of int1. The second parameter is a reference to int2, which is valid for the lifetime of int2. Variable int1 has a lifetime from the line on which it is defined, until the end of the main function. In the same way, int2 has a lifetime from the line on which it is defined, until the end of the main function. The borrow checker now knows that the lifetimes of the picked_value, which is the resulting reference from the function, will be valid until the end of the main, since both these references are valid during this period.

In the examples shown in Listing 10.11 and Listing 10.12, both int1 and int2 have basically the same lifetimes. Let's modify this code slightly so that the relationship of the returning value within the input parameters (which is the shorter of the two lifetimes of the input parameters) is clearer.

Consider the code shown in Listing 10.13.

```
fn main() {
    let int1 = 5;                                      // int1 lifetime starts
    {
        let int2 = 10;                                 // int2 lifetime starts
        let picked_value = picking_int(&int1, &int2);
```

```
        println!("{picked_value}");
    }                                          // int2 lifetime ends
}                                              // int1 lifetime ends
```

Listing 10.13 Modified Code from Listing 10.12 with int2 Now Having Shorter Lifetime

In this case, the lifetime of `int1` starts from the line on which it is defined and ends at the end of the `main` function. The variable `int2`, however, has a shorter lifetime, ending at the end of the code block. The lifetime of `picked_value`, which is the resulting reference from the function, will therefore be equal to the lifetime of `int2`, which is the shorter lifetime of the two. The reference of `picked_value` will therefore be valid in the inner scope because it corresponds to the shorter lifetime passed into the function.

Let's now slightly modify the code from Listing 10.13, as shown in Listing 10.14.


```
fn main() {
    let int1 = 5;                              // int1 lifetime starts
    let picked_value;                          // defined here now
    {
        let int2 = 10;                         // int2 lifetime starts
        picked_value = picking_int(&int1, &int2); // Error
    }                                          // int2 lifetime ends
    println!("{picked_value}");
}                                              // int1 lifetime ends
```

Listing 10.14 Modified Code from Listing 10.13 with picked_value Printed Outside the Inner Scope

The `picked_value` is now defined and printed in the `main` scope. The compiler is not happy with this code and throws an error, "`int2` does not live long enough, borrowed value does not live long enough." The shortest lifetime passed into the function in this case was `int2`. The lifetime of `int2` ends when the scope finishes. Thus, the lifetime of `picked_value` is tied to the lifetime of `int2`, which is shorter than the other value passed in, which ends at the end of `main`. As a result, after the inner scope ends, `picked_value` becomes invalid, and we cannot print it afterward. Specifically, `int2` does not live long enough for `picked_value` to still be valid at the point where we attempt to print it. If the function returns a reference to `int2`, it would result in a dangling reference.

Lifetimes with Multiple Relationships

In the previous code examples, we only considered one type of relationship between the lifetimes of inputs and the returned value. Different types of relationships do exist depending on what your function is doing. For instance, your function might always return a reference to `i`, regardless of any other parameters. The following code shows such a function:

```
fn picking_int<'a>(i: &'a i32, j: &'a i32) -> &'a i32 {
    i
}
```

In this case, we would like to establish the relationship that the return value must have the same lifetime as that of the variable i. Then, we can remove the lifetime specifier from variable j, and the code will still compile, as follows:

```
fn picking_int<'a>(i: &'a i32, j: &i32) -> &'a i32 {
    i
}
```

Notice that no errors arise in main (shown in Listing 10.14) when considered with the new definition of the picking_int function because the returning reference has a lifetime equal to the first parameter, which is int1. The lifetime of int1 ends at the end of the main function. The picked_value is therefore valid after the scope ends in the print statement.

10.1.3 Static Lifetimes

An important point to emphasize is that, typically, the lifetime of a returned value should be tied to its input parameters. After all, when a function returns a reference, that reference must point to something that was provided as an argument. If a function returns a reference to something created within the function, that reference becomes invalid as soon as the function ends. For example, if we create a local variable inside the function and try to return a reference to it, the reference would be invalid once the function exits, leading to a dangling reference issue. Consider the following example:

```
fn picking_int<'a>(i: &'a i32, j: &i32) -> &'a i32 {
    let x  = 6;
    &x // Error
}
```

The variable x is created inside the function. When the function ends, that variable will be cleaned up and therefore any reference to it will be invalid.

If you really intend to return something that has been created inside the function, consider using a *static lifetime*. This special lifetime defines a reference that can live for the entire duration of the program. Static lifetimes provide a straightforward way to manage data that must persist for the entire runtime of a program. They eliminate the complexity of tracking lifetimes manually and facilitate the easy sharing of data safely across different parts of a program. This approach can be particularly useful for ensuring certain values remain accessible without worrying about their validity.

However, static lifetimes also come with drawbacks. You code can become inflexible code by locking data into a lifetime longer than necessary. Additionally, improper use can result in resource mismanagement, such as memory that is never released. Therefore, while useful in some cases, they should be applied thoughtfully to avoid unintended consequences.

To correct the code, let's introduce static lifetimes in the following example:

```
fn picking_int<'a>(i: &'a i32, j: &i32) -> &'a i32 {
    let y: &'static i32 = &6;
    y
}
```

The variable y has a static lifetime equal to the entire duration of the program. The code will compile with no issues although the returning reference and the return type mentioned in the function signatures do not match. Let's look at why this code is correct in detail next.

The function signature, as we understood earlier, conveys that the returned reference should have the same lifetime as the first parameter, since there is no lifetime specifier with the second reference passed in. While correct, in this case, the interpretation is that the returned reference must have a lifetime at least as long as the first parameter. Since y has a 'static lifetime, which lasts for the entire program, its lifetime satisfies this requirement. Therefore, the function can safely return a reference to y without violating the borrow checker's rules.

The function signature, however, is a bit confusing in this case. Since we aren't using the generic lifetime specifier, let's remove it from the function signature. Consider the following simplified code:

```
fn picking_int (i: &i32, j: &i32) -> &'static i32 {
    let y: &'static i32 = &6;
    y
}
```

The return type indicates that the function should return something that will have a static lifetime.

10.1.4 Lifetime Elision

Before we dive into our next topic, let's consider the program shown in Listing 10.15.

```
fn main() {
    let str_1 = "some str";
    let recieved_str = return_str(&str_1);
}
```

```
fn return_str(s_1: &str) -> &str {
    s_1
}
```

Listing 10.15 A Simple Function That Accepts and Returns a String Reference

The function `return_str` takes in a string slice and then simply returns it. In `main`, we are calling this function and storing the string slice in the variable `received_str`. Notice that, even though we are taking in a reference and returning a reference, we still do not need to add the lifetime annotation. The compiler is ok with it and throws no error. What could be the possible reason? The answer is lifetime elision.

Lifetime elision is a feature in Rust that allows the compiler to automatically infer the lifetimes of references in function and method signatures, making the code more concise and readable. The Rust compiler follows three lifetime elision rules:

1. Each parameter that is a reference gets its own lifetime parameter annotation.
2. If there is exactly one input lifetime parameter, that lifetime is assigned to all output lifetimes.
3. If there are multiple input lifetime parameters, but one of them is a reference to `self` or a mutable reference to `self`, the lifetime of `self` is assigned to all output lifetimes.

If, after applying these rules, the lifetimes are still ambiguous, the compiler will throw an error and require that you use explicit lifetime annotations.

Now, let's apply these rules to the function shown in Listing 10.15. We have one input parameter (`str_1`) to the `return_str` function. Therefore, according to the first rule, that function will get its own lifetime parameter. After adding the lifetime parameter, the function should look like the following code:

```
fn return_str<'a>(s_1: &'a str) -> &str {
    s_1
}
```

The `'a` is a generic lifetime parameter associated with `s_1`. Next, according to the second rule, if there is exactly one input lifetime parameter, which is in case of the function `return_str`, that lifetime is assigned to the output lifetime. Let's add the lifetime of the input parameter to the output. The updated function definition should look like the following code:

```
fn return_str<'a>(s_1: &'a str) -> &'a str {
    s_1
}
```

By applying the first two rules, the Rust compiler can determine the lifetimes of the references used in the function signature. Generally speaking, when you have only a single input lifetime parameter in the function signature, you do not need to explicitly annotate the lifetimes. Table 10.1 shows the function signatures with and without lifetime elision.

With Lifetime Elision	Without Lifetime Elision
`fn return_str(s_1: &str) -> &str`	`fn return_str<'a>(s_1: &'a str) -> &'a str`

Table 10.1 Function Signatures with and without Lifetime Elision

Let's explore another example by adding one more input parameter to the function, but without lifetime elision. The updated code should look like the following code:

```
fn return_str<'a>(s_1: &'a str, s_2: &str) -> &'a str {
    s_1
}
```


Considering the rules again, the first rule states that each parameter that is a reference gets its own lifetime parameter. So, let's add a lifetime to `s_2` in the following way:

```
fn return_str<'a, 'b>(s_1: &'a str, s_2: &'b str) -> &'a str {
    s_1
}
```

Now, let's apply the second rule according to which, if there is exactly one input lifetime parameter, that lifetime is assigned to all output lifetime parameters. However, we have two input lifetime parameters, not one. Therefore, the second rule does not apply, and we cannot assign a lifetime to the output. Let's now remove the lifetime from the output:

```
fn return_str<'a, 'b>(s_1: &'a str, s_2: &'b str) -> &str { // Error
    s_1
}
```

The third rule only applies to methods and is not for functions. Therefore, the third rule is not applicable. We'll cover the details of this rule further in Section 10.1.5.

Since the compiler could not determine the lifetime using the elision rules, it throws an error because the lifetime of the output is not clear. Thus, you must explicitly add lifetime annotations. Let's set the lifetime to be equal to the lifetime of the first parameter since that first parameter is the returning value from the function. The updated code should look like the following code:

```
fn return_str<'a, 'b>(s_1: &'a str, s_2: &'b str) -> &'a str {
    s_1
}
```

Let's also fix the code in `main` so that it works with the new definition of the function `return_str`, as shown in Listing 10.16.

```
fn main() {
    let str_1 = "some str";
    let str_2 = "other str";
    let recieved_str = return_str(&str_1, &str_2);
}
```

Listing 10.16 Code in main from Listing 10.15 Updated According to New Definition of return_str

The code in main is now using the updated definition of the return_str and calls it with a couple of &str variables. The code compiles with no issues.

10.1.5 Lifetimes and Structs

So far, we've only seen structs containing owned fields. However, struct fields may also contain references. For instance, consider the following struct:

```
struct ArrayProcessor {
    data: &[i32], // Error
}
```

This struct has a data field that is actually a reference. While there does not seem to be any error, the compiler does not like it and throws an error, "missing lifetime specifier." The reason for this error is that the data reference could become invalid during the execution of the program, while the instance of the struct is still alive. In that situation, the struct field is a dangling reference.

To properly handle references inside a struct, all the struct fields that contain a reference to a value must be annotated with generic lifetime parameters. Just like in functions, simply mention the lifetime annotations inside angle brackets (<>) and then mention it with the field. Consider the updated struct definition after adding lifetime annotations:

```
struct ArrayProcessor<'a> {
    data: &'a [i32],
}
```

Lifetime Elisions for Structs

Unlike functions, lifetime elisions are not defined for structs. In this case, obviously, the reference to which the field is pointing must have a lifetime that is at least as long as the struct itself. You would expect the compiler to add this information automatically. However, the compiler will not.

This problem is essentially a language design issue. One possible reason functions have lifetime elision rules while structs do not is that we tend to write functions that use references much more frequently than we define data structures with references.

Next, let's create an implementation block for the `ArrayProcessor` struct:

```
Impl ArrayProcessor {} // Error
```

This code throws an error, "implicit elided lifetime not allowed here, expected lifetime parameter." Just like with generics, we must include generic lifetime annotations inside definition of our `impl` block. Therefore, the correct syntax is the following code:

```
Impl<'a> ArrayProcessor<'a> {}
```


Let's add an `update_data` method to the implementation to accept a new array, update the reference `data` field that `ArrayProcessor` is pointing at, and return a reference to the array previously pointed to by the `data` field of `ArrayProcessor`. This updated code is shown in Listing 10.17.

```
impl<'a> ArrayProcessor<'a> {
    fn update_data(&mut self, new_data: &[i32]) -> &[i32] {
        let previous_data = self.data;
        self.data = new_data; // Error
        previous_data
    }
}
```

Listing 10.17 updated_data Method Added to the Implementation of ArrayProcessor

Another error is thrown in this case, "explicit lifetime required, in the type of `new_data`." Further, the compiler offers a nice suggestion, telling us what exactly we need to do: "add explicit lifetime `'a`, to the type of `new_data`." This problem makes sense because `new_data` is assigned to the field of the struct. To ensure that the field of the struct always points to valid data, `new_data` should have a lifetime compatible with that of the struct field. Let's add compatible lifetime annotations to the `new_data` variable, as shown in Listing 10.18.

```
impl<'a> ArrayProcessor<'a> {
    fn update_data(&mut self, new_data: &'a [i32]) -> &[i32] {
        ...
    }
}
```

Listing 10.18 Fixing the Error in Listing 10.17 by Annotating Lifetime with new_array

In this code, notice how the output of the function contains a reference with no explicit lifetime annotation, but the compiler has no issues. The reason for this success is lifetime elision, as explained in Section 10.1.4.

According to the first lifetime elision rule, each parameter that is a reference gets its own lifetime parameter. The updated function signature after applying rule 1 will look as follows:

```
fn update_data<'b>(&'b mut self, new_data: &'a [i32]) -> &[i32] { ... }
```

Since 'a is already used, we used 'b. The second rule only applies when there is a single input lifetime parameter. Next, let's look at the third rule according to which, if there are multiple input lifetime parameters, but one of them is a reference to self (&self) or a mutable reference to self (&mut self), the lifetime of self is assigned to all output lifetime parameters. The rule is applicable in this case since we have multiple input lifetime parameters and one of them is a mutable reference to self (&mut self). Therefore, according to this rule, the lifetime of self is assigned to all output lifetime parameters. After applying the third rule, we'll get the following function signature:

```
impl<'a> ArrayProcessor<'a> {
    fn update_data<'b>(&'b mut self, new_data: &'a [i32]) -> &'b [i32] {
        ...
    }
}
```

Table 10.2 shows the function signature with and without lifetime elision rules.

With Lifetime Elision	Without Lifetime Elision
fn update_data(&mut self, new_data: &'a [i32]) -> &[i32]	fn update_data<'b>(&'b mut self, new_data: &'a [i32]) -> &'b [i32]

Table 10.2 Method updated_data Signature with and without Lifetime Elision

You can now use the update_data method in main to update the data field, as shown in Listing 10.19.

```
fn main() {
    let mut some_data = ArrayProcessor { data: &[4, 5, 6] };
    let previous_data = some_data.update_data(&[5, 8, 10]);
    println!("Previous data: {:?}", previous_data);
    println!("New data: {:?}", some_data.data);
}
```

Listing 10.19 Using the update_data Method in main

The main function initializes an ArrayProcessor struct with some initial data. The update_data method is then called to replace the current data with new data. This method returns the previous data before the update, which is stored in the previous_data variable. Finally, the program prints both the previous data and the updated data stored in the ArrayProcessor.

10.2 Smart Pointers

Smart pointers in Rust are a powerful concept, going beyond simple references by allowing for more advanced memory management capabilities.

Before getting started, an essential distinction to make is the key difference between a pointer and a smart pointer. A simple *pointer* variable stores the memory address of some value. We've been using such pointers throughout this book, indicated by an ampersand (&); they are also referred to as a "reference." Other than referring to or pointing to some value, these references don't have any other additional capabilities. In contrast to simple pointers, *smart pointers* are not just simple references. They have special capabilities and also metadata. We'll start with the simplest smart pointer, called a Box smart pointer. We'll then move on to two more smart pointers, Rc and RefCell.

10.2.1 Box Smart Pointer

By default, Rust allocates everything on the stack memory. For instance, consider the following line of code:

```
Let x = 0.625;
```

In this case, the variable x will be stored on the stack. If you want to store the same value on the heap (for example, in scenarios where you need to transfer ownership of the data or share data across multiple parts of your program), then you'll use the box smart pointer. An instance of the Box smart pointer is created using the new constructor function:

```
Let y = Box::new(x);
```

This code will make a new heap allocation, containing the value of 0.625. The variable y is now the Box pointer, which is pointing to some heap memory containing the value. Note that the variable x will remain in the stack. In Rust, the Box type is part of the *prelude*, which is a collection of commonly used types and traits that are automatically available in every Rust program without needing explicit imports. Since Box is included in the prelude, you don't need to use a use statement to bring it into scope.

Next, consider a reference to variable x:

```
Let z = &x;
```

Like variable y, variable z is also a pointer; however, it is pointing to some memory location on the stack, while y is pointing to some memory on the heap.

Note

Box smart pointers are similar to "unique pointers" in C++.

Let's explore some use cases for box pointers.

Using a Box Pointer to Create a Recursive Type

Consider an enum called List with two variants: Cons and Nil. Listing 10.20 shows the definition of the enum.

```
#[derive(Debug)]
enum List {
    Cons(i32, List),
    Nil,
}
```

Listing 10.20 Definition of enum List

The Cons variant has a couple of items with it (i.e., an i32 value and another List). The Nil variant has nothing associated with it. The idea in this case is to have lists that may contain other lists.

Let's create an instance of the enum in main in the following way:

```
fn main() {
    let list = List::Cons(1, (List::Cons(2, (List::Cons(3, List::Nil)))));
}
```

We begin by defining a Cons variant with a value of 1. Next, we chain it to another List that holds a Cons variant with a value of 2, followed by another List that contains a Cons variant with a value of 3. Finally, we terminate the list by specifying the Nil variant, indicating the end of the sequence. Unfortunately, we get an error in the enum definition shown in Listing 10.20, "recursive type List has infinite size."

Let's explain the reason for the error in this case. According to the definition of the enum, we must mention two things for the Cons variant, that is, an i32 value and another List. The mentioned List may be a Cons type or Nil. If the list is a Cons type, then again, we'll mention two things. This process continues until we finally mention the Nil type, which leads to a *recursive type*, meaning a type that contains repeated instances of itself.

The problem with recursive types is that the Rust compiler doesn't know the exact size of the instance of such types at compile time. The Rust complier needs to know the sizes of our variables at compile time. However, in this case, the size of the variable list created in main cannot be determined at compile time. To further understand this problem, let's see how the Rust compiler decides the amount of space it needs to store a value of a non-recursive type enum. Consider a simple enum shown in Listing 10.21.

```
enum Conveyance {
    Car(i32),
    Train(i32),
    Air(i32),
    Walk
}
```


Listing 10.21 An Enum with Four Variants

This Conveyance enum has four variants. To determine how much space to allocate for an instance of Conveyance, Rust goes through each of its variants and tries to identify the variant that needs the most space. The first three variants of Car, Train, and Air are all associated with i32 values, so they need fixed memory space equal to the size of an i32 variable (i.e., 4 bytes). The final variant of Walk does not take any memory since it is not associated with any value. As a result, the maximum amount of memory an instance of Conveyance will occupy is equal to the memory space taken by an i32 value.

In contrast, when the Rust compiler tries to determine the size of an instance of the List, it knows that the Nil type will not take any memory space. However, the compiler cannot compute exactly how much space the Cons variant will take because Cons is associated with a tuple that contains a fixed i32 value and another List. In the list, we cannot know in advance how much space will be required because the list is recursive in nature, that is, containing repeated instances of itself.

The solution in this case is to put the Cons variant behind some type of pointer. When you hover over the error message in the code editor, the compiler tells us exactly this: "insert some indirection, for example a Box, Rc, or a simple reference." Let's fix the error using the Box pointer. The updated definition of the List is shown in Listing 10.22.

```
#[derive(Debug)]
enum List {
    Cons(i32, Box<List>),
    Nil,
}
```

Listing 10.22 Updated Definition of the enum List Defined Earlier in Listing 10.20

This definition works because the compiler now knows the exact size of the variants. The first variant will contain an i32 value and the size of the Box pointer. The size of a box pointer is equal to the size of a simple pointer because a variable of type Box is a simple pointer, pointing to some memory allocation on the heap and is of fixed size. Let's also update the code in main so that it uses the new definition of the List:

```
fn main() {
    let list = List::Cons(1, Box::new(List::Cons(2,
               Box::new(List::Cons(3, Box::new(List::Nil))))));
}
```

Now, the definition of the List shown in Listing 10.22 works but still could be improved.

We mentioned earlier that calling the new function for the Box creates a new heap allocation. Looking at the enum definition shown in Listing 10.22, notice how, in the case of Cons, we are always making a new heap allocation, irrespective of whether the next variant in the List will be a Cons or Nil. The Nil does not need a heap allocation since it has no data associated with it. It basically terminates the recursion. Let's examine the instance in main again:

```
fn main() {
    let list = List::Cons(1, Box::new(List::Cons(2,
               Box::new(List::Cons(3, Box::new(List::Nil))))));
}
```

Note that the last variant of Nil is unnecessarily being boxed; that is, it has been assigned heap space.

This problem can be fixed by revising the enum definition. You can further wrap the Box<List> inside an Option. The updated enum definition is shown in Listing 10.23.

```
#[derive(Debug)]
enum List {
    Cons(i32, Option<Box<List>>),
}
```

Listing 10.23 Updated Definition of the enum List Defined Earlier in Listing 10.20

Remember that the Option type has two variants: Some and None. When we define further Cons variants, we'll wrap them in Some, and when we want to terminate the List, we'll use the None variant. This approach eliminates the need for an explicit second variant like Nil. The end of the List will be signified by the None variant. Let's modify the code in main so that it uses the new definition of the enum:

```
fn main() {
    let list = List::Cons(1,Some(Box::new(List::Cons(2,
               Some(Box::new(List::Cons(3, None)))))));
}
```

Notice that the last variant does not need any new heap allocation. This implementation is now more precise and more efficient.

Boxing Data to Avoid Unnecessary Storage

Box pointers are also useful when copying a large amount of data during a transfer of ownership. Let's look at an example, starting with the code shown in Listing 10.24.

```
struct Huge_Data;
fn main() {
    let data_1 = Huge_Data;
    let data_2 = Box::new(Huge_Data);
    let data_3 = data_1;
    let data_4 = data_2;
}
```

Listing 10.24 A Simple Program Depicting the Transfer of Some Huge Data

The struct Huge_Data is simulating a struct containing some huge data. Next, data_1 and data_2 are instances of Huge_Data with a difference that data_2 is wrapped by the Box pointer. The ownership of data_1 is next transferred to data_3, and the ownership of data_2 is transferred to data_4. When transferring ownership, data is copied around the stack. Therefore, in the case of data_1, the entire dataset will be copied because the entire dataset resides on the stack. However, in the case of data_2, only the Box pointer will be copied, which is a small amount of data. The actual data is not copied or relocated to some other place within the heap because the assignment of heap-allocated data results in a move and not a copy (refer to Chapter 4, Section 4.1.1).

In our example, this distinction does not really matter since the struct does not contain something substantial. Imagine, however, if Huge_Data stores a lot of information, box pointers can have a significant impact.

Using Box Pointers with Trait Objects for Dynamic Dispatch

Let's look at one more example where the use of the Box is beneficial. In this example, we want to create a vector of different types that implement some trait. To simulate this scenario, let's define a Storage trait that may be extended later to provide storage-related functions. We'll implement this trait for the Huge_data struct defined in Listing 10.24 and implement another type called Small_Data. The code is shown in Listing 10.25.

```
struct Huge_Data;
struct Small_Data;
trait Storage {}
impl Storage for Huge_Data {}
impl Storage for Small_Data {}
```

Listing 10.25 Storage Trait and Its Implementation for Huge_Data and Small_Data

Next, we'll update the code in main shown earlier in Listing 10.24 by adding an instance of Small_Data, with the following code:

```
fn main() {
    ...
    let data_5 = Box::new(Small_Data);
}
```

Now, let's assume we need to define a vector of types to implement the Storage trait. The following code may be added for this purpose:

```
fn main(){
    ...
    let data = vec![Box::new(data_3), data_4, data_5]; // Error
}
```

In this code, we want to store the data of the types that implement the Storage trait. However, the compiler throws an error, "mismatched types," further elaborating, "expected struct Box<Huge_Data> however it found struct Box<Small_Data>." By considering the first instances of the vector, the Rust compiler detects that the type should be Box<Huge_Data>. However, the later instance does not conform to this type. Remember that vectors can only store values that all have the same type.

To enable the vector to store different types, you can tell the compiler to store any type in the vector that implements the Storage trait. This goal can be achieved using the trait objects (see Chapter 8, Section 8.2.6). Thus, let's add the following type annotation:

```
fn main(){
    ...
    let data: Vec<Box<dyn Storage>> = vec![Box::new(data_3), data_4, data_5];
}
```

dyn Storage is a trait object indicating that the elements stored in the vector can be of any type that implements the Storage trait. Recall that that trait objects must be stored behind some type of pointer. In this case, we used the Box smart pointer, which is typically used.

10.2.2 Rc Smart Pointer

The Rc (reference counted) smart pointer in Rust enables multiple ownership of data by keeping track of how many references point to the same value. Unlike typical ownership, where only one owner is allowed, Rc allows several parts of a program to share ownership of a value.

Let's illustrate one such case where shared ownership is required. Consider the diagram shown in Figure 10.1.

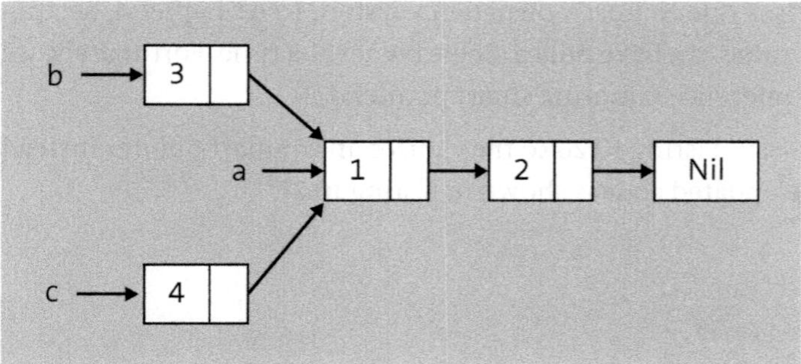

Figure 10.1 A Scenario Where Multiple Ownership Is Needed

In this scenario, you have list "a" containing two elements: The first element contains a value of 1, and the second element contains a value of 2. The last element is pointing to `Nil`, indicating the end of the list. We have another list "b," which has the first element, containing value 3, and then it points to the elements in the list "a," thereby making list "b" a larger list. In the same way, we have list "c," with the first element of 4, and then it is pointing to the elements in list "a," also making it also a larger list. Thus, list "a" is being pointed by both list "b" and list "c." Obviously, we want list "a" to remain in the memory as long as one of the lists (either "b" or "c") exists. If we delete list "a," then the pointers in lists "b" and "c" will become invalid.

Let's see what happens when we implement this scenario using the enum `List` defined earlier in Listing 10.23. The code shown in Listing 10.26 illustrates this implementation.

```
enum List {
    Cons(i32, Option<Box<List>>),
}
fn main() {
    let a = List::Cons(1, Some(Box::new(List::Cons(2, None))));
    let b = List::Cons(3, Some(Box::new(a)));
    let c = List::Cons(4, Some(Box::new(a))); // Error
}
```

Listing 10.26 Implementing the Scenario in Figure 10.1

We first created list `a` which contains the elements of `1` and `2`. List `b` next contains `3`, and then it contains list `a`. List `c` contains the element of `4` and also contains list `a`. The compiler throws an error, "use of moved value: `a`."

Let's go through this step by step. First, we declare `list a`, meaning that `a` takes ownership of the list. In the next line, we create `list b`, which includes `list a` within the `Cons` variant, effectively moving `list a` into `list b`. This transfer of ownership means that `list a` can no longer be accessed. When we attempt to access `list a` again, the compiler throws an error because its ownership has already been moved to `list b`, making it inaccessible. This issue arises due to Rust's ownership system (see Chapter 4, Section 4.1), which enforces that a value can have only a single owner at a time. Fortunately, we can resolve this using the reference-counting smart pointer, Rc.

Let's update the code shown in Listing 10.26 so that it uses an Rc smart pointer instead of a Box smart pointer. The updated code is shown in Listing 10.27.

```
use std::rc::Rc;
enum List {
    Cons(i32, Option<Rc<List>>),
}
fn main() {
    let a = List::Cons(1, Some(Rc::new(List::Cons(2, None))));
    let b = List::Cons(3, Some(Rc::clone(&a)));
    let c = List::Cons(4, Some(Rc::clone(&a)));
}
```

Listing 10.27 Updated Code from Listing 10.26 Based on the Rc Smart Pointer

To use an Rc pointer, you must bring it into the scope using the line `use std::rc::Rc`. Next, use the Rc pointer instead of the Box pointer in the Cons variant of the `enum List`. To enable `List` b to point to `List` a, the Rc pointer provides a `clone` method. The input to the method is a reference to another Rc pointer. However, list a is simply a list, which means we cannot pass it to the clone. We'll therefore change it into an Rc smart pointer, so that we can pass it to the `clone` method:

```
fn main() {
    let a = Rc::new(List::Cons(1, Some(Rc::new(List::Cons(2, None)))));
    ...
}
```

The variable a is now an Rc smart pointer holding the list. What the Rc enables is that the compiler now treats a, b, and c all as owners of the list. In other words, they share ownership of the list.

Clone in Case of Rc

An important point to highlight is that, internally, the clone doesn't make a deep copy in this case, which is unlike most implementations of clones. In this case, the clone does not store the same thing in some different memory; it only increments the reference count, which doesn't take much time and therefore is computationally efficient.

When we created variable a, the reference count to the list became 1. After creating b, the reference count became 2, and after c, it was incremented to 3. At the end of the `main` function, all the three variables will be dropped. The variables are dropped in last in, first out (LIFO) order, meaning the variable defined last is deleted first. Therefore, when main ends, first the variable c will be dropped, which will also erase one reference to list a. Therefore, the reference count of list a will be decremented by 1, to the value 2. Next, b will be dropped, which will further decrement the reference count to the list to 1. Finally, when a is dropped by itself, the reference count will reach a value of 0. At that point, the Rc pointer will drop the value it is holding, and the value is cleaned up from memory.

To clearly see the values of the reference count, you can include the variables of b and c inside a code segment to print the reference count after each created reference. Listing 10.28 shows the code in main.

```
fn main() {
    let a = Rc::new(List::Cons(1, Some(Rc::new(List::Cons(2, None)))));
    println!("Reference count after a: {}", Rc::strong_count(&a));
    {
        let b = List::Cons(3, Some(Rc::clone(&a)));
        println!("Reference count after b: {}", Rc::strong_count(&a));
        let c = List::Cons(4, Some(Rc::clone(&a)));
        println!("Reference count after c: {}", Rc::strong_count(&a));
    }
    println!("Reference count after scope: {}", Rc::strong_count(&a));
}
```

Listing 10.28 Understanding How the Reference Counts Are Updated

Executing this program results in the following output:

```
Reference count after a: 1
Reference count after b: 2
Reference count after c: 3
Reference count after scope: 1
```

The strong_count method on Rc returns the reference count of the value passed in. After creating list a, the reference count is 1. After creating list b, the reference count becomes 2, and after creating list c, the reference count becomes 3. At the end of the inner scope, b and c are dropped, and therefore, the reference count becomes 1 as indicated by the program's output.

The Rc smart pointer has many use cases. For instance, in a graph data structure, we have multiple edges that may point to the same node. Conceptually, the node is owned by all of the edges that point to it. A node shouldn't be cleaned up unless it doesn't have any edges pointing to it and so has no owners. Another use case is in a doubly linked list, which we'll cover in Chapter 11.

As a side note, you've already used a couple of smart pointers, such as String and Vec, which provide useful information about the amount of memory they occupy, like their capacity. Additionally, both of these types also own the data they reference in memory, meaning they are responsible for managing the memory and deallocating it when it is no longer needed.

10.2.3 RefCell Smart Pointer

The `RefCell` smart pointer provides for single ownership of data like the `Box` smart pointer. This smart pointer, however, offers some interesting properties, including enforcing the borrowing rules to be checked at runtime and enabling interior mutability. Let's cover each of these properties in detail.

Checking Borrowing Rules at Runtime

In Rust, borrowing rules are typically enforced at compile time, ensuring that you cannot have simultaneous mutable and immutable references that could lead to data races (refer to Chapter 4, Section 4.3.2). However, with `RefCell`, the borrowing rules are checked at runtime instead of at compile time. Let's walk through an example to understand this difference.

Consider the code shown in Listing 10.29.

```
fn main() {
    let mut x = 50;
    let ref1 = &x;
    let ref2 = &x;
    let ref3 = &mut x;
    println!("{} {} ", ref1, ref2); // Error
}
```

Listing 10.29 Illustration of Borrowing Rules Checking at Compile Time

As you may have guessed already, Rust's borrowing rules are being violated. According to these rules, we cannot have immutable and mutable references at the same time, and this rule is checked at compile time.

Let's now look at a similar example, but now using the `RefCell` smart pointer, as shown in Listing 10.30.

```
use std::cell::RefCell;
fn main() {
    let x = RefCell::new(50);
    let ref1 = x.borrow();
    let ref2 = x.borrow();
    let ref3 = x.borrow_mut();
    println!("{} {} ", ref1, ref2);
}
```

Listing 10.30 Code Similar to Listing 10.29 but Using RefCell

The line use std::cell::RefCell brings the RefCell smart pointer into scope. The new constructor function creates a new RefCell smart pointer. The borrow method allows for an immutable borrow of a value inside a RefCell, while the borrow_mut method allows for a mutable borrow of a value inside a RefCell. Note that the immutable and mutable borrows coexist, but the compiler raises no issues. You'll see the error only when you execute the program, and the compiler will throw the error "already borrowed." This error means that the borrowing rules are being checked at runtime and not at compile time.

The advantage of checking borrowing rules at compile time is that errors are caught earlier in the development cycle, with no additional performance cost at runtime. This reason is why compile-time borrowing checks are the default in Rust. On the other hand, the benefit of checking borrowing rules at runtime is that it allows for certain memory-safe scenarios that would otherwise be disallowed by compile-time checks. This flexibility is needed because some program behaviors are impossible to detect using static analysis alone (i.e., analysis of code without executing it to detect potential issues or errors), for instance, borrowing violations.

The references to RefCell do not follow the non-lexical lifetimes, which means that they are tied to the scopes in which they are defined. The variables ref1, ref2, and ref3 will remain in memory until the end of main. The print statement shown in Listing 10.30 is therefore not needed. The code shown in Listing 10.31 will still have mutable and immutable references at the same time.

```
use std::cell::RefCell;
fn main() {
    let x = RefCell::new(50);
    let ref1 = x.borrow();
    let ref2 = x.borrow();
    let ref3 = x.borrow_mut();
}
```

Listing 10.31 Print Line from Listing 10.30 Not Required Since RefCell Doesn't Follow Non-Lexical Lifetimes

For details on non-lexical lifetimes, refer to Section 10.1.1.

You can use the drop function to explicitly end the scopes of references. Let's update the code shown in Listing 10.31 and drop the immutable references before the mutable reference, as shown in Listing 10.32.

```
use std::cell::RefCell;
fn main() {
    let x = RefCell::new(50);
    let ref1 = x.borrow();
    let ref2 = x.borrow();
```

```
    drop(ref1);
    drop(ref2);
    let ref3 = x.borrow_mut();
}
```

Listing 10.32 Using drop to End the Scope of References and Clear from Memory

The references `ref1` and `ref2` are now dropped thereby ending their scope before a mutable reference. The code should execute without error since all immutable references are dropped before the mutable reference.

An alternative approach is to define the references inside a scope. For instance, consider the code shown in Listing 10.33.

```
use std::cell::RefCell;
fn main() {
    let x = RefCell::new(50);
    {
        let ref1 = x.borrow();
        let ref2 = x.borrow();
    }
    let ref3 = x.borrow_mut();
}
```

Listing 10.33 Alternative to Drop Is Scope

The references `ref1` and `ref2` are now limited to the inner scope, and therefore, they will be cleared when the scope ends. At the end of the scope, the `drop` function will be automatically called for references. Explicit calls to this method are no longer needed.

Let's add a print statement to the code shown in Listing 10.33 to display the value of the variable `x`, as follows:

```
use std::cell::RefCell;
fn main() {
   ...
   println!("x: {:?}",x);
}
```

When you execute this code, you get something strange. Instead of the value, you'll see the following output:

```
x: RefCell { value: <borrowed> }
```

This result means that when Rust tries to print the value, it was mutably borrowed, giving an indication that it may change. Therefore, the actual value is not displayed, as it might not be properly updated. Let's try dropping `ref3`, thereby ending its scope early before printing, as shown in Listing 10.34.

```
fn main() {
    ...
    let ref3 = x.borrow_mut();
    drop(ref3);
    println!("x: {:?}", x);
}
```

Listing 10.34 Calling drop before the Printing Prints the Actual Value

Now, the actual value of x can be displayed because no mutable reference to it exists anymore, and there is no chance that the value might be updated when it is printed.

Interior Mutability

The checking of borrowing rules at compile time allows for something called *interior mutability*. Rust's rules by default do not allow a mutable reference to a variable that is immutable itself. To see this rule in action, consider the following code:

```
fn main() {
    let x = 32;
    let x1 = &mut x; // Error
}
```

As expected, the code throws an error. The variable x was declared to be immutable, and therefore, a mutable reference to it is not allowed.

With the RefCell smart pointer, you can mutate a value even if it is declared as immutable. This capability allows for interior mutability, meaning the data can be changed internally without exposing that mutability to the outside. As a result, the outside world remains unaware that the data can be modified. This feature is useful when you need to modify data internally while maintaining an immutable interface to the outside world, such as in scenarios involving caching or lazily initialing data (i.e., delaying the creation or computation of a value until it is needed). For instance, consider the code shown in Listing 10.35.

```
use std::cell::RefCell;

fn main() {
    let a = RefCell::new(10);
    let mut ref1 = a.borrow_mut();
    *ref1 = 15;
    drop(ref1);
    println!("{:?}", a);
}
```

Listing 10.35 Interior Mutability with RefCell

In this case, a is a RefCell, and b mutably borrows a and then updates it. Note that, even though a is not declared to be mutable, you can still have a mutable reference to it, which mutates its internal value. Executing this code now will display the updated value of a, which is 15.

The RefCell does not implement the Deref trait. Therefore, using the dereference operator on an instance of this pointer will generate an error. For example, the following code will throw an error:

```
use std::cell::RefCell;
fn main() {
    let a = RefCell::new(10);
    let c = *a; // Error
    ...
}
```

This error arises because RefCell enforces Rust's borrowing rules at runtime, and dereferencing it directly would bypass the borrowing checks.

RefCell Combined with Rc

The RefCell smart pointer may not seem all that powerful by itself, but when combined with the Rc smart pointer, amazing things can happen. For instance, consider the code shown in Listing 10.36.

```
use std::{cell::RefCell, rc::Rc};

fn main() {
    let a = Rc::new(RefCell::new(String::from("c++")));
}
```

Listing 10.36 Combing RefCell with Rc

The Rc provides for multiple immutable ownership. With RefCell providing mutable access to internal data, multiple owners are further given the ability to mutate the internal data. Thus, with the help of an Rc pointer wrapped with RefCell, multiple owners can now mutate the internal data.

Let's create another shared owner of the data, one that can mutate the shared value, as shown in Listing 10.37.

```
use std::{cell::RefCell, rc::Rc};
fn main() {
    let a = Rc::new(RefCell::new(String::from("c++")));
    let b = Rc::clone(&a);
    *b.borrow_mut() = String::from("rust");
```

```
    println!("{:?}", a);
}
```

Listing 10.37 Code from Listing 10.36 Updated by Adding a New Owner That Mutates the Data

Both a and b are now owners of the data. The clone method on an Rc pointer creates a new owner for the data. If you execute the code, you'll see an updated value for the variable a. This technique will be especially useful when we look at the implementation of doubly linked lists in Chapter 11, where multiple owners should have the ability to modify the data.

10.3 Deref Coercion

Deref coercion is a feature that lets you automatically treat a smart pointer (or a reference) as if it's the actual value without needing to manually dereference it. For instance, if you have a Box<T> or String, the compiler can automatically dereference these types when necessary to access the underlying data.

The Deref trait is defined in the following way:

```
trait Deref {
    type Target: ?Sized;
    fn deref(&self) -> &Self::Target;
}
```

When deref coercion is applied, the compiler repeatedly calls the deref method until the required reference type is obtained.

Smart pointers such as Box<T> are common use cases for deref coercion because they let you automatically access the data inside them, making the code simpler. The code shown in Listing 10.38 is an example of deref coercion in Box.

```
use std::ops::Deref;
struct MyBox<T>(T);
impl<T> MyBox<T> {
    fn new(value: T) -> MyBox<T> {
        MyBox(value)
    }
}
impl<T> Deref for MyBox<T> {
    type Target = T;
    fn deref(&self) -> &Self::Target {
        &self.0
    }
}
```

```
fn greet(name: &str) {
    println!("Hello, {}!", name);
}
fn main() {
    let my_box = MyBox::new(String::from("Rust"));
    greet(&my_box); // Deref coercion
}
```

Listing 10.38 Using the Deref Trait to Enable Automatic Coercion of MyBox to a String Slice

In this case, `MyBox<T>` implements `Deref`, where `T` is its target type. Note that the `&self.0` is the inner value contained within the type of `MyBox`, which is `T`. The `greet` function expects a `&str`, but we pass `&my_box`. The compiler automatically dereferences `my_box` to `&String` and then further to `&str`.

Deref coercion can also work recursively. Consider the code shown in Listing 10.39.

```
use std::ops::Deref;
struct InnerBox<T>(T);
struct OuterBox<T>(T);
impl<T> Deref for InnerBox<T> {
    type Target = T;
    fn deref(&self) -> &Self::Target {
        &self.0
    }
}
impl<T> Deref for OuterBox<T> {
    type Target = InnerBox<T>;
    fn deref(&self) -> &Self::Target {
        &self.0
    }
}
fn print_value(value: &str) {
    println!("Value: {}", value);
}
fn main() {
    let nested = OuterBox(InnerBox(String::from("Nested Rust")));
    print_value(&nested); // Multiple levels of deref coercion
}
```

Listing 10.39 Demonstrating Deref Coercion with Nested Smart Pointers

Now, `OuterBox` dereferences to `InnerBox`, and `InnerBox` dereferences to `String`. Finally, the `String` is dereferenced to `&str` for the `print_value` function. This approach simplifies the code and makes it more intuitive by automatically converting types to the values they contain, so you don't have to manually dereference them each time you want to access the value. The end result is cleaner, more readable code with less chance of errors.

Deref coercion can chain multiple levels of dereferencing as long as each intermediate type implements `Deref`. Moreover, the coercion only works for immutable references (`&T`) and not for mutable references (`&mut T`).

In summary, deref coercion is a vital Rust feature that makes working with references and smart pointers seamless. By understanding and implementing the `Deref` trait, developers can leverage this feature to write cleaner and more intuitive code. In scenarios involving custom types, implementing `Deref` enables compatibility with Rust's existing application programming interfaces (APIs), improving code reuse and maintainability.

10.4 Practice Exercises

In this section, we'll provide some practice questions to help reinforce your learning. The exercise questions cover almost all the concepts introduced in the chapter. Solutions to each exercise will follow in Section 10.5.

1. **Resolving borrowing conflicts in mutable and immutable references**
   Reorganize the following code so that it adheres to Rust's borrowing rules while still preserving the functionality of the program.

```
fn main() {
    let mut some_str = String::from("I am String");
    let ref1 = &some_str;
    let ref2 = &mut some_str;
    ref2.push_str(" additional information");
    println!("{ref1}"); // move this line only
    println!("{ref2}");
}
```

2. **Resolving lifetimes in function references**
   Modify the following code by addressing the lifetime issue, ensuring the reference is valid when used in the assertion.

```
fn identity(a: &i32) -> &i32 {
    a
}

fn main() {
    let mut x_ref: Option<&i32> = None;
    {
        let x = 7;
        x_ref = Some(identity(&x));
```

```
    }
    assert_eq!(*x_ref.unwrap(), 7); // Issue at this line
}
```

3. **Correcting scope and lifetimes in option handling**
   Rearrange the code to ensure the variable y stays in scope when used by the option function and the program compiles and runs successfully.

```
fn option(opt: Option<&i32>) -> &i32 {
    opt.unwrap()
}
fn main() {
    let answer = {
        let y = 4; // move this line only
        option(Some(&y))
    };
    assert_eq!(answer, &4);
}
```

4. **Adding lifetime annotations for function references**
   Modify the function signature of some_if_greater to include the necessary lifetime annotations, ensuring that the returned reference is valid.

```
fn some_if_greater(number: &i32, greater_than: &i32) -> Option<&i32> {
    if number > greater_than {
        Some(number)
    } else {
        None
    }
}
fn main() {
    let num_1 = 7;
    let greater_val = 4;
    let test = some_if_greater(&num_1, &greater_val);
}
```

5. **Identifying and expanding lifetime parameters in function signatures**
   Identify the function signature that needs explicit lifetime parameters. For functions that do not require explicit lifetime parameters, write their expanded code to be generated by the compiler.

   Note: Do not compile this code; it is not a program.

```
fn print(s: &str) {}
fn debug(v: usize, s: &str) {}
fn substr(s: &str, until: usize) -> &str {}
fn get_str() -> &str {}
fn frob(s: &str, t: &str) -> &str{}
```

```
fn get_mut(&mut self) -> &mut T;
fn new(buf: &mut [u8]) -> BufWriter;
```

6. **Fixing compilation errors in a binary tree enum definition**
   The following code defines an enum for a binary tree structure, but it fails to compile. Your task is to identify and fix the issues in the code to ensure it compiles successfully.

```
enum BinaryTree {
    Leaf,
    Node(i32, BinaryTree, BinaryTree),
}
fn main() {}
```

10

7. **Fixing function signature in data modification**
   The following code defines a structure Wrapper and a function modify_data that is intended to modify the data contained in a Wrapper. However, the function signature is incomplete, leading to compilation errors. Your task is to complete the function signature so that the code compiles correctly.

```
struct Wrapper {
    data: String,
}

fn modify_data(mut wrapper: Box<Wrapper>) -> ? {
    wrapper.data = String::from("Modified");
    wrapper
}

fn main() {
    let original_wrapper = Box::new(Wrapper {
        data: String::from("Original"),
    });
    let modified_wrapper = modify_data(original_wrapper);
}
```

8. **Completing the linked list enum definition**
   The following code defines an enum for a generic linked list, but it is incomplete. You need to declare the enum variants properly to create a functional linked list. Specifically, you need to add a Node variant that includes a Box pointer to the next node, and another variant that acts as a placeholder for the end of the list. Complete the code so it represents a valid linked list structure.

```
#[derive(Debug)]
enum ListNode<T> {
    /*TODO: Declare an enum variant called Node, with Box pointer for the
            next node of type 'T' */
```

```
    /*TODO: Another variant for the placeholder for the end of the list */
}

fn main() {
    // Create a linked list representing: Node(1, Node(2, Node(3,
    // Node(4, None))))
    let list = ListNode::Node(1, /* TODO: Box pointer for the next node */);
    println!("{:?}", list);
}
```

9. **Implementing shared ownership of a file**
   The idea is to simulate multiple users sharing ownership of a file. The Rc smart pointer can be used to allow multiple users to own the same file, ensuring shared access without ownership conflicts. Your task is to complete the User struct and the main function so that multiple users can own a single file.

```
use std::rc::Rc;
struct File {}

struct User {
    file: /*You code here*/
}

fn main() {
    let txt_file = Rc::new(File {});
    let'ser_1 = User {
        file: /*You code here*/
    };
    let'ser_2 = User {
        file: /*You code here*/
    };
}
```

10. **Safely modifying and checking values within a RefCell**
    In the following code, you need to complete the TODOs to modify the value inside a RefCell and check its contents. The goal is to use borrow_mut to change the Some value inside the RefCell to None and then check whether the RefCell contains a Some variant before printing the final value.

```
use std::cell::RefCell;
fn main() {
    let data: RefCell<Option<i32>> = RefCell::new(Some(42));

    /* TODO: Use borrow_mut to safely modify the value inside the
             RefCell to None. */
```

```
    if /* TODO: add code to check if data contains the some variant */ {
        println!("Final value: {:?}", data.borrow());
    } else {
        println!("No value present.");
    }
}
```

11. **Resolving borrowing conflicts and mutating values in `RefCell`**
    The code provided will compile but will panic at runtime due to borrowing conflicts. Your tasks are:
    - Task 1: Add some code so that there is no panic at execution time.
    - Task 2: The value at the last print line will not be displayed. Instead of value, `<Borrowed>` will be displayed. Add appropriate code so that the value of x is displayed.

```
use std::cell::RefCell;

fn main() {
    let x = RefCell::new(5);
    let x_ref1 = x.borrow();
    let x_ref2 = x.borrow();
    println!("x_ref1: {}, x_ref2: {}", x_ref1, x_ref2);

    /* Code for Task 1 */

    let mut x_ref3 = x.borrow_mut();
    *x_ref3 = 6;

    /* Code for Task 2 */

    println!("Stored value: {:?}", x);
}
```

10.5 Solutions

This section provides the code solutions for the practice exercises in Section 10.4. The code is largely self-explanatory; however, we have included comments and additional explanations wherever necessary to enhance understanding.

1. **Resolving borrowing conflicts in mutable and immutable references**

```
fn main() {
    let mut some_str = String::from("I am String");
    let ref1 = &some_str;
    println!("{ref1}");
```

```
    let ref2 = &mut some_str;
    ref2.push_str(" additional information");
    println!("{ref2}");
}
```

2. **Resolving lifetimes in function references**

```
fn identity(a: &i32) -> &i32 {
    a
}
fn main() {
    let mut x_ref: Option<&i32> = None;
    {
        let x = 7;
        x_ref = Some(identity(&x));
            assert_eq!(*x_ref.unwrap(), 7);
    }
}
```

3. **Correcting scope and lifetimes in option handling**

```
fn option(opt: Option<&i32>) -> &i32 {
    opt.unwrap()
}
fn main() {
    let y = 4;
    let answer = {
        option(Some(&y))
    };
    assert_eq!(answer, &4);
}
```

4. **Adding lifetime annotations for function references**

```
fn some_if_greater<'a>(number: &'a i32, greater_than: &'a i32) -> Option<&'a
i32> {
    if number > greater_than {
        Some(number)
    } else {
        None
    }
}
fn main() {
    let num_1 = 7;
    let greater_val = 4;
    let test = some_if_greater(&num_1, &greater_val);
}
```

5. **Identifying and expanding lifetime parameters in function signatures**

```
fn print(s: &str) {}                          // does not need explicit
                                              // lifetime
fn print<'a>(s: &'a str) {}                   // expands code by compiler

fn debug(v: usize, s: &str) {}                // does not need explicit
                                              // lifetime
fn debug<'a>(v: usize, s: &'a str) {}         // expands code by compiler

fn substr(s: &str, until: usize) -> &str {}   // does not need explicit
                                              // lifetime
fn substr<'a>(s: &'a str, until: usize) -> &'a str; // expands code by
                                                    // compiler

fn get_str() -> &str {}                       // needs explicit lifetimes

fn frob(s: &str, t: &str) -> &str{}           // needs explicit lifetimes

fn get_mut(&mut self) -> &mut T;              // does not need explicit
                                              // lifetime
fn get_mut<'a>(&'a mut self) -> &'a mut T;    // expands code by compiler

fn new(buf: &mut [u8]) -> BufWriter;          // does not need explicit
                                              // lifetime
fn new<'a>(buf: &'a mut [u8]) -> BufWriter<'a>; // expanded
```

6. **Fixing compilation errors in a binary tree enum definition**

```
enum BinaryTree {
    Leaf,
    Node(i32, Box<BinaryTree>, Box<BinaryTree>),
}
fn main() {}
```

7. **Fixing function signature in data modification**

```
struct Wrapper {
    data: String,
}
fn modify_data(mut wrapper: Box<Wrapper>) -> Box<Wrapper> {
    wrapper.data = String::from("Modified");
    wrapper
}
fn main() {
    let original_wrapper = Box::new(Wrapper {
        data: String::from("Original"),
```

10

```
    });
    let modified_wrapper = modify_data(original_wrapper);
}
```

8. **Completing the linked list enum definition**

```
#[derive(Debug)]
enum ListNode<T> {
    Node(T, Box<ListNode<T>>),
    None,
}
fn main() {
    // Create a linked list representing: Node(1, Node(2, Node(3,
    // Node(4, None))))
    let list = ListNode::Node(
        1,
        Box::new(ListNode::Node(
            2,
            Box::new(ListNode::Node(
                3,
                Box::new(ListNode::Node(4, Box::new(ListNode::None))),
            )),
        )),
    );
    println!("{:?}", list);
}
```

9. **Implementing shared ownership of a file**

```
use std::rc::Rc;
struct File {}
struct User {
    file: Rc<File>,
}
fn main() {
    let txt_file = Rc::new(File {});
    let user_1 = User {
        file: Rc::clone(&txt_file),
    };
    let user_2 = User {
        file: Rc::clone(&txt_file),
    };
}
```

10. **Safely modifying and checking values within a `RefCell`**

```
use std::cell::RefCell;
fn main() {
    let data: RefCell<Option<i32>> = RefCell::new(Some(42));
    *data.borrow_mut() = None;

    if  data.borrow().is_some() {
        println!("Final value: {:?}", data.borrow());
    } else {
        println!("No value present.");
    }
}
```


11. **Resolving borrowing conflicts and mutating values in `RefCell`**

```
use std::cell::RefCell;
fn main() {
    let x = RefCell::new(5);
    let x_ref1 = x.borrow();
    let x_ref2 = x.borrow();
    println!("x_ref1: {}, x_ref2: {}", x_ref1, x_ref2);
    drop(x_ref1);
    drop(x_ref2);
    let mut x_ref3 = x.borrow_mut();
    *x_ref3 = 6;
    drop(x_ref3);
    println!("Stored value: {:?}", x);
}
```

10.6 Summary

This chapter explored Rust's robust memory management features, starting with lifetimes, which ensure references remain valid for the duration in which they are needed. We also discussed lifetime elision, a feature that simplifies code by allowing the Rust compiler to automatically infer lifetimes in certain cases. Thus chapter then delved into the use of lifetimes in structs, demonstrating how they help you manage relationships between data. Moving into advanced topics, we introduced you to smart pointers, such as `Box` for heap allocation, `Rc` for reference counting, and `RefCell` for interior mutability. These tools are essential for writing safe, efficient Rust code, especially when managing memory and ownership in more complex scenarios.

In the next chapter, you'll learn how to implement common data structures such as singly linked lists and doubly linked lists.

Chapter 11
Implementing Typical Data Structures

Data structures form the backbone of all effective algorithms. In this chapter, you'll implement the foundational structures that empower your programming endeavors.

In this chapter, you'll learn to implement common data structures, beginning with singly and doubly linked lists. We'll provide detailed explanations and examples of how to build these lists, highlighting their differences and use cases. We'll also discuss reference cycles, a common issue in linked structures, and how they can lead to memory leaks. By understanding these concepts, you'll be equipped to create efficient data structures and handle potential pitfalls in memory management. 11

11.1 Singly Linked List

A *linked list* is a data structure used to organize and store data. Unlike arrays, which have a fixed size, linked lists grow and shrink as needed. This list consists of a sequence of elements called *nodes*, where each node contains a reference or link to the next node in the sequence. To begin, let's understand the basics with the help of some visuals. Figure 11.1 shows the typical structure of a singly linked list.

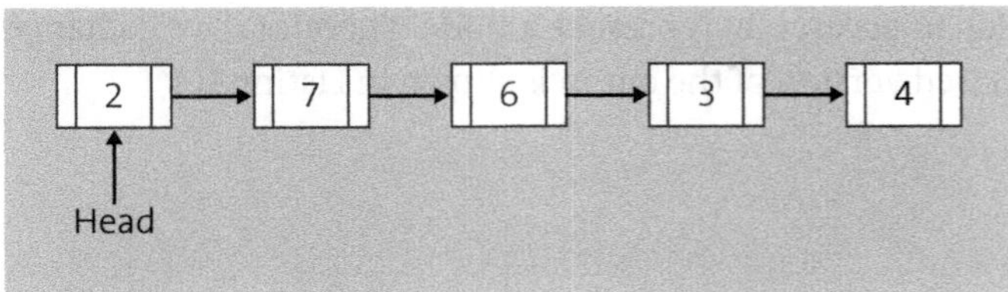

Figure 11.1 A Typical Singly Linked List

The square boxes shown in Figure 11.1 represent the nodes, with each node pointing to the next node in the sequence, except for the last node, which points to nothing. The first node is known as the *head* of the list, while the last node is known as the *tail*. Each node contains a value. Elements can be added or removed either from the head or from the tail of the linked list.

Let's dive into implementing linked lists, starting with refining the List enum; resolving issues to make it compatible with linked lists; and finally exploring other functionalities, such as adding, removing, and printing the linked list.

11.1.1 Implementation by Modifying the List Enum

Let's now turn to the implementation. We'll start with the code for the List enum from Chapter 10, Section 10.2.1. In this section, you'll modify the List enum to structure it in such a way that allows the definition and implementation of a linked list.

```
#[derive(Debug)]
enum List {
    Cons(i32, Option<Box<List>>),
}
```

First, we'll change to using terminology that is compatible with linked lists. The basic ingredient of a linked list is a node, so we'll change List to Node, as in the following example:

```
#[derive(Debug)]
enum Node {
    Cons(i32, Option<Box<Node>>),
}
```

Next, in the current implementation, we have the value and the link to the next node bound together in a single variant. Let's separate the two, that is, separate the value from the pointer to next, which we'll name them element and next, respectively. Moreover, enums may not be sufficient in this case because, with the enum, we can create an instance of the enum that's either the element variant or the next variant, but not both. However, we need both fields to accurately represent a node. Therefore, we'll change the enum to a struct. The revised version of the enum is shown in Listing 11.1.

```
#[derive(Debug)]
struct Node {
    element: i32,
    next: Option<Box<Node>>,
}
```

Listing 11.1 Definition of the Node Struct

next is basically an optional pointer to the next node. We can now use the Node struct in main. Listing 11.2 shows the code for creating a list containing one element.

```
fn main() {
    let list_1 = Node {
```

```
        element: 1,
        next: None,
    };
}
```

Listing 11.2 Creating a List Containing a Single Node

The code creates an instance of Node containing both the fields of elements and next, which was not possible with the List enum definition.

Longer lists can also be created by nesting one node within another node, as shown in Listing 11.3.

```
fn main() {
    ...
    let list_2 = Node {
        element: 1,
        next: Some(Box::new(Node {
            element: 2,
            next: Some(Box::new(Node {
                element: 3,
                next: None,
            })),
        })),
    };
}
```

Listing 11.3 Creating a Longer List Containing More Nodes

In this case, the main function creates a linked list by initializing a node where list_2 represents a chain of nodes: The first node contains the value 1 and points to the second node with the value 2, which in turn points to the third node with the value 3, marking the end of the list with None.

11.1.2 Resolving Issues with the Implementation

A couple of issues exist in the implementation shown earlier in Listing 11.1. First, this implementation will not allow you to have explicit information about the head, or the starting node. Second, this implementation will also not allow for an empty list. In other words, the first node must contain something for the element part. Let's refine the implementation to overcome these limitations.

To have explicit information regarding the head of the list, we can define another wrapper struct, with only one field called head, which will be of type Node. We'll name this struct linkedlist in the following way:

```
#[derive(Debug)]
struct Linkedlist {
    head: Node,
}
```

The head serves as the starting point to traverse or manipulate the entire list, and therefore, its information is necessary. We can now define an instance of the Linkedlist, instead of a simple node. The instance will have an explicit head field, which represents the starting node. Let's use it in main, as shown in Listing 11.4.

```
fn main() {
    ...
     let list_3 = Linkedlist {
        head: Node {
            element: 1,
            next: Some(Box::new(Node {
                element: 2,
                next: Some(Box::new(Node {
                    element: 3,
                    next: None,
                })),
            })),
        },
    };
}
```

Listing 11.4 Using the Linkedlist Wrapper Struct to Create a List

This code now provides explicit information regarding the head.

The introduction of the wrapper struct solved the first problem because explicit information regarding the head of the list is now available. However, we still cannot have any list with an empty head; in other words, we cannot have an empty list. Unfortunately, an empty list may be required in scenarios for initialization or in scenarios where no data is available yet.

This problem arises because the head field is defined as Node, which cannot be empty. A node must have two fields: element and next. For instance, we cannot have a list with an empty head, for example:

```
fn main() {
    ...
    let list_4 = Linkedlist { head: None } // Error
}
```

This problem can be fixed by revising the definition of the Linkedlist. In particular, we can redefine the head as an Option node. The updated definition of the Linkedlist is as follows:

```
#[derive(Debug)]
struct Linkedlist {
    head: Option<Node>,
}
```

This change will cause a few errors in main. The head node must be wrapped by the Some. The updated code in main is shown in Listing 11.5.

```
fn main() {
    ...
    let list_3 = Linkedlist {
        head: Some(Node { // The head is wrapped by Some
            element: 1,
            next: Some(Box::new(Node {
                element: 2,
                next: Some(Box::new(Node {
                    element: 3,
                    next: None,
                })),
            })),
        }),
    };

    let list_4 = Linkedlist { head: None };  // This now compiles
}
```

Listing 11.5 Code in main Fixed Based on the New Definition of Linkedlist

Note that the list_4 which is a Linklist with no head now compiles. Therefore, we can now create empty lists.

11.1.3 Refining the Next Field

Let's inspect the Node struct shown earlier in Listing 11.1 once more:

```
#[derive(Debug)]
struct Node {
    element: i32,
    next: Option<Box<Node>>,
}
```

Reading the type of the next field can be difficult; therefore, a beneficial task is to simplify complex types by creating custom types, which helps improve the clarity and maintainability of your code. In this case, define a pointer type with the same type as the next field and update the Node definition. As a result, the new pointer type can be used to declare the type of the next field in the Node struct definition. The updated struct definition is shown in Listing 11.6.

```
#[derive(Debug)]
struct Node {
    element: i32,
    next: pointer,
}
type pointer = Option<Box<Node>>;
```

Listing 11.6 Updated Node Definition and the New Pointer Type

Notice how the head in the Linkedlist has a similar type to that of the pointer type, with one minor difference: it is not boxing the Node. If we change the type of head from Option<Node> to that of the Option<Box<Node>>, then its type is the same as the pointer. Now, we can also use the pointer type for the head. The updated definition of the Linkedlist looks as follows:

```
#[derive(Debug)]
struct Linklist {
    head: pointer,
}
```

This now reads better. In main, we'll get rid of all the previously created lists and create a new list that conforms to our new definitions of the Node and the LinkedList. Listing 11.7 shows the code in main for creating a new list using the revised definitions of the structs.

```
fn main() {
    let list = Linkedlist {
        head: Some(Box::new(Node {
            element: 100,
            next: Some(Box::new(Node {
                element: 200,
                next: None,
            })),
        })),
    };
}
```

Listing 11.7 List Created with the Revised Definitions of Node and Linkedlist

To access the element part of the head for printing or updating, we'll use a line similar to the following:

```
println!("{:?}", &list.head.unwrap().element);
```

The println! statement in the main function retrieves the head of the list variable, which represents the starting point of the linked list. This statement uses unwrap() to extract the Some value from the head and then accesses the element field of the first node, printing its value, which is 100. To access the second Node element part, we'll use the following line of code:

```
println!("{:?}", &list.head.unwrap().next.unwrap().element);
```

First, we access the next field after unwrapping the head, which provides access to the second node. Afterwards, we access the element part. Let's now try displaying only the head using the following code:

```
println!("{:?}", &list.head);
```

The execution of this line should result in the following output:

```
Some(Node { element: 100, next: Some(Node { element: 200, next: None }) })
```

What we see is basically the whole of the list. The reason why the head displays the whole list is that the Node struct is a recursive type. The head is basically wrapping all the stuff inside it, and then the next of head is wrapping all the stuff that comes afterwards, and so on.

An important point to emphasize is that inside the memory, the layout will be the same as shown earlier in Figure 11.1. The head will be the first node containing a pointer to the next node, and each subsequent node points to the next one, continuing in this manner.

11.1.4 Adding Elements

Now that our basic data structures are defined, let's add some functionality to the Linkedlist. First, we'll add a new constructor function to initialize an empty Linkedlist, as shown in Listing 11.8.

```
impl Linkedlist {
    fn new() -> Self {
        Linkedlist { head: None }
    }
}
```

Listing 11.8 Definition of new Constructor Function for Linkedlist

The function returns a Linkedlist with the head set to None, meaning an empty Linkedlist. Next, we'll include an add method for adding an element to the Linkedlist at the beginning. The skeleton of the method will resemble the following example:

```
fn add(&mut self, element: i32) {}
```

The input is a mutable reference to self since we'll be updating the Linkedlist. The second argument is the element that we wish to add. Let's now focus on the logic this method needs to implement.

The method must first check whether the head contains Some Node or None. If None, meaning that no head exists in the node, then we'll simply make a new node and set it equal to the head. However, if a head already exists, then we'll make a new node, and the current head will be replaced with the new node. The next of the new node will point to the old head. Let's summarize these two cases:

- **Case 1: head is None**
  Make a new node and set it equal to the head.
- **Case 2: head is Some**
  In this case, head already exists. Make a new node. Set the next of the new node to point to old head. The newly created node will be the new head.

Listing 11.9 shows the code for these two cases.

```
impl Linkedlist {
    ...
    fn add(&mut self, element: i32) {
        match self.head { // Error
            None => {
                let new_node = Some(Box::new(Node {
                    element: element,
                    next: None,
                }));
                self.head = new_node
            }
            Some(previous_head) => {
                let new_node = Some(Box::new(Node {
                    element: element,
                    next: Some(previous_head),
                }));
                self.head = new_node;
            }
        }
    }
}
```

Listing 11.9 Logic for Adding a New Element

The compiler is not happy with the code, telling us we "cannot move out of self.head as enum variant Some, which is behind a mutable reference." Note that the variable inside the Some (i.e., previous_head in this case) will capture the unwrapped value of the Option we are matching on. Furthermore, the variable takes over the value by ownership. However, with a mutable reference (self is passed as &mut self to the method), we cannot take ownership of something.

Another way of thinking about this scenario is that, if we take ownership of the head, then the list will no longer have access to the head. Since the remaining nodes are only accessible through head, they will be basically unusable.

Let's look at it from another perspective. When we take ownership of the head, in some sense, we are trying to rip off the head from a data structure we do not own. If we rip off the head, then we won't have a pointer that we can use to point to the remaining nodes.

To fix this problem, we can leave something temporarily in the place of the head while we are updating it. This technique can be performed using a special memory handling function called take, which replaces a value in memory by returning the original value and replacing it with a default value. The take function has the following syntax:

```
fn take<T>(dest: &mut T) -> T
```

Specifically, in this case, the function takes a mutable reference to some value T, replaces the destination value with a default value of T, and returns T.

Let's look at the semantics of the take function when called on the head. Consider the following code:

```
let previous_head = self.head.take();
```

This line will replace the head with the default value. Since the head is an Option, the default value of Option, which is None, will be assigned to head. Furthermore, the value of head will be returned and stored in the previous_head. The previous_head now contains the old_head.

We can now use the result of calling the take on the head to complete the definition of the add method. Listing 11.10 shows the new definition of the add method based on the take method.

```
impl Linkedlist {
    ...
    fn add(&mut self, element: i32) {
        let previous_head = self.head.take();
        let new_node = Some(Box::new(Node {
            element: element,
            next: previous_head,
        }));
        self.head = new_head;
```

```
    }
}
```

Listing 11.10 Updated Definition of the add Method

As explained earlier, the first step will take the head, replace it with a default value of Option (i.e., None), and return the old head in the previous_head. Next, we make a new node as before, with the next field set to the previous_head. Finally, we update the head to that of the new node.

Notice that we don't need explicit cases for checking whether a node is empty. Irrespective of whether the node is empty or not, we'll always update the head with a newly created node. A more appropriate approach is to name the variable new_node as new_head in this case. Let's update the code accordingly:

```
fn add(&mut self, element: i32) {
    ...
    let new_head = Some(Box::new(Node { // variable name changed
    ...
}
```

You can now use the implementation in main, as shown in Listing 11.11.

```
fn main() {
    let mut list = Linkedlist::new();
    list.add(5);
    list.add(7);
    list.add(10);
    list.add(15);
    list.add(20);
    println!("List: {:?}", list);
}
```

Listing 11.11 Using the add Method in main

This function first creates a new Linkedlist using the new constructor function and then adds a few elements. When executing the code, notice how the elements are displayed in reverse order. This order is followed because our function is inserting elements at the beginning. The last element added, therefore, appears at the front of the list.

11.1.5 Removing Elements

Similar to the add method, you can define a remove method. This method will remove an element from the beginning of the Linkedlist and will return that element. The skeleton of this method will resemble the following example:

```
fn remove(&mut self) -> Option<i32> {}
```

The input is a mutable reference to self. The output is an Option<i32> value. The method will return a None if the list is empty and there is nothing to remove.

Let's now focus on the logic this method must implement. If the head is empty, we'll simply return None since the list is empty and there is nothing to remove. If there is Some existing head, we'll remove the head, return its element, and update the head to the next of the existing head. Listing 11.12 shows the code for the remove method.

```
impl Linkedlist {
    ...
    fn remove(&mut self) -> Option<i32> {
        match self.head.take() {
            Some(previous_head) => {
                self.head = previous_head.next;
                Some(previous_head.element)
            }
            None => None,
        }
    }
}
```

Listing 11.12 Definition of the remove Method

We first take out the head using the take method and match on it. As explained earlier, the take method will put a default value in place of the head which is None (for an Option) and returns the original head which is moved to the variable used in the arm corresponding to Some variant. Next, if there is Some existing head, then we update the head to the next of previous_head and returns the element field of the previous_head, wrapped around by the Some variant. In case of None, we simply return nothing. Notice that inside the Some arm, we update the head with the next of the previous_head. This implicitly means that the old head is being removed.

We can now use the implementation. Consider adding the following line to the code shown earlier in Listing 11.11:

```
fn main() {
    ...
    println!("{}", list.remove().unwrap());
}
```

The call to remove() on the list will invoke the remove method defined in Listing 11.12. Since the remove method returns an Option<i32> value, to print the actual value, use the unwrap() method to access the inner value, which is behind the Some variant. In summary, this method will remove the first value in the list, which in this case should be the value 20.

11.1.6 Printing Singly Linked Lists

Printing a Linkedlist inside a print statement works but has strange formatting. Let's define a proper printing method for printing the element part of the nodes included in the Linkedlist. The skeleton of this method will resemble the following example:

```
fn print(&self) {}
```

The method will not modify the Linkedlist; therefore, it takes an immutable reference to self. The logic this method needs to implement is fairly simple. Starting from the head of the Linkedlist, the method then follows the next fields to traverse through all the nodes. While passing through each node, the method will display its element part. The code for this method is shown in Listing 11.13.

```
impl Linkedlist {
    ...
    fn printing(&self) {
        let mut list_traversal = &self.head;
        while !list_traversal.is_none() {
            println!("{:?}", list_traversal.as_ref().unwrap().element);
            list_traversal = &list_traversal.as_ref().unwrap().next;
        }
    }
}
```

Listing 11.13 Definition of the printing Method

The method first defines a variable list_traversal, which is a reference to the head. Next, it iterates as long as list_traversal is not None. During each iteration, it updates the list_traversal to the next node and displays the element part.

The semantics for printing the element part of a node inside the print statement, i.e., list_traversal.as_ref().unwrap().element, may be a bit hard to comprehend. So let's break it down. The list_traversal has a type &Option<Box<Node>>. If we call unwrap directly on it, we'll get an owned value, which we don't want. In particular, calling unwrap on list_traversal will result in a type of Box<Node>, which is a type indicating an owned value. To ensure that ownership doesn't change, use the as_ref method. Calling the as_ref on list_traversal returns a type Option<&Box<Node>>. Next, calling unwrap on the result of as_ref will have a type of &Box<Node>, which uses a reference and not an owned value. The element part is finally accessed. Table 11.1 shows the variables involved and their respective types.

Variable	Type
list_traversal	&Option<Box<Node>>
list_traversal.unwrap()	Box<Node>
list_traversal.as_ref()	Option<&Box<Node>>
list_traversal.as_ref().unwrap()	&Box<Node>

Table 11.1 Variables and Their Types in the print Statement from Listing 11.13

In the last line in Listing 11.13, we used a reference to list_traversal on the right side during assignment. This is because the next field gives Option<Box<Node>> and not &Option<Box<Node>>. In this way, the next field is defined in the struct Node. Since the variable list_traveral is defined as a &Option<Box<Node>> and not as Option<Box<Node>>, to avoid type mismatch errors (since types must be assigned values of the same type), we'll use a reference before the variable list_traversal during assignment.

The method can now be used in main for printing the list in the following way:

```
fn main() {
    ...
    list.printing();
}
```

11.2 Doubly Linked List

Singly linked lists allow navigation in only one direction (forward). In contrast, doubly linked lists allow navigation in both forward and backward directions.

Let's visualize this capability through the structure of a typical doubly linked list, as shown in Figure 11.2.

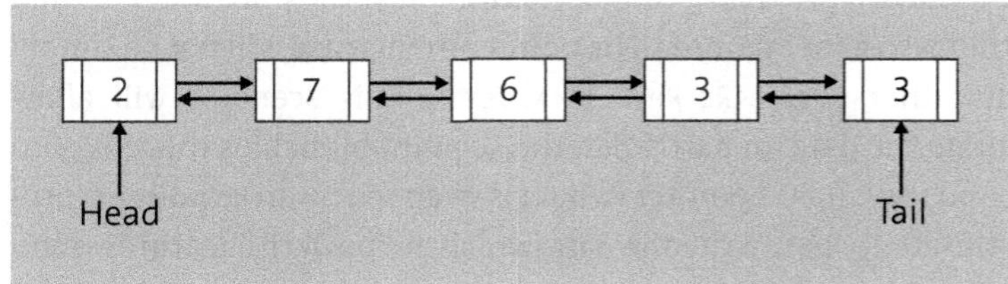

Figure 11.2 A Typical Doubly Linked List

Unlike a singly linked list, we have explicit information about both the head and tail of the list. Additionally, each node contains two pointers: one to the next node and one to the previous. An important observation from a programming perspective is that each node is pointed to by more than one pointer. For example, a single node will have two pointers referencing it, meaning you must enable multiple ownership for a node.

A simple box pointer, which only allows single ownership, won't be sufficient in this case. You must use an Rc pointer, which allows multiple owners to reference the same data (refer to Chapter 10, Section 10.2.2). Furthermore, you may need to modify the data by accessing nodes through either the next or previous pointers. This setup requires not only multiple owners but also the ability to mutate the data. This functionality is provided by the RefCell pointer when wrapped around by an Rc pointer (refer to Chapter 10, Section 10.2.3). With this understanding, let's move on to the implementation.

11.2.1 Setting Up the Basic Data Structure

Start with the struct definitions from the previous section for implementing the singly linked list. For the sake of convenience, this code is shown in Listing 11.14.

```
#[derive(Debug)]
struct Linkedlist {
    head: pointer,
}

#[derive(Debug)]
struct Node {
    element: i32,
    next: pointer,
}
type pointer = Option<Box<Node>>;
```

Listing 11.14 Basic Struct Definitions Used for Implementing Singly Linked List

To be considered a doubly linked list, the node not only needs to contain the next pointer but also needs to have a previous pointer. Additionally, the Linkedlist should not only have the head but also a tail pointer. The pointer in this case will not be simply `Option<Box<Node>>`, but rather `Option<Rc<RefCell<Node>>>`. This scenario will allow multiple owners to mutate the data. In particular, the Rc pointer enables the ability to have multiple owners, and the RefCell pointer (which is wrapped by an Rc pointer) provides each owner with the ability to mutate the data, which are powerful features when combined. Finally, the Linkedlist wrapper struct name shown in Listing 11.14 should be changed to something more meaningful. The updated struct definitions are shown in Listing 11.15.

```
use std::{cell::RefCell, rc::Rc};
#[derive(Debug)]
```

```
struct DoublyLinkedlist {
    head: pointer,
    tail: pointer,
}

#[derive(Debug)]
struct Node {
    element: i32,
    next: pointer,
    previous: pointer,
}
type pointer = Option<Rc<RefCell<Node>>>;
```

Listing 11.15 Basic Struct Definitions for Doubly Linked List Implementation

11.2.2 Adding Elements

Now, we can add functionality by defining a new constructor function. This definition is similar to the new constructor function defined for the singly linked list earlier in Section 11.1.4. Listing 11.16 shows the implementation of a doubly linked list.

```
impl DoublyLinkedlist {
    fn new() -> Self {
        DoublyLinkedlist {
            head: None,
            tail: None,
        }
    }
}
```

Listing 11.16 Definition of new Constructor Function for the DoublyLinkedlist

The constructor simply returns a `DoublyLinkedlist` with `head` and `tail` set to `None`.

Next, create an `add` method for inserting a node at the beginning of the list. When trying to insert a new node, two things might happen. The list may contain some nodes, meaning that a `head` will exist, or the list is empty, meaning that there is no `head`. If a `head` exists, the method must do two things: first, set the `previous` pointer of the old `head` to that of the new `head`. Then, the `next` of the new `head` should be set as equal to the old `head`. This step will ensure that the new node is inserted at the beginning of the list. If no `head` exists in the list, then the node will be the first node in the list. In this case, we'll set the `tail` to point to this node and also make it the `head` of the list. Let's summarize the two cases along with the actions that must be taken corresponding to each case:

- **Case 1: head is Some: Head already exists**
  Set the previous pointer of the old head = new head and the next of the new head = to old head. Finally we update the head to new_head.
- **Case 2: head is None: Head is empty**
  Set the tail to point to the new node. The new node is also the head of the list.

Now that you understand the logic, let's look at the implementation. Listing 11.17 shows the definition of the add method.

```
impl Doubly_Linklist {
    ...
    fn add(&mut self, element: i32) {
        let new_head = Rc::new(RefCell::new(Node {
            element: element,
            next: None,
            previous: None,
        }));
        match self.head.take() {
            Some(old_head) => {
                old_head.borrow_mut().previous = Some(new_head.clone());
                new_head.borrow_mut().next = Some(old_head.clone());
                self.head = Some(new_head);
            }
            None => {
                self.tail = Some(new_head.clone());
                self.head = Some(new_head);
            }
        }
    }
}
```

Listing 11.17 Definition of the add Method for the DoublyLinkedlist

Let's go over the method line by line. The first thing to notice is that the signature of the method looks the same as that of the add method for the singly linked list (Section 11.1.4). Inside the method, we first create a new node and name it new_head. The name of the variable reflects the fact that the add method will always add the new node at the start of the list and therefore is named as new_head. The new_head has the element field set to the element passed into the method, while the next and previous are set to None. Note that we need to wrap the node by Rc and RefCell so that it is compatible with the pointer type defined previously in Listing 11.15.

Next, we match on the self.head.take. As explained in the previous section, the take will return the value of the head, replacing it with default value which is None in this case.

The None value for the head, will be replaced inside the match arms. The returned value from the self.head.take is being moved to the variable in the Some arm, in this case the variable of old_head.

The arm corresponding to the Some variant represents the case when we have some existing head. As explained under case 1, we need to do two things: First, we'll set the previous old_head to that of the new_head, and second, the next of the new_head is set equal to old_head. These modifications are applied using the following statements:

```
old_head.borrow_mut().previous = Some(new_head.clone());
new_head.borrow_mut().next = Some(old_head.clone());
```

To mutate old_head, we'll call the borrow_mut method, which allows mutation of data behind an Rc pointer. The clone method on new_head will increment the number of pointers to the new_head. This will ensure that the node remains in memory as long as there are pointers to it.

Remember that the take function on the head puts a None in place of head. So we need to properly refill the head so that the list remains usable. The head is therefore updated to new_head at the end of the arm. In the Some arm the tail is not affected. Also note that previous of the new_head is not being set because previous of the head is always None. We explicitly set this during initialization of the new_head at the start of the method in Listing 11.17.

Now let's look at the second arm of Listing 11.17, which corresponds to None or no head in the list. When there is no existing head in the list, then it means that the node we are inserting is the first node in the list. In this case, we'll set the tail to point to this new_head and also make it the head of the list. The following two lines are used to code this:

```
self.tail = Some(new_head.clone());
self.head = Some(new_head);
```

Notice that we did not call the clone on the head in the second line. The clone increments the owners to the data, which are essentially pointers to the data. When we do not call the clone and assign Rc pointer to something, then ownership is transferred to the new variable and simple ownership rules apply. For instance, in Listing 11.18, we'll have an error due to change of ownership.

```
use std::rc::{self, Rc};
fn main() {
    let x = Rc::new(50);
    let y = x;
    println!("{}", x); // Error
}
```

Listing 11.18 Error from Change of Ownership

If we rather called the clone as follows,

```
self.head = Some(new_head.clone());
```

then we'll have three owners: the new_head, the self.tail, and self.head. However without the clone, we have two owners, the self.tail and the self.head, with the self.head holding the original resource. In particular, the statement without the clone, i.e., self.head = Some(new_head), transfers the ownership of new_head to that of self.head. This completes the add method.

11.2.3 Adding a New Constructor Function for Node

Let's simplify the add method slightly. At the beginning of this method (refer to Listing 11.17), we created a new node. During implementation of different methods, nodes may be created many times, and therefore it is not very efficient to write this long code every time. Instead, let's define a new constructor function for creating a new node. The new constructor function will be defined inside an implementation block for a Node. Listing 11.19 shows the code for adding a new constructor function.

```
impl Node {
    fn new(element: i32) -> Rc<RefCell<Node>> {
        Rc::new(RefCell::new(Node {
            element: element,
            next: None,
            previous: None,
        }))
    }
}
```

Listing 11.19 Definition of Constructor Function new for a Node

The function accepts an element as input and returns a Rc<RefCell<Node>>>. The function simply returns a new node with element set to the value passed in and the previous and next set to None.

We can now revise the add function to use the constructor instead of manually defining the new_head. The updated definition of the add method is shown in Listing 11.20.

```
impl Doubly_Linklist {
    ...
    fn add(&mut self, element: i32) {
        let new_head = Node::new(element);

        match self.head.take() {
            ...
```

```
        }
    }
}
```

Listing 11.20 Updated Definition of the add Method from Listing 11.17

Following this code, you can add a similar method for adding an element at the end of the doubly_Linkedlist. The code will be quite similar but with few changes, and this scenario appears as an exercise in Section 11.4.

11.2.4 Removing Elements

When managing doubly linked lists, removing elements can be just as crucial as adding them. Let's look at the implementation, starting with the signature of the remove method:

```
fn remove(&mut self) -> Option<i32> {}
```

The input is a mutable reference to self, and the return value will be an Option<i32> value. We'll need to handle two cases. First, the list may be empty which means that the head will be None. In this case, we'll print that the list is empty and return a None value. Listing 11.21 shows the code for handling this case.

```
impl Doubly_Linklist {
    ...
    fn remove(&mut self) -> Option<i32> {
        if self.head.is_none() {
            println!("List is empty so we can not remove");
            None
        } else {
            ...
        }
    }
}
```

Listing 11.21 Partial Implementation of the remove Method

The else part will correspond to the second case of when the head is not empty. There are two scenarios within this second case.

The first scenario is when there are two or more nodes in the list, as shown in Listing 11.22.

```
// Possibility: 1
// ----------------------
//          Head        Tail
```

```
// None <-- 1 --> 2 --> 3 --> None
// None     1 <-- 2 <-- 3     None
// -----------------------
```

Listing 11.22 First Possibility in the else Part of the remove Method

In this case, the head is pointing to the first node with value 1. The next of head is also not None and contains a node with a value of 2. We need to update the head to point to its next node, delete the previous head, and set the previous pointer of the new head to None. The list after removal is shown in Listing 11.23.

```
// Possibility: 1 (After Removal)
// -----------------------
//       Head  Tail
// None <-- 2 --> 3 --> None
// None     2 <-- 3     None
// -----------------------
```

Listing 11.23 Updated List after Removal of the First Node

The second scenario is when we only have a head node. This possibility is shown in Listing 11.24.

```
// Possibility: 2
// -----------------------
//        Head
//        Tail
// None <-- 1 --> None
// -----------------------
```

Listing 11.24 Second Possibility in the else Part of the remove Method

The next of head is empty, meaning that we only have a single node in the list. Therefore, we'll delete that node and make the list empty. We'll also make sure that the tail becomes None in this case, since there is nothing left in the list. The updated list after removal is shown in Listing 11.25.

```
// Possibility: 2 (After Removal)
// -----------------------
//       Head = None
//       Tail = None
// -----------------------
```

Listing 11.25 Updated List after Removal of the First Node

Let's now add some suitable code for the two cases in the else part of the remove method. First, take out the head because we are updating the head and will call the map function on it. The map accepts a closure as an input and maps the input to something else represented by the body of the closure. Listing 11.26 shows the partial code in the else part.

```
fn remove(&mut self) -> Option<i32> {
        if self.head.is_none() {
            ...
        } else {
            self.head
                .take()
                .map(|old_head| {});
        }
    }
}
```

Listing 11.26 Partial Implementation of the else Part in the remove Method

The call to the take function will return the head, which is transferred to the old_head inside the map closure. The map will swap the old_head, which is the input to the closure, with some new_head which, will be returned from the body of the closure. (Currently, we haven't added any code yet.)

Inside the body of the closure, we'll add a match for handling the two scenarios. In particular, we want a match on the next node following old_head. The code after the match inside the closure body in the map is shown in Listing 11.27.

```
impl Doubly_Linklist {
    ...
    fn remove(&mut self) -> Option<i32> {
        ...
        } else {
            self.head
                .take()
                .map(|old_head| match old_head.borrow_mut().next.take() { });
        }
    }
}
```

Listing 11.27 Partial Implementation of the else Part with Code Added in the Closure inside map

The take method is used because the next of the node, if it exists, will be replaced. Recall that the take method replaces the variable with a default value and returns its original value. In this case, the original value will be transferred to the variable used inside the Some arm. If the next of the old_head is Some node, then it means that we've encountered the first scenario, and the next of the node will be replaced with a new value. On the other hand, if the next of the old_head is None, then it means we've encountered the second scenario. Listing 11.28 shows the code with the two arms in detail.

```
impl Doubly_Linklist {
    ...
    fn remove(&mut self) -> Option<i32> {
        ...
          else {
            self.head
                .take()
                .map(|old_head| match old_head.borrow_mut().next.take() {
                    Some(new_head) => {
                        new_head.borrow_mut().previous = None;
                        self.head = Some(new_head);
                        self.head.clone()
                    }
                    None => {
                        self.tail = None;
                        println!("List is empty after removal");
                        None
                    }
                });
        }
    }
}
```

Listing 11.28 remove Method after Adding Code to the Match Body

When the next of the old_head is Some variant, then we have two or more already existing nodes. The next of the old_head on which the take function is called is first transferred to the variable new_head. Next, we make the previous of the new_head as None and update the list head to the new_head. Finally, we'll return a pointer to the head.

In the case of None, which means that the next of the old_head is empty, we make sure that the tail also becomes empty and will display a suitable message and return None. Overall, the match inside the map method shown previously in Listing 11.27 will either return new_head or None. The map will further use the returned value to fill in the old_head, which was temporarily filled with None due to the call to the take function.

Let's also return the removed value, which is the element part of the head. Returning the removed value allows the caller to know what was removed or if the removal was unsuccessful (indicated by None). We'll grab it at the start of the else branch since the head is updated later on and return it after the end of the else branch. This completes the remove method given in Listing 11.29.

```
impl Doubly_Linklist {
    ...
    fn remove(&mut self) -> Option<i32> {
        if self.head.is_none() {
            println!("List is empty so we can not remove");
            None
        } else {
            let removed_val = self.head.as_ref().unwrap().borrow().element;
            self.head
                .take()
                .map(|old_head| match old_head.borrow_mut().next.take() {
                    Some(new_head) => {
                        new_head.borrow_mut().previous = None;
                        self.head = Some(new_head);
                        self.head.clone()
                    }
                    None => {
                        self.tail = None;
                        println!("List is empty after removal");
                        None
                    }
                });
            Some(removed_val)
        }
    }
}
```

Listing 11.29 Definition of the Completed remove Method

Let's focus on the statement `let removed_val = self.head.as_ref().unwrap().borrow().element;`, where we first access the head and then call `as_ref`. This approach ensures that the value doesn't move out of the `head`. Since `self.head.as_ref()` returns an option, we `unwrap` it, which is followed by an immutable access to the `head` since we are not interested in updating the data. Finally, we access the `element` part.

The removal of a node from the end can also be implemented in the same way. You'll perform this task in one of the exercises in Section 11.4.

11.2.5 Printing Doubly Linked Lists

The function for printing the `DoublyLinkedlist` will have a similar definition to that of a singly linked list, as discussed in Section 11.1.6. Listing 11.30 shows the code for the method.

```
impl Doubly_Linklist {
    ...
    fn printing(&self) {
        let mut traversal = self.head.clone();
        while !traversal.is_none() {
            println!("{}", traversal.as_ref().unwrap().borrow().element);
            traversal = traversal.unwrap().borrow().next.clone();
        }
    }
}
```

Listing 11.30 Definition of the printing Method for Doubly_Linkedlist

We first initialize a variable that points to the head. Note that simple references can also be used instead of using the clone as we did in the Linkedlist implementation. However, the clone serves the same purpose. The cloned copy will be cleaned up at the end of the function, and therefore, we avoid issues regarding ownership management of the head (which should ideally have one owner). Next, we iterate as long as the variable traversal is not None, and during each iteration, we print the element part of a node and update the traversal variable.

This step completes the implementation of basic methods for the DoublyLinkedlist. You can now use the implementation in main, as shown in Listing 11.31.

```
fn main() {
    let mut list1 = Doubly_Linklist::new();
    list1.add(30);
    list1.add(32);
    list1.add(34);
    list1.add(36);
    list1.printing();
    list1.remove();
    println!("After Removal");
    list1.printing();
}
```

Listing 11.31 Sample Usage of the Methods Defined for DoublyLinkedlist

11.3 Reference Cycles Creating Memory Leakage

Rust is renowned for being a memory-safe language, offering strong guarantees against data races. However, Rust does not provide equally strict guarantees when it comes to memory leaks, that is, situations where memory is not properly freed up. While challenging in Rust, you can create memory that is never deallocated. Memory leaks can occur, particularly when using Rc and RefCell smart pointers, the latter of which enables mutability.

To demonstrate, let's define a Node struct and its Drop implementation, as shown in Listing 11.32.

```
use std::cell::RefCell;
use std::rc::Rc;
#[derive(Debug)]
struct Node {
    next: Option<Rc<RefCell<Node>>>,
}

impl Drop for Node {
    fn drop(&mut self) {
        println!("Dropping {:?}", self);
    }
}
```

Listing 11.32 Definition of Node Struct and Implementation of Drop for Node

In this scenario, the Node struct contains a next field, which is Option<Rc<Refcell<Node>>. Next, we implemented the Drop trait for Node. Defined in Rust's standard library, the Drop trait runs cleanup code when a value goes out of scope. This trait provides a way to specify what should happen when an object is destroyed, such as releasing resources like memory, files, or network connections. The drop method is automatically called when a value is dropped. Types can redefine the specific behavior to execute at the drop time by providing a custom defined implementation, as we provided in the code shown in Listing 11.32. The drop method that we've defined will execute the code inside it when Node goes out of scope.

In the main function, let's create some nodes to later use to create a reference cycle that leads to memory leakage. Listing 11.33 shows the code.

```
fn main() {
    let a = Rc::new(RefCell::new(Node { next: None }));
    println!("a strong count: {:?}", Rc::strong_count(&a));
    let b = Rc::new(RefCell::new(Node {
        next: Some(Rc::clone(&a)),
    }));

    println!("\nB is created:");
    println!("a strong count: {:?}, ", Rc::strong_count(&a));
    println!("b strong count: {:?}", Rc::strong_count(&b));
    let c = Rc::new(RefCell::new(Node {
        next: Some(Rc::clone(&b)),
```

```
    }));
    println!("\nC is created:");
    println!("a strong count: {:?}", Rc::strong_count(&a));
    println!("b strong count: {:?}", Rc::strong_count(&b));
    println!("c strong count: {:?}", Rc::strong_count(&c));
}
```

Listing 11.33 Creating a Few Nodes, Linking Them to Each Other, and Displaying Their strong_counts

First, we created node a with its next field set to None. Next, node b is defined, which points to a. Finally, we have node c pointing to node b. After creating each node, we display its respective reference count using the strong_count method. After creating a, we only print the reference count of a; after creating b, we print the reference count of both a and b. After creating c, we print the reference count of all a, b, and c. Everything is fine until now, and we don't have any cycles. Node b is pointing to node a, and node c is pointing to node b, while node a is not pointing to anything:

```
c -> b -> a
```

A cycle will be created if the next field of node a is set to point to node c, as shown in Listing 11.34.

```
fn main() {
    ...
    (*a).borrow_mut().next = Some(Rc::clone(&c));
    println!("\nAfter creating cycle:");
    println!("a strong count: {:?}", Rc::strong_count(&a));
    println!("b strong count: {:?}", Rc::strong_count(&b));
    println!("c strong count: {:?}", Rc::strong_count(&c));
}
```

Listing 11.34 Code from Listing 11.33 Updated by Adding a Cycle

Now, to mutate the internal value of a variable of type Rc, you must use the deref operator (*) and then call the borrow_mut on it. The next field of node a is set to point to node c, essentially creating a cycle. Node b is pointing to node a, node c is pointing to node b, and node a is pointing back to node c, creating a non-ending cycle, as follows:

```
c -> b -> a -> c -> ...
```

Creating reference cycles can be useful in scenarios like modeling bidirectional relationships or graph-like structures where nodes need to reference each other.

If you execute the code shown in Listing 11.34, you should get the output shown in Listing 11.35.

```
a strong count: 1

B is created:
a strong count: 2,
b strong count: 1

C is created:
a strong count: 2,
b strong count: 2
c strong count: 1

After creating cycle:
a strong count: 2
b strong count: 2
c strong count: 2
```

Listing 11.35 Output of the Code from Listing 11.34

Recall that the `clone` function increments the reference count. The initial count of node a is 1 because only a points to itself. When node b is created, the count of node a is incremented to 2, while b's count is set to 1. After creating node c, the reference count of node b is incremented to 2 while node a's count remains at 2. Moreover, node c's count is set to 1. When a cycle is introduced, which means that node a points to node c, the count of node a and node b remain at 2, while the count of c is incremented to 2.

When `main` ends, the destructor attempts to clean up memory held by a variable, which is only possible if the reference count for a variable is exactly 1. This specific value indicates that only a single reference to the variable exists, which is the variable itself. However, if the reference count is greater than 1, meaning there are multiple pointers sharing the data, the memory cannot be freed. Cleaning up in such a case would invalidate those remaining pointers, leading to potential issues like dangling references. The nodes a, b, and c are therefore not cleared from the memory since their reference counts are greater than 1, which therefore leads to a memory leak.

Let's add a print statement to the code shown earlier in Listing 11.34 for printing node a, as shown in Listing 11.36.

```
fn main() {
    ...
    println!("a {:?}", a);
}
```

Listing 11.36 Print Statement Adding to the Code from Listing 11.34 for Printing Node a

After executing the code, note how the display is never ending and ultimately results in an error "`main` has overflowed stack." This error occurs because, when you try to print a, Rust will also attempt to print b since a contains b. However, since b contains c, it will attempt to print c as well. The problem arises when printing c requires printing a again, which creates an infinite loop, ultimately leading to an error in the display process.

Let's understand the reason for this memory leakage from another perspective. Recall that we added an implementation for the Drop trait in the code shown earlier in Listing 11.32. The drop method will be automatically called when an instance of a node goes out of scope. However, the code shown in Listing 11.34, when executed, does not make a call to the drop method for any node since we do not have any output in Listing 11.35 corresponding to the drop method.

However, let's see what happens if we comment out the line that creates the cycles in Listing 11.34. In other words, comment out the following line:

```
(*a).borrow_mut().next = Some(Rc::clone(&c));
```

Now, when you rerun the program, notice how drop is called three times at the end of the main. This behavior occurs because we do not have cycles anymore, and therefore, drop is called three times, one time for each node, and everything is cleaned up nicely. Throughout execution, you'll see reference counts for the nodes, as shown in Listing 11.37.

```
a strong count: 1

B is created:
a strong count: 2,
b strong count: 1,

C is created:
a strong count: 2,
b strong count: 2,
c strong count: 1

After creating cycle:
a strong count: 2,
b strong count: 2,
c strong count: 1
```

Listing 11.37 Reference Counts of Nodes without Cycles

Variables are dropped in the reverse order of their creation. Thus, when the main function ends, Rust will first try to clean up node c. Since c has a reference count of 1, it can be safely dropped. After c is cleaned up, the reference count of b will be decremented to 1, as c was pointing to b. This step allows b to be cleaned up safely as well since its reference count is 1. Finally, the reference count of a will also be decremented to 1, since b was pointing to it, and it was cleaned up. The end result is safe cleaning of a from memory also, and in this way, everything is cleared from memory.

Let's reconsider the code containing the cycles shown earlier in Listing 11.34. In some situations, some behavior involving reference cycles might be required, such as in bidirectional relationships or graph-like structures. We have a nice solution for handling such cases. Instead of using the Rc, Rust has a Weak Rc smart pointer. Listing 11.38 shows the definition of Node with a weak Rc smart pointer.

```
use std::cell::RefCell;
use std::rc::{Rc, Weak};
#[derive(Debug)]
struct Node {
    next: Option<Weak<RefCell<Node>>>,
}
```


Listing 11.38 Revised Definition of Node Using Weak Rc Smart Pointer

weak is a special type of Rc with two key functions named upgrade and downgrade. Calling upgrade on a weak Rc will attempt to convert a weak Rc into an Rc pointer, thereby incrementing its strong count.

Calling the downgrade creates a new weak pointer to an Rc pointer. This new pointer will hold a non-owning reference to a managed allocation. As a result, this pointer does not share ownership over its underlying data. Moreover, this step increases the weak count by 1 and does not change the strong count. In summary, strong references allow you to share ownership of an Rc smart pointer instance. On the other hand, weak references do not signify ownership, meaning they do not prevent the data from being dropped. You can simply access the data without owning it.

Let's change the code shown earlier in Listing 11.34 to use a weak Rc instead. Listing 11.39 shows the updated code based on the new Node definition shown in Listing 11.38.

```
fn main() {
    let a = Rc::new(RefCell::new(Node { next: None }));
    println!("a strong count: {:?}", Rc::strong_count(&a),);

    let b = Rc::new(RefCell::new(Node {
        next: Some(Rc::downgrade(&a)),          // Using downgrade now
    }));

    println!("\nB is created:");
    println!("a strong count: {:?},", Rc::strong_count(&a),);
    println!("b strong count: {:?}", Rc::strong_count(&b),);

    let c = Rc::new(RefCell::new(Node {
```

```
        next: Some(Rc::downgrade(&b)),          // using downgrade now

    }));
    println!("\nC is created:");
    println!("a strong count: {:?}, ", Rc::strong_count(&a),);
    println!("b strong count: {:?},", Rc::strong_count(&b),);
    println!("c strong count: {:?}", Rc::strong_count(&c),);

    (*a).borrow_mut().next = Some(Rc::downgrade(&c)); // using downgrade now

    println!("\nAfter creating cycle:");
    println!("a strong count: {:?},", Rc::strong_count(&a),);
    println!("b strong count: {:?},", Rc::strong_count(&b),);
    println!("c strong count: {:?},", Rc::strong_count(&c),);
}
```

Listing 11.39 Updated Code from Listing 11.34 Based on a Weak Rc Pointer

Instead of using the clone method in the next fields of b, c, and a, you can now use the downgrade method. This method will essentially update the weak count of these nodes to which they are pointing without updating the respective strong counts. Moreover, they will also not share ownership over the resources they are holding. As a result, the strong counts of all the nodes remain at a value of 1 and, therefore, can be safely deleted from the memory at the end of main. Since everything from the memory is being wiped out, no memory leakages are created.

11.4 Practice Exercises

In this section, we'll provide some practice questions to help reinforce your learning. The exercise questions cover almost all the concepts introduced in the chapter. Solutions to each exercise will follow in Section 11.5.

1. **Implementing the peek method in a linked list**
   Your task is to complete the peek function for the Linklist. This method should return an Option<i32> representing the value at the head of the list without removing it.

2. **Making a linked list generic in Rust**
   Your task is to modify the current linked list implementation to make the element field of each Node generic, rather than the concrete type i32. This task involves updating the Linklist and Node structs to accept a generic type T. Make necessary changes in the following code:

```
#[derive(Debug)]
struct Linklist<?> {  // This line needs a fix
    head: pointer<?>, // This line needs a fix
}
#[derive(Debug)]
struct Node<?> {
    element: T,
    next: pointer<?>,  // This line needs a fix
}
type pointer<?> = Option<Box<Node<?>>>;  // This line needs a fix

impl<T: ?> Linklist<?> {  // This line needs a fix
    fn new() -> Linklist<?> { // This line needs a fix
        Linklist { head: None }
    }

    fn add(&mut self, element: i32) { // This line needs a fix
        let previous_head = self.head.take();
        let new_head = Some(Box::new(Node {
            element: element,
            next: previous_head,
        }));
        self.head = new_head;
    }

    fn remove(&mut self) -> Option<i32> {  // This line needs a fix
        match self.head.take() {
            Some(previous_head) => {
                self.head = previous_head.next;
                Some(previous_head.element)
            }
            None => None,
        }
    }

    fn peek(&self) -> Option<i32> {  // This line needs a fix
        match &self.head {
            Some(H) => Some(H.element),
            None => None,
        }
    }

    fn print(&self) {
        let mut list_traversal = &self.head;
```

```
        while !list_traversal.is_none() {
            println!("{:?}", list_traversal.as_ref().unwrap().element);
            list_traversal = &list_traversal.as_ref().unwrap().next;
        }
    }
}
fn main() {
    let mut list = Linklist::new();
    list.add(5);
    list.add(7);
    list.add(10);
    list.add(15);
    list.add(20);

    println!("{:?}", list.peek());
}
```

3. **Adding elements to the end of a doubly linked list**
   Your task is to implement the `add_back()` method for a doubly linked list. This method should add a new element at the tail (or the end) of the list. The doubly linked list already tracks both the head and the tail, so when adding an element to the back, the method should properly update the tail pointer to point to the new node. Additionally, ensure that the previous pointer of the new node links back to the old tail and that the next pointer of the old tail points to the new node.
4. **Removing elements from the end of a doubly linked list**
   The goal is to implement the `remove_back()` method for a doubly linked list. This method should remove the element at the tail (or the end) of the list. The doubly linked list tracks both the head and the tail, so the method should properly update the tail pointer to point to the previous node. Additionally, ensure that the `next` pointer of the new tail is set to `None`.

11.5 Solutions

This section provides the code solutions for the practice exercises in Section 11.4. The code is largely self-explanatory; however, we have included comments and additional explanations wherever necessary to enhance understanding.

1. **Implementing the `peek` method in a linked list**

```
#[derive(Debug)]
struct Linklist {
    head: pointer,
}
#[derive(Debug)]
```

```
struct Node {
    element: i32,
    next: pointer,
}
type pointer = Option<Box<Node>>;

impl Linklist {
    fn peek(&self) -> Option<i32> {
        match &self.head {
            Some(H) => Some(H.element),
            None => None,
        }
    }
}
fn main() {}
```

2. **Making a linked list generic in Rust**

```
#[derive(Debug)]
struct Linklist<T> {
    head: pointer<T>,
}

#[derive(Debug)]
struct Node<T> {
    element: T,
    next: pointer<T>,
}
type pointer<T> = Option<Box<Node<T>>>;

impl<T: std::fmt::Debug + std::marker::Copy> Linklist<T> {
    fn new() -> Linklist<T> {
        Linklist { head: None }
    }

    fn add(&mut self, element: T) {
        let previous_head = self.head.take();
        let new_head = Some(Box::new(Node {
            element: element,
            next: previous_head,
        }));
        self.head = new_head;
    }

    fn remove(&mut self) -> Option<T> {
```

```
        match self.head.take() {
            Some(previous_head) => {
                self.head = previous_head.next;
                Some(previous_head.element)
            }
            None => None,
        }
    }

    fn peek(&self) -> Option<T> {
        match &self.head {
            Some(H) => Some(H.element),
            None => None,
        }
    }

    fn print(&self) {
        let mut list_traversal = &self.head;
        while !list_traversal.is_none() {
            println!("{:?}", list_traversal.as_ref().unwrap().element);
            list_traversal = &list_traversal.as_ref().unwrap().next;
        }
    }
}
fn main() {
    let mut list = Linklist::new();
    list.add(5);
    list.add(7);
    list.add(10);
    list.add(15);
    list.add(20);

    println!("{:?}", list.peek());
}
```

3. **Adding elements to the end of a doubly linked list**

```
use std::{cell::RefCell, rc::Rc};
#[derive(Debug)]
struct Doubly_Linklist {
    head: pointer,
    tail: pointer,
}
```

```
#[derive(Debug)]
struct Node {
    element: i32,
    next: pointer,
    prev: pointer,
}
type pointer = Option<Rc<RefCell<Node>>>;
impl Doubly_Linklist {
    fn add_back(&mut self, element: i32) {
        let new_tail = Node::new(element);
        match self.tail.take() {
            Some(old_tail) => {
                old_tail.borrow_mut().next = Some(new_tail.clone());
                new_tail.borrow_mut().prev = Some(old_tail.clone());
                self.tail = Some(new_tail);
            }
            None => {
                self.head = Some(new_tail.clone());
                self.tail = Some(new_tail);
            }
        }
    }
}
impl Node {
    fn new(element: i32) -> Rc<RefCell<Node>> {
        Rc::new(RefCell::new(Node {
            element: element,
            next: None,
            prev: None,
        }))
    }
}
fn main() {}
```


4. **Removing elements from the end of a doubly linked list**

```
use std::{cell::RefCell, rc::Rc};
#[derive(Debug)]
struct Doubly_Linklist {
    head: pointer,
    tail: pointer,
}

#[derive(Debug)]
struct Node {
```

```
    element: i32,
    next: pointer,
    prev: pointer,
}
type pointer = Option<Rc<RefCell<Node>>>;
impl Doubly_Linklist {

    fn remove_back(&mut self) {
        if self.tail.is_none() {
            println!("list is emtpy so we can not remove");
        } else {
            self.tail
                .take()
                .map(|old_tail| match old_tail.borrow_mut().prev.take() {
                    Some(new_tail) => {
                        new_tail.borrow_mut().next.take();
                        self.tail = Some(new_tail);
                        self.tail.clone()
                    }
                    None => {
                        self.head.take();
                        println!("List is empty after removal");
                        None
                    }
                });
        }
    }

}
impl Node {
    fn new(element: i32) -> Rc<RefCell<Node>> {
        Rc::new(RefCell::new(Node {
            element: element,
            next: None,
            prev: None,
        }))
    }
}
fn main() {}
```

11.6 Summary

This chapter focused on implementing typical data structures in Rust, starting with detailed implementations of both singly linked lists and doubly linked lists. You learned how these lists are constructed, the key differences between them, and the use cases where each excels. In this chapter, we also introduced you to the concept of reference cycles, a common issue in linked structures that can lead to memory leaks. Through practical examples and explanations, we provided insights into how to avoid such pitfalls when working with references in Rust. By the end of this chapter, we hope you are better equipped to build efficient data structures and manage memory correctly.

Next, we'll discuss useful patterns for handling structs, such as the builder pattern, to make your Rust code more flexible and maintainable.

Chapter 12
Useful Patterns for Handling Structs

Patterns in programming provide a roadmap to efficiency. This chapter explores design patterns that simplify the way we work with data structures.

In this chapter, we'll walk you through various patterns for managing and initializing struct instances. You'll learn about the builder pattern, which simplifies the creation of complex objects by providing a step-by-step construction process. This chapter also covers techniques for simplifying structs, making them more maintainable and easier to use. By mastering these patterns, you'll be able to design robust and flexible data structures in Rust.

12.1 Initializing Struct Instances

We discussed the basics of defining structs in Chapter 5, Section 5.1. In this section, we'll dive more deeply into the nuances of struct initialization. We'll focus on the usage of the `new` constructor, beginning with an example, followed by some default constructors.

12.1.1 New Constructors

In programming, a new instance of a struct is typically created using a constructor. Rust, however, does not provide built-in constructors. The only way to create a new instance of a struct is by specifying the struct's name and manually initializing all of its fields.

For instance, let's add a `Student` struct in the library crate of a new package called `using_patterns`. The definition of the struct is shown in Listing 12.1.

```
#[derive(Debug)]
pub struct Student {
    pub age: u8,
    pub name: String,
}
```

Listing 12.1 Definition of Student Struct inside a Library Crate

In main, only one way to create a new instance of the struct exists: by writing the name of the struct and then initializing all of its fields. Listing 12.2 shows how you can create an instance of the Student struct in main.

```
use using_patterns::Student;
fn main() {
    let std_1 = Student {
        age: 20,
        name: "Joseph".to_string(),
    };
}
```

Listing 12.2 Initializing Student Struct Instance in main

Since the Student struct is defined in the library, we first bring it into scope with use using_patterns::Student; where using_patterns is the crate-level module or root module.

The initialization shown in Listing 12.2 works well for simple cases, but what happens if we add a private field Student struct? The revised definition of the Student struct is shown in Listing 12.3.

```
#[derive(Debug, Default)]
pub struct Student {
    id: u8,
    pub age: u8,
    pub name: String,
}
```

Listing 12.3 Student Definition from Listing 12.1 Revised with Private Field of id

The id field is not prefixed with the pub keyword, which means that it is a private field. The addition of the private field will now raise an error in main during struct instance initialization. The error is "missing structure fields: id." When creating a new instance of a struct, you must initialize all fields. Let's try to initialize all the fields, as shown in Listing 12.4.

```
fn main() {
    let std_1 = Student {
        id: 11, // Error
        age: 20,
        name: "Joseph".to_string(),
    };
}
```

Listing 12.4 Initializing the Struct Instance with a Private Field Gives an Error

Another error arises now, "field id of struct Student is private." We keep the id private because we do not want anyone to set it manually; instead, it should be automatically generated.

To fix this problem, follow the Rust convention of creating an associated function for initializing struct instances (as introduced in Chapter 5, Section 5.1.2). To recap, this associated function is typically called new. This function works like a constructor. Let's add it to the implementation Student, as shown in Listing 12.5.

```
impl Student {
    pub fn new(std_name: String) -> Self {
        Self {
            id: 0,
            age: 20,
            name: std_name,
        }
    }
}
```

Listing 12.5 Definition of the Constructor Function new for Student

The new constructor function must be declared as public. The function takes student name (std_name) as an input and returns a Student instance. The function sets the name field to the name passed in while the remaining fields are some sort of default values. This function works because it has access to the private fields of the struct, since it belongs to the same top-level module (which is crate module or root module in this case) containing the Student struct. For more on modules and the visibility of items inside modules, refer to Chapter 6.

In the main function, instead of manually creating an instance, we can call the new function with the following line of code:

```
let std_1 = Student::new("Joseph".to_string());
```

One advantage of using the new function is that you can check in advance for some precondition, only creating a new instance if the condition holds. For instance, let's say we want to check before creating a new instance that the name passed in contains all characters. Listing 12.6 shows this logic in coded form.

```
impl Student {
    pub fn new(std_name: String) -> Result<Self, String> {
        if std_name.chars().all(|x| matches!(x, 'a'..='z')) {
            Ok(Self {
                id: 0,
                age: 20,
                name: std_name,
            })
```

```
        } else {
            Err("The name is invalid.".to_string())
        }
    }
}
```

Listing 12.6 Revised Definition of the new Function for Checking If name Contains All Characters

The `chars` function on the `std_name` returns an `iterator` over all the characters in the `String`. Next, we use the `all` combinator, which tests if every element of the `iterator` matches a predicate. We are checking that all the values passed in must match on characters from `a` to `z`. The `match` macro checks one thing against the other. In this case, the macro will check each character in the `std_name` against letters from `a` to `z`. If the condition is true, we return a new `Student` by wrapping it around an `Ok` variant. In any other case, we return an `Err`.

12.1.2 Default Constructors

The Rust standard library has a `Default` trait. When implemented on a type, this trait allows that type to provide useful default values via the default function. The default function acts as a *default constructor*, enabling the initialization of instances of the type with predefined values.

To demonstrate, let's implement thus trait for `Student`. The trait has a single method called `default` with no inputs and returns an instance of `Self`. Listing 12.7 shows an implementation of the trait for `Student`.

```
impl Default for Student {
    fn default() -> Self {
        Self {
            id: 0,
            age: 20,
            name: "Unknown".to_string(),
        }
    }
}
```

Listing 12.7 Implementation of the Default Trait for Student

The fields are initialized from some default values. A key advantage of the `Default` implementation is that it can be used anywhere default implementations are required. For example, consider the following line of code:

```
let std_1 = Student::new("Joseph123".to_string());
```

The new function defined earlier in Listing 12.6 returns a Result, which will either return an Ok variant containing a Student or an Err. A useful method is available from the Result enum, called unwrap_or_default. When called on Result, this method will first attempt to unwrap the Result if it contains an Ok variant. However, in case of Err, it will try to attempt calling the default method. Let's call it on the new function in the following way:

```
let std_1 = Student::new("Joseph123".to_string()).unwrap_or_default();
```

Unfortunately, we have an incorrect student name in this case. Therefore, the default method on the Student instance will be called, returning a default instance.

The Default trait can also be automatically implemented via the derive macro (see Chapter 15 for more information on macros). For instance, in the code shown earlier in Listing 12.1, we can derive the trait for the Student struct in the following way:

```
#[derive(Debug, Default)]
pub struct Student {
    ...
}
```


The derive attribute in this code enables the compiler to insert an implementation of Default for the Student struct, which will initialize all the variables of the Student struct from their default values. As a result, we now have two implementations for Default: The first implementation is defined in Listing 12.7, and the second implementation is automatically inserted by the compiler due to the use of the derive attribute. The compiler therefore throws an error "conflicting implementations, of trait Default for type Student." This error arises because the compiler has already inserted the implementation, so we don't need to add it explicitly. To fix this problem, remove the implementation for Default in Listing 12.7. In main, we can use Default trait implementation to create a default Student instance as follows:

```
Let std_2 = Student::default();
```

Note that calling the default function will assign default values to the fields. Strings defaults to an empty string, integers defaults to 0, and the Booleans defaults to false. For custom types, you must define manual default values.

12.2 Builder Pattern

Some data structures are complicated to construct because they require a large number of inputs and optional configuration choices. This flexibility can easily lead to a large number of distinct constructors, each having many arguments. We suggest the *builder pattern* make it manageable.

For instance, consider the Customer struct shown in Listing 12.8.

```
#[derive(Debug, Default, Clone)]
struct Customer {
    name: String,
    username: String,
    membership: Membershiptype,
    gender: char,
    country: String,
    age: u8,
}

#[derive(Debug, Clone)]
enum Membershiptype {
    new,
    causual,
    loyal,
}

impl Default for Membershiptype {
    fn default() -> Self {
        Membershiptype::new
    }
}
```

Listing 12.8 Definition of Customer Struct Simulating a Complicated Data Structure

The Customer struct contains several fields. The membership field is an enum, which is defined below the struct separately. Although this struct is not too complicated, for the sake of illustration, we assume that it represents some type that may become complicated based on future additions. We also have a default implementation for membership type, which returns a new variant.

Let's see how the builder pattern can help in this scenario.

12.2.1 Motivating Example for the Builder Pattern

Let's first examine the motivation for the use of builder pattern. In this case, we'll add a new constructor function to an implementation block of Customer. The constructor will create a basic type of Customer about whom we do not know much; we only have their name information. Listing 12.9 shows the definition of the constructor function.

```
impl Customer {
    fn new(name: String) -> Self {
        Customer {
```

```
            name: name,
            ..Default::default()
        }
    }
}
```

Listing 12.9 Definition of the new Constructor Function for Customer

This function just sets the `name` field while the remaining fields are set to the default values. For the `membership`, the default value will be taken from the `Default` trait implementation for `membership`, shown earlier in Listing 12.8.

The constructor function in Listing 12.9 creates a basic `Customer`. Let's consider a slightly more advanced level of `Customer`, who is not simply a guest `Customer` but is interested in a logon account with us.

To create an instance of such a `Customer`, you'll need another constructor. Rust does not allow overloading; therefore, we'll define another constructor function called `new_2` and pass in the `name` and the `username` to create a new `Customer`. Listing 12.10 shows the definition of the `new_2` constructor function.


```
impl Customer {
    ...
    fn new_2(name: String, username: String) -> Self {
        Customer {
            name: name,
            username: username,
            ..Default::default()
        }
    }
}
```

Listing 12.10 Definition of the new_2 Constructor Function for Customer

No Overloading in Rust

Overloading refers to the ability to define multiple functions, methods, or operators with the same name but using different parameters or types. In Rust, overloading is not directly supported for functions or methods. A function or method with the same name but different parameters within a certain scope is simply not allowed.

Our function now sets the `name` and `username` fields while keeping the remaining fields to their default values.

Next, say we have a customer who, in addition to having an account, also holds membership with us. To create an instance of such a user, define yet another constructor function called new_3. Listing 12.11 shows the definition of the new_3 constructor function.

```
impl Customer {
    ...
    fn new_3(name: String, username: String, membership: Membershiptype) ->
Self {
        Customer {
            name: name,
            username: username,
            membership: membership,
            ..Default::default()
        }
    }
}
```

Listing 12.11 Definition of the new_3 Constructor Function for Customer

We'll set the fields according to the inputs and set the remaining to defaults.

The three constructor functions can now be used in main, as shown in Listing 12.12.

```
fn main() {
    let new_user = Customer::new("AliceNouman".to_string());
    let'ser_with_login = Customer::new_2("Joseph".to_string(), "joe123".to_
string());
    let'ser_with_membership = Customer::new_3(
        "Micheal".to_string(),
        "micheal2000".to_string(),
        Membershiptype::loyal,
    );
}
```

Listing 12.12 Using the Three Constructors in main

This code compiles, but it could be improved. Currently, we have three different constructor functions, each with different names, and the list of arguments for each function keeps growing. If we add more fields to the Customer, we might need even more constructor functions with longer argument lists. To avoid this increase in the number of constructor functions, we recommend using the builder pattern.

12.2.2 Solving the Proliferation of Constructors

Let's see how the builder pattern can prevent the increase in the number of constructors.

The first ingredient of the builder pattern is defining a new struct containing the same fields as that of the struct for which we are implementing the builder pattern. To proceed, introduce a new struct called CustomerBuilder. This struct will have the same fields as the Customer struct, but all fields are optional, with the exception of the mandatory name field. Listing 12.13 shows the definition of the struct.

```
#[derive(Default)]
struct CustomerBuilder {
    name: String,
    username: Option<String>,
    membership: Option<Membershiptype>,
    gender: Option<char>,
    country: Option<String>,
    age: Option<u8>,
}
```

Listing 12.13 Definition of the CustomerBuilder Struct

The next part of the builder pattern is the definition of individual methods for setting the values of the individual fields. You'll have as many methods as there are optional fields. Let's add these fields to the implementation of CustomerBuilder, as shown in Listing 12.14.

```
impl CustomerBuilder {
    fn username(&mut self, username: String) -> &mut Self {
        self.username = Some(username);
        self
    }

    fn membership(&mut self, membership: Membershiptype) -> &mut Self {
        self.membership = Some(membership);
        self
    }

    fn gender(&mut self, gender: char) -> &mut Self {
        self.gender = Some(gender);
        self
    }
    fn country(&mut self, country: String) -> &mut Self {
        self.country = Some(country);
        self
    }
    fn age(&mut self, age: u8) -> &mut Self {
        self.age = Some(age);
```

```
        self
    }
}
```

Listing 12.14 Definition of Individual Methods for Setting the Values of Fields

Perhaps you've noticed a common pattern in all these functions. They are named after the fields of the `Customer` struct, take a mutable reference to `self`, and return a mutable reference to `Self`. The syntax `&mut Self` ensures that subsequent method calls can continue to mutate the same instance without needing to re-borrow it.

Finally, the third ingredient of the builder pattern is a `build` method. This method will finalize the instance and will return a completed `Customer` instance. Listing 12.15 shows the code for the `build` method.

```
impl CustomerBuilder {
    ...
    fn build(&mut self) -> Customer {
        Customer {
            name: self.name.clone(),
            username: self.username.clone().unwrap_or_default(),
            membership: self.membership.clone().unwrap_or_default(),
            gender: self.gender.unwrap_or_default(),
            country: self.country.clone().unwrap_or_default(),
            age: self.age.unwrap_or_default(),
        }
    }
}
```

Listing 12.15 Definition of the build Method

The input to the `build` method is a mutable reference to `self`, and the method returns a `Customer` instance. Inside the method, use a `clone` of the fields of `CustomerBuilder`. Note that `self` in this `impl` block refers to the `CustomerBuilder` instance. Since we don't want to move the values from the `CustomerBuilder`, we'll `clone` them. Lastly, since only the first field is mandatory, the rest of the values may not be provided during calls to the constructor. The `unwrap_or_default` in this case will either attempt to `unwrap` the value if a value is set by the respective method, or in case of `None`, the method will attempt to assign a default value.

In the implementation of the `Customer`, you don't need the individual constructor functions now. Instead, let's add one constructor function, which will take the mandatory `name` value and return a `CustomerBuilder` instance. Listing 12.16 shows the new implementation for the `Customer` containing a single constructor.

```
impl Customer {
    fn new(name: String) -> CustomerBuilder {
        CustomerBuilder {
            name: name,
            username: None,
            membership: None,
            gender: None,
            country: None,
            age: None,
        }
    }
}
```

Listing 12.16 A Single Constructor Function Is Now Required for Customer

In the body of the function, we only set the name field, and the remaining values are set to None. Alternatively, we can derive the Default for the struct and assign default values to all the fields, as shown in Listing 12.17.

```
impl Customer {
    fn new(name: String) -> CustomerBuilder {
        CustomerBuilder {
            name: name,
            ..Default::default()
        }
    }
}
```

Listing 12.17 Alternate Definition of the Constructor Function Using the Default Trait

Let's now use the implementation in main. First, we'll attempt to create a new_user about whom we do not have sufficient information; we only have their name information, as follows:

```
fn main() {
    let new_user = Customer::new("Alice".to_string());
}
```

First, we make a call to the constructor function defined in Listing 12.17. This constructor will return an instance of CutomerBuilder, with the name field set to name and the remaining fields set to default value of None. Next, the build is called on the CustomerBuilder instance. If you look at the definition of the build pattern shown earlier in Listing 12.15, notice how it returns a Customer, with the values being copied from the CustomerBuilder. In particular, the new function changes the type from Customer to CustomerBuilder, while the build changes it back from CustomerBuilder to that of Customer.

Listing 12.18 shows how users with more advanced levels of information can be constructed using the builder pattern.

```
fn main() {
        ...
    let'ser_with_login = Customer::new("Joseph".to_string())
        .username("joe123".to_string())
        .build();

    let'ser_with_membership = Customer::new("Micheal".to_string())
        .username("micheal2000".to_string())
        .membership(Membershiptype::loyal)
        .build();
}
```

Listing 12.18 Creating Users with More Advanced Level of Information

Notice how the methods are nicely chained together, and the instances are created in a clean and clear way, step by step. We also have a nice single interface for constructing new instances.

This pattern is seen more frequently in Rust compared to many other languages because Rust lacks overloading.

12.3 Simplifying Structs

Sometimes, a large, poorly designed struct can lead to borrowing issues. While individual fields might be borrowable independently, borrowing the struct as a whole can block further borrows or block usage of the struct, ultimately preventing other parts of the code from accessing it.

Consider `struct A` defined in Listing 12.19.

```
struct A {
    f1: u32,
    f2: u32,
    f3: u32,
}
```

Listing 12.19 Definition of Struct A Containing Three Fields

Let's define a few functions that will use the struct instance. Listing 12.20 shows the definitions of these functions.

```
fn fn1(a: &mut A) -> &u32 {
    &a.f2
}
fn fn2(a: &mut A) -> u32 {
    a.f1 + a.f3
}

fn fn3(a: &mut A) {
    let x = fn1(a);
    let y = fn2(a);
}
```

Listing 12.20 Some Functions Defined on Struct A

All three functions take a mutable reference to struct A. However, function f1 returns &i32, and fn2 returns u32, while fn3 does not return anything. The code compiles nicely. However, let's modify the definition of the fn3 by adding a print statement at the end of the function, as shown in Listing 12.21.


```
fn fn3(a: &mut A) {
    let x = fn1(a);
    let y = fn2(a);
    println!("{x}"); // Error
}
```

Listing 12.21 Revised Definition of fn3

The compiler throws an error, "cannot borrow *a as mutable more than once at a time."

Let's walk through the code of fn3 step by step to understand what happened. Before we dive in, recall the borrowing rules, discussed in Chapter 4, Section 4.3.2. According to borrowing rules, multiple mutable references to the same data are not permitted at the same time. In function fn3, we first call fn1, which returns a reference to a value that is a field of the struct. The variable x is therefore tied to the struct field. Moreover, since the function is using the struct as a mutable borrow, as long as the variable x is alive or in scope, the Rust compiler assumes that the whole of the struct is being borrowed as mutable. Therefore, no other reference to the struct is allowed.

We can also look at the code from the perspective of lifetime elision rules (see Chapter 10, Section 10.1.4). If you have exactly one input lifetime parameter, that lifetime is assigned to all output lifetime parameters. Look at the signature of fn1, i.e., fn1(a: &mut A) -> &u32, and note that the output will be assigned the lifetime of the input struct a. Thus, the return value from fn1 is captured in variable x inside fn3. As long as x is alive, and the struct instance indicated by variable a will be alive as a mutable reference.

Although we know from the inside code of the function that we are not updating the struct fields, the compiler thinks otherwise. It assumes that, since the function is using the struct as mutable reference, some chance exists that the variable that points to a resource inside the structure may update some of its fields.

In summary, the problem in this case is that Rust does not allow fields to be borrowed independently. As a result, borrowing a field enforces the borrowing of the whole struct. The question now is how can we allow the borrowing of the individual fields independently of each other, when they reside inside the same struct?

Passing Immutable Struct Instances

No issues will arise if we pass the struct immutably to functions. For instance, the code shown in Listing 12.22 will compile.

```
fn fn1(a: &A) -> &u32 {
    &a.f2
}
fn fn2(a: &A) -> u32 {
    a.f1 + a.f3
}
fn fn3(a: &mut A) {
    let x = fn1(a);
    let y = fn2(a);
    println!("{x}");
}
```

Listing 12.22 Using an Immutable Reference to Fix the Error in fn3

The definitions of `fn1` and `fn2` are slightly modified and now accept an immutable reference instead of a mutable reference. The compiler in this case is now sure that the struct instance will not be updated inside these functions, and therefore, it can safely immutably borrow it multiple times inside `fn3`. The `print` statement is therefore not causing any issues inside `fn3`.

Alternatively, we can change the definition of `fn1` (shown earlier in Listing 12.20) so that it returns an owned copy, as in the following example:

```
fn fn1(a: &mut A) -> u32 {
    &a.f2
}
```

The code will still compile. In this case, the return value is an owned copy and not a reference. Since there are no references involved in the return value, the compiler has no issues. The variable x inside `fn3` in is now an owned copy and is not referring to something inside the struct. The reference to a field of a struct, when the struct is borrowed mutably, leads the compiler to believe that the structure may be updated.

Let's refocus on the issue shown earlier in Listing 12.20 and Listing 12.21. The key issue is that Rust doesn't allow the borrowing of individual fields independently of each other.

One solution to this limitation might be to decompose the struct into several smaller structs and then combine them back again into the original struct. Each decomposed struct can be borrowed separately, allowing for more flexible behavior. To decompose the struct, let's now examine how the struct is being used inside the individual functions.

The function `fn1` is using field `f2`, and `fn2` is using the fields `f1` and `f3`. We'll decompose the struct into `B` and `C`, with `B` containing `f2` and `C` containing `f1` and `f3`. Finally, we'll create these new structs as fields of `struct A`. Listing 12.23 shows the updated definition of the `struct A` and its decomposed structs.

```
struct A {
    b: B,
    c: C,
}
struct B {
    f2: u32,
}
struct C {
    f1: u32,
    f3: u32,
}
```

Listing 12.23 Updated Definitions of Struct A Based on Decomposition

This decomposition will now allow us to mutably borrow the structures separately and therefore have more flexible behavior. For instance, in `fn1`, we'll use `struct B` since it only requires the field of `f2`. In the same way, the function `fn2` will use the `struct C`. The revised definitions of the functions are shown in Listing 12.24.

```
fn fn1(a: &mut B) -> &u32 {
    &a.f2
}
fn fn2(a: &mut C) -> u32 {
    a.f1 + a.f3
}
```

Listing 12.24 Revised Definitions of fn1 and fn2 Based on Decomposed Struct A

The signature of `fn3` will remain the same. Inside the function body of `fn3`, instead of passing in the entirety of `struct a` in the calls to `fn1` and `fn2`, we'll pass in the fields `b` and `c`, respectively. Listing 12.25 shows the updated `fn3`.

```
fn fn3(a: &mut A) {
    let x = fn1(&a.b); // Error
    let y = fn2(&a.c);
    println!("{x}");
}
```

Listing 12.25 Revised fn3 Based on the New Definitions of fn1 and fn2

This version throws an error, "expected `&mut B`." The variable `a` is a mutable reference, but the compiler wants us to explicitly set `b` as mutable also. The call to `fn2` will also require a similar change. The updated code for `fn3` is shown in Listing 12.26.

```
fn fn3(a: &mut A) {
    let x = fn1(&mut a.b);
    let y = fn2(&mut a.c);
    println!("{x}");
}
```

Listing 12.26 Finalized Code for fn3, Which Does Compile

The compiler has no issues now. The decomposition of a struct enables us to borrow fields independently of each other. In this example, the borrow checker knows that `a.b` and `a.c` are distinct and can be borrowed independently. As a result, it does not try to borrow all of `struct a`, which was the case previously.

Other Solutions for Borrowing of Individual Fields

Other solutions may include the following:

- **Avoid borrowing the entire struct**
  Instead of borrowing the entire struct, pass only the fields that are needed for each function. This focus allows you to separate mutable borrows and immutable borrows.
- **Use interior mutability**
  Interior mutability allows you to mutate a value even if it's behind an immutable reference. You can use `RefCell` pointers to make these fields mutable even when they are part of an immutable reference.

The decomposition of structs often leads to a better design with smaller and therefore easily manageable units of functionality.

12.4 Practice Exercises

In this section, we'll provide some practice questions to help reinforce your learning. The exercise questions cover almost all the concepts introduced in the chapter. Solutions to each exercise will follow in Section 12.5.

1. **Initialize a struct with public and private fields**
   Add the implementation for the new constructor function so that the following program compiles.

```
#[derive(Debug)]
struct Employee {
    employee_id: u32,
    pub age: u8,
    pub name: String,
}

impl Employee {
   // Add constructor function here
}

fn main() {
    let emp_1 = Employee::new("Alice".to_string());
}
```

2. **Validate struct fields using custom validation logic**
   Consider the following code. Add the body for the new constructor to validate the year field to ensure the car is not older than 1990. In case the car is older, the constructor should return an Err and should not create an instance of Car.

```
#[derive(Default)]
struct Car {
    car_id: u32,
    model_name: String,
    year: u16,
}

impl Car {
    pub fn new(car_model_name: String, car_year: u16) -> Result<Self, String>
{
        // Validation: year must be 1990 or later
    }
}

fn main() {
    let car_1 = Car::new("MODEL_X".to_string(), 1985);
}
```

3. **Creating a flexible employee struct using the builder pattern**
   Consider the following Employee struct definition. Implement the builder pattern to allow for step-by-step construction of an Employee instance where name is a mandatory field and the remaining fields are optional.

```
#[derive(Debug, Default)]
struct Employee {
    name: String,
    position: String,
    salary: f64,
    years_of_service: u32,
}
```

4. **Solving borrowing conflicts by decomposing structs**
   In this question, apply the principles of struct decomposition to resolve borrowing conflicts. You're provided with a single struct that causes borrowing issues. Refactor the struct into smaller structs and adjust the function calls to borrow individual fields mutably without causing conflicts.

```
struct Library {
    books: u32,
    shelves: u32,
    readers: u32,
}
fn manage_books(lib: &mut Library) -> &u32 {
    &lib.books
}
fn calculate_shelf_space(lib: &mut Library) -> u32 {
    lib.shelves * 5
}
fn manage_library(lib: &mut Library) {
    let book_count = manage_books(lib);
    let shelf_space = calculate_shelf_space(lib);
    println!("{book_count}");
}
```

12.5 Solutions

This section provides the code solutions for the practice exercises in Section 12.4. The code is largely self-explanatory; however, we have included comments and additional explanations wherever necessary to enhance understanding.

1. **Initialize a struct with public and private fields**

```
#[derive(Debug)]
struct Employee {
    employee_id: u32,
    pub age: u8,
```

```
    pub name: String,
}
impl Employee {
    pub fn new(emp_name: String) -> Self {
        Self {
            employee_id: 0,
            age: 30,
            name: emp_name,
        }
    }
}
fn main() {
    let emp_1 = Employee::new("Alice".to_string());
}
```

2. **Validate struct fields using custom validation logic**

```
#[derive(Default)]
struct Car {
    car_id: u32,
    model_name: String,
    year: u16,
}

impl Car {
    pub fn new(car_model_name: String, car_year: u16) -> Result<Self, String>
{
        if car_year < 1990 {
            return Err("The car year must not be older than 1990.".to_
string());
        }

        Ok(Self {
            car_id: 12345, // Auto-assigned ID
            model_name: car_model_name,
            year: car_year,
        })
    }
}

fn main() {
    let car_1 = Car::new("MODEL_X".to_string(), 1985);
}
```

3. **Creating a flexible employee struct using the builder pattern**

```
#[derive(Debug, Default)]
struct Employee {
    name: String,
    position: String,
    salary: f64,
    years_of_service: u32,
}

struct EmployeeBuilder {
    name: String,
    position: Option<String>,
    salary: Option<f64>,
    years_of_service: Option<u32>,
}

impl EmployeeBuilder {
    fn new(name: String) -> Self {
        EmployeeBuilder {
            name,
            position: None,
            salary: None,
            years_of_service: None,
        }
    }

    fn position(&mut self, position: String) -> &mut Self {
        self.position = Some(position);
        self
    }

    fn salary(&mut self, salary: f64) -> &mut Self {
        self.salary = Some(salary);
        self
    }

    fn years_of_service(&mut self, years: u32) -> &mut Self {
        self.years_of_service = Some(years);
        self
    }

    fn build(&self) -> Employee {
        Employee {
            name: self.name.clone(),
```

```
            position: self.position.clone().unwrap_or_else(|| "Unknown".to_
string()),
            salary: self.salary.unwrap_or(0.0),
            years_of_service: self.years_of_service.unwrap_or(0),
        }
    }
}

fn main() {
    let new_employee = EmployeeBuilder::new("John".to_string())
        .position("Manager".to_string())
        .salary(50000.0)
        .years_of_service(5)
        .build();

    println!("{:?}", new_employee);

    let entry_level_employee = EmployeeBuilder::new("Alice".to_string())
        .build();

    println!("{:?}", entry_level_employee);
}
```

4. **Solving borrowing conflicts by decomposing structs**

```
struct Books {
    books: u32,
}

struct Shelves {
    shelves: u32,
}

struct Library {
    books: Books,
    shelves: Shelves,
}

fn manage_books(b: &mut Books) -> &u32 {
    &b.books
}

fn calculate_shelf_space(s: &mut Shelves) -> u32 {
    s.shelves * 5
}
```

```
fn manage_library(lib: &mut Library) {
    let book_count = manage_books(&mut lib.books);
    let shelf_space = calculate_shelf_space(&mut lib.shelves);
    println!("{book_count}");
}

fn main() {
    let mut lib = Library {
        books: Books { books: 150 },
        shelves: Shelves { shelves: 20 },
    };

    manage_library(&mut lib);
}
```

12.6 Summary

This chapter tackled useful patterns for handling structs, offering practical ways to streamline struct management in Rust. We began with approaches for initializing struct instances, ensuring clarity and simplicity in object creation. Next, you learned all about the builder pattern, a powerful technique for constructing complex objects step by step, allowing for greater flexibility and maintainability. This chapter also emphasized simplifying structs, focusing on reducing complexity to make them easier to use and modify. Mastering these patterns will give you the ability to design robust, adaptable data structures that scale well with complexity.

In the next chapter, we'll dive into Rust's handling of size, covering sized types versus unsized types and the impact of zero-sized types on memory and performance.

PART III

Advanced Language Concepts

Chapter 13
Understanding Size in Rust

Understanding how size affects your code can lead to better design choices. In this chapter, we'll unravel the complexities of sized and unsized types in Rust.

13

In the programming context, *size* typically refers to the amount of memory that a data type occupies. This can include the space required for variables, data structures, or any allocated resources and is often measured in bytes. Understanding the size of types is crucial for efficient memory management and performance optimization in programs. This chapter delves into the nuances of size in Rust, starting with the distinction between sized and unsized types. We explain how to work with references to unsized types and the implications for memory management. You'll also learn about optionally sized traits and generic parameters, as well as unsized coercion. This chapter also covers zero-sized types, including the `never` type, `unit` type, `unit` structs, and `PhantomData`, providing a deep understanding of Rust's type system and its impact on performance and memory usage.

13.1 Sized and Unsized Types

Understanding size-related issues in Rust is integral to writing safe, performant, and efficient code. With this knowledge, you can make informed decisions, design better data structures, and ensure memory safety throughout your programs.

From the size perspective, Rust categorizes types into two groups:

- **Sized types**
  A type is considered *sized* if its size in bytes can be determined at compile time.
- **Unsized types**
  A type is *unsized* if its size cannot be determined at compile time, also referred to as *dynamically sized* types.

We'll explore examples of sized and unsized types in the following sections.

13.1.1 Examples of Sized Types

Let's start with some examples of sized types. All primitive types are sized. Use the `size_of` function defined in the memory module of Rust's standard library to display the size in bytes for the specified type. For instance, the code shown in Listing 13.1 will display the size of an integer.

```
use std::mem::size_of;
fn main() {
    println!("i32 size is: {}", size_of::<i32>());
}
```

Listing 13.1 Displaying the Size of i32

The `print` statement will display 4 bytes. In this case, the size is known at compile time. Tuples constructed from primitives are also sized. Consider the following statement:

```
println!("(i32,i32) size is: {}", size_of::<(i32, i32)>());
```

This statement will `print` a size of 8 bytes.

Arrays are also sized. For instance, consider the following statement:

```
println!("[i32: 3] size is: {}", size_of::<[i32; 3]>());
```

This statement prints the size of an `i32` array containing three elements. In this case, the size of the array size is `12` bytes.

> **Note**
>
> Generally speaking, all types that are constructed from primitive types, including struct, enums, hash maps, and arrays, are also sized.

Let's go over an example of a struct. Consider a struct with a few fields defined, as shown in Listing 13.2.

```
struct Point {
    x: bool,
    y: i64,
}
```

Listing 13.2 Definition of a Point Struct

The size of a struct is determined by the sizes of its individual fields, plus any padding added by the compiler for alignment purposes. In this case, field x is a `bool`, which takes 1 byte, plus field y, which takes 8 bytes, resulting in a total size of 9 bytes. The compiler may add additional bytes for alignment purposes. The following statement displays the size of the struct:

```
println!("Struct size is: {}", size_of::<Point>());
```

This statement will display a size of 16 bytes, which means that 7 bytes are being used as padding for alignment.

Pointers or references are also of fixed size regardless of whether they're mutable or immutable and regardless of the value they are pointing to. The following `print` lines will print the size of an immutable and a mutable reference:

```
println!("i32 reference is: {}", size_of::<&i32>());
println!("i32 mutable reference is: {}", size_of::<&mut i32>());
```

The sizes are `8` bytes each. Pointers or references are one machine word in size. A *machine word* refers to the unit of data that a particular computer architecture or processor can handle in one operation. The size of a machine word varies depending on the computer architecture. Common machine word sizes are `32` bits (or `4` bytes), and `64` bits (or `8` bytes). To see how many bytes a word is on your machine, `print` a reference to a `unit` type, for example, in the following way:


```
println!("Machine word size is: {}", size_of::<&()>());
```

Smart pointers such as `Box`, `Rc`, or `RefCell` and function pointers are also sized. Use the following statements to see their respective sizes:

```
println!("Box<i32> is: {}", size_of::<Box<i32>>());
println!("fn(i32) -> i32 is: {}", size_of::<fn(i32) -> i32>());
```

These statements will display the size of the box and the functional pointers that are equal to the size of pointer, which is typically 8 bytes on most machines.

13.1.2 Examples of Unsized Types

Let's go over some unsized types now. Slices such as arrays and string slices are unsized. Let's try printing the size of an `i32` array slice:

```
println!("[i32] size is: {}", size_of::<[i32]>()); // Error
```

The compiler is not happy and throws an error, "the size for values of type `[i32]`, cannot be known at compilation time." A *slice* is just a reference to a contiguous sequence of elements in a collection; we don't know its size. We just know that they are next to each other, and we have a pointer to it. However, the Rust compiler needs to know the sizes of variables at compile time. Therefore, the compiler is not happy. To solve this issue, you can use a pointer to a raw array slice, which has a fixed size:

```
println!("[i32] size is: {}", size_of::<&[i32]>());
```

Array Slice versus Raw Array Slice

Let's clarify terminology about slices before proceeding further. In literature and the community in general, the type [i32] is not typically called an "array slice"; usually what is meant is type &[i32], which is a reference to a slice. To differentiate between the two, use some the name *array slice* to refer to &[i32] and *raw array slice* for referring to [i32].

Now, when we try to declare a variable of type raw array slice, the compiler will have similar issues of unknown sizes at compilation time. For instance, consider the following code:

```
Let a: [i32];  // Error
```

The number of elements that may be part of the array cannot be determined right now. We only know that it will be some contiguous i32 values in memory, but we do not know how many of them will exist. If we mention the size information, as follows,

```
Let a: [i32; 3];
```

then the error disappears because it is now not a raw slice but instead an i32 array with fixed size. To properly use a raw slice, we must always use a reference to it.

Similar to the raw array slices, there are also raw string slices whose size is also not known at compile time. Let's try to display the size of a raw string slice in a print statement:

```
println!("str size is: {}", size_of::<str>()); // Error
```

As expected, you'll see similar error message. By using a reference at the start of the raw string slice (i.e., &str), the error will go away, as in the following example:

```
println!("str size is: {}", size_of::<&str>());
```

Finally, trait objects also do not have a size known at compile time. For instance, consider the code shown in Listing 13.3.

```
trait Some_trait {}
fn main() {
        println!(
        "The size of the trait object is: {}",
        size_of::<dyn Some_trait>()
    ); // Error
}
```

Listing 13.3 Displaying the Size of a Trait

The compiler again throws an unknown size error. This error makes sense because trait objects can be implemented by any number of structs or enums and thus can also be of any size at runtime. Therefore, the size cannot be determined at compile time. As in the previous cases, fix this problem by adding a reference in the print statement before dyn Some_trait, as in the following example:

```
println!("The size of the trait object is: {}",size_of::<&dyn Some_trait>());
```

All types which are constructed from the unsized types are also unsized.

13.2 References to Unsized Types

In this section, we'll consider the size of a reference to an array and an array slice. Consider the code shown in Listing 13.4.

```
fn main() {
    println!(
        "Size of a reference to sized type: {}",
        size_of::<&[i32; 3]>()
    );
    println!(
        "Size of a reference to unsized type: {}",
        size_of::<&[i32]>()
    );
}
```

Listing 13.4 Displaying the Size of a Reference to an Array and Reference to an Array Slice

In the previous section, you learned that an array is a sized type, while an array slice is unsized. We also discussed how a simple reference is one machine word in size, which is typically 8 bytes on most machines. If you execute this code, you may notice the following output:

```
Size of a reference to sized type: 8
Size of a reference to unsized type: 16
```

The size of a reference to &[i32; 3] (which is a sized) is 8 bytes, while the size of a reference to &[i32] (which is unsized) is 16 bytes. Simple references to sized types are one machine word in size. However, references to unsized types such as raw slices or trait objects are two machine words in size. Let's explore the reason.

Size versus Data Length Information

Size refers to the total amount of memory that a type occupies in memory. This value tells you how much space a particular variable or type requires. For example, if you have an array with three integers, the size of the array is the total memory used by those three integers combined.

Data length information, on the other hand, tells you how many elements are in a collection, like an array or a vector. This value is the count of items in the collection, such as how many numbers are stored in an array or how many elements a slice contains. For instance, in the expression `let num_1: &[i32; 3] = &[10, 12, 30];`, the data length information refers to the fixed size `3`, which indicates that the array `num_1` contains exactly three `i32` elements.

Consider the following lines of code:

```
let num_1: &[i32; 3]  = &[10, 12, 30];
let num_2: &[i32]  = &[10, 12, 30];
```

In the first line, we have an array containing three elements, and in the next line, we have an array slice also containing three elements.

In the first case, however, the data length information is embedded in the type information itself and is therefore not required to be stored again. However, in the second case, the data length information is not attached to the type itself, and therefore the pointer or reference must be added with some additional information. That added information is stored in the next 8 bytes or one machine word.

References that are 2 words are also referred to as *fat pointers*. On the other hand, references that are one word are referred to as *thin pointers*. To confirm that the size information is really stored inside the pointer, we'll iterate through the two types, as shown in Listing 13.5.

```
fn main() {
    ...
    for num in num_1 {
        sum += num;
    }

    for num in num_2 {
        sum += num;
    }
}
```

Listing 13.5 Iterating through the Array and Array Slice

In the first `for` loop, the length information is already embedded in the type information, so Rust knows it's exactly 3 elements to iterate over. In the second case, one would expect that the Rust compiler should complain since the size is not known. However, the code works because the Rust compiler is still able to know the number of times it will iterate by inspecting the pointer, which contains this information in the form of data length information. The compiler therefore has no issues, and the code of both loops should compile.

References to trait objects are also two machine words in size. Let's look at the example shown in Listing 13.6.

```
#[derive(Debug)]
struct Circle;

#[derive(Debug)]
struct Rectangle;

impl Shape for Circle {
    fn print(&self) {
        println!("{:?}", self);
    }
}

impl Shape for Rectangle {
    fn print(&self) {
        println!("{:?}", self);
    }
}
fn main() {
    println!("Size of &Cricle is: {}", size_of::<&Circle>());
    println!("Size of &Rectangle is: {}", size_of::<&Rectangle>());
    println!("Size of &dyn Shape: {}", size_of::<&dyn Shape>());
}
```

Listing 13.6 Definition of a Shape Trait and Its Implementation for Circle and Rectangle

In this case, a `Shape` trait has a `print` function, and then we have its implementation for the `Circle` and `Rectangle`. In `main`, we display the sizes of references to the two structs and a reference to the `trait` object. (For more on trait objects and dynamic dispatch, see Chapter 8, Section 8.2.6.) If you execute this code, you'll see the following output:

```
Size of &Cricle is: 8
Size of &Rectangle is: 8
Size of &dyn Shape: 16
```

The size of references to `Circle` and `Rectangle` are one word each while the size of the reference to the `trait` object is `2` words or `16` bytes. The extra `8` bytes, in this case, will be used to point to the *virtual table (vtable)*, which is a table of function pointers associated with a trait. When a trait object is created, Rust generates a vtable that contains the pointers to the actual implementations of the trait methods for the specific types.

13.3 Sized and Optionally Sized Traits

In this section, we'll explore the Sized trait, which is automatically implemented at compile time for types of a known size. We'll also look at the optional Sized trait, which is useful for handling types that may or may not have a fixed size.

The first point to note about the Sized trait is that it serves as both an auto trait and a marker trait. Let's quickly recap auto traits and marker traits, which we introduced in Chapter 8, Section 8.2.7 and Section 8.2.8, respectively.

Auto traits are traits that are automatically implemented for a type if certain conditions are met. *Marker traits*, on the other hand, indicate that a type has a specific property. All auto traits are marker traits, which means that we can mention them with a type and the compiler will provide automatic default implementations for them. For instance, for an i32 type, some of the auto traits are Copy, Clone, Default, Send, and Sync. You can call the methods of these traits. For instance, to directly call the default method from the Default trait on an i32 type to set its value to a default value, use the following code:

```
fn main() {
    let x: i32 = Default::default();
}
```

Derived versus Auto Traits

In Rust, derived and auto traits are both used to automatically implement certain behaviors for types, but they work in different ways.

Derive is a mechanism that allows Rust to automatically implement common traits for your types. By using the #[derive(...)] attribute, you can tell the compiler to generate implementations for traits like Debug, Clone, or Default.

Auto traits, on the other hand, are traits that the Rust compiler automatically applies to certain types based on specific conditions.

Let's return to our discussion of Sized traits. The auto implementation of the Sized trait occurs if all members of a type are also Sized. The definition of members varies depending on the containing type, such as fields in a struct, variants in an enum, elements in an array, or items in a tuple. Once a type receives the auto implementation of Sized trait, its size in bytes is known at compile time.

In the following sections, we'll delve more deeply into the Sized trait and its role in Rust's type system. You'll learn how to opt out of the Sized trait for certain types, explore the use of generics with explicit Sized trait bounds, and discover how to design flexible generic functions to handle both sized and unsized types effectively.

13.3.1 Opting Out of Sized Trait

There's something special about the Sized trait: It's impossible to opt out of this trait's implementation, unlike some other auto marker traits. To explain this limitation, we'll use a *negative implementation* crate. Add the following line to the *Cargo.toml* file under the dependencies section:

```
negative-impl = "0.1.4"
```

This crate enables us to impose restrictions so that a specific trait cannot be implemented for a particular type. In other words, you can opt out of the implementation of a specific auto trait for a type. For instance, consider the code shown in Listing 13.7.

```
use negative_impl::negative_impl;
struct ABC;
#[negative_impl]
impl !Send for ABC {}
#[negative_impl]
impl !Sync for ABC {}
```

Listing 13.7 Opting Out of the Auto Traits Implementation of Send and Syn for ABC

In this case, we declared the struct ABC. Then, using the negative_impl crate to derive an attribute macro with the same name of negative_impl, we opted out of the auto traits Send and Syn for ABC. An *attribute macro* in Rust is a procedural macro that operates on attributes applied to items like functions, structs, or modules. With this macro, you can generate or modify code based on the annotated item's metadata, thus enabling custom behaviors and simplifying repetitive patterns in your code.

Send and Sync Traits

The Send trait indicates whether a type can be safely transferred across threads, while the Sync trait signifies whether references of a type can be safely shared between threads. We'll cover threads in Chapter 14.

Now the special thing about the Sized trait is that you cannot opt out of it. For instance, adding the following line to the code shown in Listing 13.7 will throw an error:

```
#[negative_impl]
impl !Sized for ABC {} // Error
```

This error arises because it is illogical to imagine a situation where you actually want the compiler to lose awareness of a type's size and regard it as unsized. Treating a sized type like an unsized type yields no advantages and only complicates interactions with the type.

On the other hand, the negative implementations for the Send and Sync appear to be logical. For instance, in some specific scenarios, you may prefer to not allow our type to be transmitted or exchanged between threads.

13.3.2 Generic Bound of Sized Traits

In this section, we'll explore the ?Sized syntax (also referred to as *optionally sized traits*), which allows for flexibility when working with generics by accommodating both sized and unsized types. We'll also dive deeply into how ?Sized interacts with unsized structs, and you'll learn how to design and handle such structures effectively in Rust.

?Sized Syntax

By default, the Sized trait is automatically applied as a bound to every generic type parameter. For instance, let's consider the following function signature containing a generic:

```
fn some_fn<T>(t: T) {},
```

This function signature will be desugared (i.e., expanded) into the following syntax:

```
fn some_fn<T: Sized>(t: T) {}
```

As a result, by default, the parameter T is expected to have a known size at compile time, ensuring memory allocation predictability and simplifying memory management.

In Rust, the ?Sized syntax indicates that a type parameter might be either sized or unsized. Let's add this syntax as a bound in the function signature of some_fn:

```
fn some_fn<T: ?Sized>(t: T) {} // Error
```

We'll handle the error shortly, but first, note that this approach is often used in trait bounds to allow the implementation of a trait for both sized and unsized types. Its advantage is flexibility since this approach allows for generic code that can work with various types regardless of whether or not they have a known size at compile time.

Let's take care of the error now, which is similar to the errors we've seen in earlier sections, "the size for values of type T, cannot be known at compilation time. It doesn't have a size, known at compile-time." Since there is a possibility that type T may be unsized, Rust's rule of having known sizes for variables at compile time is being violated. To compile the code, we'll pass a reference to the type T instead of the type itself:

```
fn some_fn<T: ?Sized>(t: &T) {}
```

This change resolves the error because now, regardless of T being a sized type or an unsized type, it's behind a reference of a fixed size.

The ?Sized bound is also referred to as a *widening*, *expanded*, or *relaxed bound* because it loosens the limitations on the type parameter rather than restricting it. What makes the optional size bound unique is that, among Rust's bounds, it is the only example of a relaxed constraint.

?Sized with Unsized Struct

An unsized struct is a struct that contains an unsized field. An unsized field, as the name suggests, will have an unsized type such as a raw slice or a trait object. Listing 13.8 shows the definition of an unsized struct.

```
struct UnSizedStruct {
    sized_field: i32,
    unsized_field: [i32],
}
```

Listing 13.8 Definition of an Unsized Struct

Rust enforces two requirements for an unsized struct:

- First, the struct must have exactly one unsized field.
- Second, the unsized field must be the last field of the struct.

The struct shown in Listing 13.8 satisfies these two requirements. It contains exactly one unsized field, which is the last field of the struct. If we try to add one more field to the struct, the compiler will generate an error, as shown in Listing 13.9.

```
struct UnSizedStruct {
    sized_field_1: i32,
    unsized_field: [i32], // Error
    sized_field_2: i32,
}
```

Listing 13.9 Revised Definition of the Unsized Struct by Including Another Sized Field

The error exactly indicates the problem, in this case, "only the last field of a struct may have a dynamically sized type (or unsized type) change the field's type to have a statically known size." In this example, the second requirement is being violated.

If we change the last field type shown in Listing 13.9 from sized to unsized, we'll get another error, as shown in Listing 13.10.

```
struct UnSizedStruct {
    sized_field_1: i32,
    unsized_field: [i32],
    unsized_field_2: [i32], // Error
}
```

Listing 13.10 Changing the Last Field from Sized to Unsized

In Listing 13.10, the error arises due to the violation of the first requirement for unsized structs. According to the first requirement, we must have exactly one unsized field.

The requirements for an unsized struct are related to Rust memory management and safety features. Rust must know the sizes of all the fields in a struct at compile time to correctly calculate memory offsets and create the necessary allocations required to store an instance of a struct. This behavior prevents scenarios where it might be unclear how much memory should be allocated for an instance of the struct. As shown in Listing 13.10, `unsized_field` is unsized because its size isn't known at compile time. When another unsized field is added after an existing unsized field in Listing 13.10, Rust simply cannot calculate memory offsets or manage memory correctly, which can lead to potential safety issues.

Let's refocus on the correct implementation of the unsized struct shown earlier in Listing 13.8. Let's try to make an instance of the `UnSizedStruct` in `main`, as shown in Listing 13.11.

```
fn main() {
    let x = UnSizedStruct { // Error
        sized_field_1: 3,
        unsized_field: [3],
    };
}
```

Listing 13.11 Creating an Instance of the UnSizedStruct

Surprisingly, the Rust compiler does not allow this change, throwing an error of "the size for values of type array slice containing `i32` values, cannot be known at compilation time." This problem is a bit irritating, as Rust allows the struct to be defined, but you cannot create an instance of it. The reason for this error is that every type has the `Sized` trait since `Sized` is an auto trait, from which you cannot opt out. However, the struct is not sized, so the created instance of the struct will also be unsized.

The optionally sized trait (`?Sized`) can be used to fix this problem. Recall from the previous section that optionally sized traits can be used as a bound on generics. Let's redefine the struct to use a generic field that is an optionally sized trait bound, as shown in Listing 13.12.

```
struct UnSizedStruct<T: ?Sized>{
    sized_field_1: i32,
    unsized_field: T,
}
```

Listing 13.12 Revised Definition of the UnSizedStruct to Fix the Error in main

The revised definition fixes the error in `main`. The generic `T` can now be any type that is optionally sized. The reason this approach works is because the `?Sized` trait made the field `unsized_field` optionally sized, thereby relaxing Rust's size constraint.

13.3.3 Flexible Generic Function with Optionally Sized Trait

In this section, we'll focus on creating flexible generic functions that can handle both sized and unsized types using the *optionally sized* trait. You'll learn how to leverage Rust's type system to write more versatile and efficient code that can adapt to different kinds of data.

Consider a `print_fn` function that takes in a generic parameter `T`, as shown in Listing 13.13.

```
fn print_fn<T: Debug>(t: T) {
    println!("{:?}", t);
}
```

Listing 13.13 Definition of a Function print_fn

As pointed out earlier, the function signature will be desugared to something like the following:

```
fn print_fn<T: Debug + Sized>(t: T)
```

This desugaring occurs because the `generic` parameters are auto bound by the `Sized` trait. Let's use this function in `main`, as shown in Listing 13.14.

```
fn main() {
    let x = "my name";
    print_fn(x);
}
```

Listing 13.14 Calling the Function print_fn in main

The variable `t` in this case will be treated as a string slice, with two traits (`Debug` and `Sized`). Everything is fine up to this point. However, the function assumes ownership of any values given to it (indicated by the parameter type `T` without a reference), which can be somewhat inconvenient if we pass on non-copy types (i.e., heap-allocated types). Let's modify the function to only accept references, as shown in Listing 13.15.

```
fn print_fn<T: Debug + Sized>(t: &T) {
    println!("{:?}", t);
}
```

Listing 13.15 Revised Definition of print_fn Gives an Error in main

Notice a change in the input parameter type from T to &T. Surprisingly, the compiler is still not happy and throws an error in main on the line in which the call to the print_fn is made, as shown in Listing 13.16.

```
fn main() {
    let x = "my name";
    print_fn(x); // Error
}
```

Listing 13.16 Revised print_fn Definition Leads to an Error in main

This error is the same as we've been seeing throughout this chapter, "the size for values of type str cannot be known at compilation time." At compile time, Rust engages in pattern matching when resolving the actual concrete types that need to be substituted for T. Table 13.1 shows how T is resolved.

Parameter Type	T	&T	&T
Function Call Input	&str	&str	&&str
Resolves To	T = &str	T = str	T = &str

Table 13.1 Generic Resolution Table Based on the Input to the Function print_fn

When the parameter is defined as a simple T without any &, as shown in Listing 13.13, and the function is called with a &str, as shown in Listing 13.14, then T resolves to &str. However, when we change the type from T in the function parameter to that of &T (as shown in the revised print_fn definition in Listing 13.15), the compiler resolves T to simple str due to pattern matching. In this case, the & used in the &T and the & of the String slice will match; therefore, T will be resolved to a raw string slice. This resolution basically causes the error in main shown in Listing 13.16.

We know that the size of the raw string slice (str) is not known at compile time; therefore, it cannot have the Size bound. The error will go away if we add one more reference to the string slice. To fix the error in main, call print_fn with a reference to the string slice:

```
print_fn(&x);
```

In this case, we are passing in the type &&str to the function. The first & will match with the reference & to T, and therefore, T will be resolved to a &str. The compiler has no issues with string slices since they are located behind a reference.

Working with references in this manner is rather inconvenient, and always remembering the correct conversion is hard to manage. The optionally sized trait can really help in such situations. Listing 13.17 shows the revised definition of `print_fn` with an optionally sized trait as a bound on `T`.

```
fn print_fn<T: Debug + ?Sized>(t: &T) {
    println!("{:?}", t);
}
```

Listing 13.17 Revised Definition of the print_fn with Optionally Sized Trait

Now, we can pass in a string slice or a reference to string slice without having to worry how the `T` will be actually resolved. The code shown in Listing 13.18 therefore compiles.

```
fn main() {
    let x = "my name";
    print_fn(x);
    print_fn(&x);
}
```

13

Listing 13.18 Function Now Works with Both Sized and Unsized Types

13.4 Unsized Coercion

Rust provides a feature called *unsized coercion*, which allows a sized type to be transformed into an unsized type. This mechanism is somewhat similar to deref coercion, which we covered in Chapter 10, Section 10.3. Let's first revisit deref coercion, and then we'll dive into how unsized coercion differs.

13.4.1 Deref Coercion

Deref coercion enables the automatic conversion of a reference from one type to another, particularly when interacting with methods or functions expecting a certain type. Consider the following function that accepts a string slice as an input:

```
fn str_slice_fn(s: &str) {}
```

You can call this function in `main` with inputs that can be dereferenced into a `String` slice, as shown in Listing 13.19.

```
fn main() {
    let some_string = String::from("String");
    str_slice_fn(&some_string);
}
```

Listing 13.19 Deref Coercion with String Slices

Although the function `str_slice_fn` accepts a `&str`, we could pass a reference to an owned `String` because `String` can be coerced into `&str` through deref coercion. In general, deref coercion allows the function to accept any type that can ultimately be dereferenced into a `String` slice, including custom-defined types.

13.4.2 Unsized Coercion

Now, let's look at how unsized coercion differs from deref coercion. Consider the following function, which accepts an array slice:

```
fn array_slice_fn<T>(s: &[T]) {}
```

This function accepts any input that can be coerced into an array slice. Let's define some variables in `main` and call the function, as shown in Listing 13.20.

```
fn main() {
    let slice: &[i32] = &[1];
    let vec = vec![1];
    let array = [1, 2, 3];
    array_slice_fn(slice);
    array_slice_fn(&vec);  // deref coercion
    array_slice_fn(&array); // Unsized coercion
}
```

Listing 13.20 Deref and Unsized Coercion with Array Slices

In the first call to the function, we provided the anticipated input of array `slice`. In the second call to the function, we provided a reference to a vector (&vec). Since vectors can be coerced into array slices, no problem arose, and a `deref` coercion took place. In the last case, however, we passed a reference to an `array` `&[i32; 3]`, but the function accepts an `array` slice `&[i32]`. In this case, the array that has a known size has been coerced into an array slice, which does not have a known size. So, an unsized coercion took place.

The key difference between deref coercion and unsized coercion is that, in deref coercion, the type changes, while in unsized coercion, the type does not change. Instead, the property of the type changes from sized to unsized, which impacts the way the compiler handles the type. For example, a fixed-size array could be coerced into a slice, which doesn't have a fixed size. The impact of this change is that the code becomes more flexible, allowing operations on types for which sizes are unknown at compile time.

13.4.3 Unsized Coercion with Traits

Consider a trait with a single method and its implementation for a raw array slice, as shown in Listing 13.21.

```
trait Some_Trait {
    fn method(&self);
}
impl<T> Some_Trait for [T] {
    fn method(&self) {}
}
```

Listing 13.21 Definition of Some_Trait and Its Implementation for [T]

The trait implementation for the type `[T]` allows for calling it on different types due to deref coercion and unsized coercion. More specifically, it can be called for an array slice `&[T]`, any type that can be coerced into the array slice such as vectors, and also for simple arrays with size information. In summary, the following three types can be used to call the method:

- any `&[T]`
- `Vec<T>`
- `[T; N]`

Let's call the method in `main` using the three types, as shown in Listing 13.22.

```
fn main() {
    let slice: &[i32] = &[1];
    let vec = vec![1];
    let array = [1, 2, 3];

    slice.method();
    vec.method();  // deref coercion
    array.method(); // Unsized coercion
}
```

Listing 13.22 Calling the Method Using Three Different Types

In the first call of `slice.method()`, the method is expecting a reference to array slice. We've provided one, so nothing special happens. In the second call of `vec.method()`, a `deref coercion` will take place since the vector can be dereferenced into an array slice. In the last call of `array.method`, an `Unsized coercion` will take place. In this case, before the call to the method, the size information was embedded into the type itself, but inside the method, this information will be lost, which leads to a change from sized to unsized type.

You may have noticed that the way we use the functions and methods is now more flexible, enhances the usability of the code, and avoids unnecessary duplication of code.

13

13.5 Zero-Sized Types

Zero-sized types play an essential role in Rust's type system because they developers to define types that occupy no memory at runtime. These types often serve as markers or are used to indicate certain behaviors without requiring additional resources. In this section, we'll explore various forms of zero-sized types, including the `never` type, `unit` type, `unit` struct, and `PhantomData`.

13.5.1 Never Type

The `never` type represents computations that never resolve to any value. In other words, these computations will always panic or will always exit the program.

> **Release Notes**
>
> This type is only available in the nightly version of Rust as of the time of writing (spring 2025). According to the documentation, this type was supposed to be stabilized in version 1.41, but some last-minute regressions were detected, and the stabilization was temporarily reverted. It may be available in future releases. For now, you must switch to the nightly version. However, you must be sure you have the nightly version installed.
>
> Use the following command to switch to the nightly version of Rust:
>
> ```
> C:\> rustup override set nightly
> ```

Let's see how we can work with the `never` type and then walk through a few use cases.

Using the Never Type

We first must indicate to the compiler that we want to use the `never` type. Include the following line at the start of the code file:

```
#![feature(never_type)]
```

The first use case of the `never` type is when we want to signify that a function will never return normally. Consider the following code:

```
fn unrecoverable_state() -> ! {
    panic!("This function will never return normally with something valid");
}
```

The `never` type is indicated by the exclamation point (`!`). Calling this function in `main` will not generate any compile-time errors:

```
fn main() {
    unrecoverable_state();
}
```

However, when you execute this code, the compiler will panic. Functions returning the never type are also known as *diverging functions*. These functions can only panic, exit the program, or result in an infinite loop. In summary, use the never type when a function is guaranteed to never return normally.

You cannot create variables initialized from a never type, as shown in the following example:

```
let x = !; // Error
```

The never type means no value, and no value can ever not be assigned to something that needs a value. The compiler throws an error of "expected expression" in this case. The assignment of a function returning a never type is currently not a compile time error. For instance, consider the following line:

```
let x = unrecoverable_state();
```

This line will compile with no errors. You can create variables of the never type in the following way:

```
let x: !;
```

13

Never Type with Match

The never type can be used in match arms that are guaranteed to be unreachable. For instance, consider the code shown in Listing 13.23.

```
fn main(){
    let x = match "123".parse::<i32>() {
        Ok(num) => num,
        Err(_) => panic!(),
    };
}
```

Listing 13.23 never Type in the match Arm

The match in this case is matching on the Result of parsing a String into an i32. The parse function returns a Result. Since the parse method will always be successful resulting in a valid value, the arm corresponding to Err is never reached. Macros such as panic!, unimplemented!, and unreachable! all resolved to never type. These macros are used to indicate situations where the program will either panic, has not yet been implemented, or will never reach certain code, respectively. (We'll cover these macros in more detail in Chapter 15.)

You can confirm that panic! returns a never type by assigning panic to a variable in the following way:

```
let x  = panic();
```

Then, you can look at the type of x in the editor.

We know that returning values from the match arms should be of the same type. However, in Listing 13.23, note that the second arm returns a never type that is different from the returning value from the first arm. The compiler, however, has no issues because the never type can be coerced to any other type. In the case of the code shown in Listing 13.23, the type was coerced implicitly to an i32 type because the compiler was expecting an i32. Although the compiler found a never type, it automatically converted it into an i32 type.

Never Type with Return, Break, and Continue

The return statement also results in a never type and therefore can coerced to any type. For instance, the following statement will compile with no issues:

```
Let x: String = return;
```

In addition to the return, break and continue also result in never types. Consider the code shown in Listing 13.24.

```
fn main() {
  let result = loop {
        counter += 1;
        if counter == 10 {
            break;
        }
    };
}
```

Listing 13.24 Returning from a Loop Using break and Assigning the Loop to a Variable

This code simply increments a counter by 1, until the counter value reaches a value of 10. The variable result will assume a value that is returned from the loop. As pointed out before, the break and continue return a never type; however, when you inspect the type of the variable result in the editor, notice it is a unit value.

This result is likely due to reasons rooted in history. The never type wasn't present in the earlier days of Rust. As time passed, it gained prominence as a fundamental component, although specific situations arose to ensure compatibility with older versions. In these cases, the never type is treated as equivalent to unit. Although they are not the same thing, as we'll see in later sections.

Never Type for Representing States of Failure

Another use case of never type is that to permit the designation of particular states as unachievable at the type level. Consider a function that returns a Result, such as the following:

```
fn function() -> Result<i32, String> {}
```

If the function completes successfully and returns, a Result will have an instance of the Success type. Conversely, if it encounters an error, the Result will encompass an instance of the Err type. Now, let's see another similar function but with the never type:

```
fn function_1() -> Result<i32, !> {}
```

The success case is the same as before. However, in cases of error, there's a catch. The function cannot actually encounter errors, as creating instances of never type is impossible. With the function signature given, you can confidently affirm that this function will never experience errors. Let's also look at a similar example:

```
fn function_2() -> Result<!, i32> {}
```

Now, the opposite of the previous case holds. If this function returns something, it's clear that it must have encountered an error because achieving success is unattainable.

Custom Never Types

An alternative to simulating the behavior of the never type on the stable version is to use our own custom defined never type. For instance, consider the following enum definition:

```
enum NeverType{}
```

Like the never type, we can create a variable of type NeverType; however, we cannot assign it to some variable. Consider the following examples:

```
Let x: NeverType; // this compiles
Let x = NeverTypes // Does not compile since enum has no variant
```

We can now use this custom type in our functions, like function_1 and function_2 defined earlier, to indicate that success is not possible or alternatively to indicate that error cannot occur, as in the following examples:

```
fn function() -> Result<NeverType, String> {}
fn function_1() -> Result<i32, NeverType> {}
```

However, unlike the never type (!), our custom defined NeverType cannot be coerced into any other type.

13.5.2 Unit Type

A unit type represents a lack of meaningful data or information and is denoted by parentheses. This type has only one possible value, called unit value, which is also denoted as parentheses. The following function is an example of a unit variable, initialized from a unit:

```
fn main() {
    let x = ();
}
```

Unit values are useful when you have some piece of code that does something, but it doesn't return anything meaningful. Let's look at this scenario and then dive more deeply into how the unit type works and how it differs from the never type.

Functions with No Meaningful Value

A common scenario where you'll encounter a unit value is when a function doesn't return a meaningful value. These functions are typically used for operations that don't produce a result, such as logging, modifying state, or handling errors.

Consider the code shown in Listing 13.25.

```
fn f1() {}
fn main() {
    let y = f1();
}
```

Listing 13.25 Function That Doesn't Return Anything and Its Usage in main

Functions that do not return anything explicitly return a unit value. In this case, variable y stores the returning unit value. The function can be desugared to the following form:

```
fn f1() -> () {}
```

In many cases, we want to know if the function has successfully done some processing or if it failed for a variety of reasons. In case of success, we are not interested in the details. Listing 13.26 shows an example of this check using a division function.

```
fn division_status(divident: f64, divisor: f64) -> Result<(), String> {
    let answer = match divisor {
        0.0 => Err(String::from("Error: Division by zero")),
        _ => {
            println!("The division is invalid");
            Ok(())
        }
    };
    answer
}
```

Listing 13.26 Definition of division_status Simulating a Division Function

This function returns a unit value in case of success and throws an error otherwise. This use case of the unit value, together with the Result, can be used to check whether the status of some operation is valid or not.

Statements and Unit Type

All statements return a unit value. The following statement will return a unit value:

```
let z = println!("Hello, world!");
```

The variable z is therefore of type unit.

Vector of Unit

Vectors with zero capacity are treated as vector of unit types, as follows:

```
let mut vec: Vec<()> = Vec::with_capacity(0);
```

Let's push a few instances a vector:


```
vec.push(());
vec.push(());
vec.push(());
```

The compiler recognizes that unit type has no size and optimizes its interactions with instances of a unit type. Therefore, pushing unit values will only update the length of the vector and will not lead to any heap allocations or change in the capacity of the vector. Let's check if the size is 3 with the following code:

```
assert_eq!(3, vec.len());
```

If we execute, the assertion will pass, meaning that the length of the vector is 3.

The capacity function is used to specify an initial capacity for the vector. The capacity refers to the number of elements the Vec can hold before needing to resize its internal memory allocation due to the addition of further elements into the vector. You may be expecting that the capacity should be zero because there is no allocation needed. However, consider the following statement:

```
println!("{}", vec.capacity());
```

This statement will produce some strange output when executed, in particular, a huge number. The compiler basically sets this value to be the highest possible value that a vector can be allocated. Why is this?

Whenever the length of a vector exceeds its allocated capacity, the capacity is increased by pinging the allocator, and a new allocation takes place. Since zero-sized types do not take any memory, we should not be pinging the allocator at all. To rule out the possibility of pinging the allocator, the vector has been allocated the highest possible value that a vector can take.

Unit Type versus Never Type

A particularly important distinction to understand clearly is the difference between a `unit` type and the `never` type.

The `never` type represents computations that never produce a value. The `unit` type, on the other hand, represents computations with no meaningful value. The functions that return the `never` type are guaranteed to never return normally. In contrast, functions that return the `unit` type always return normally. Finally, the `never` type has no associated value and can be coerced to all other types. The `unit` type has a single value called `unit` and cannot be coerced into any other type. These differences are summarized in Table 13.2.

Never Type	Unit Type
Never produces a value	No meaningful value
Function returning `never` will never returns normally	Function returning `unit` always returns normally
No associated value and can be coerced into all other types	Single value which cannot be coerced to any other type

Table 13.2 Differences between the Never and Unit Types

13.5.3 Unit Structs

A `Unit` struct is a struct with no fields. Due to absence of associated fields, it has zero size. You can use `Unit` structs as a marker type to ensure certain concepts or properties are enforced. Let's go over a login authentication example to explain this use.

Let's say we want a different login method invoked depending on the role of the user. A user can be either an `Admin` or a simple `User`. The code shown in Listing 13.27 defines the necessary marker types of `Admin` and `User` for representing the roles in a login system, along with an `Authenticate` trait.

```
struct Admin;
struct User;
trait Authenticate {
    fn authenticate(&self, username: &str, password: &str) -> bool;
}
```

Listing 13.27 Definitions of Admin and User and the Authenticate Trait

The Authenticate trait defines a method named authenticate, which will essentially contain the authentication logic for each implementing role. Next, we'll add the implementation of the Authenticate trait for the two roles. Listing 13.28 shows the implementation.

```
impl Authenticate for Admin {
    fn authenticate(&self, username: &str, password: &str) -> bool {
        username == "admin" && password == "adminpass"
    }
}

impl Authenticate for User {
    fn authenticate(&self, username: &str, password: &str) -> bool {
        username == "user" && password == "userpass"
    }
}
```

Listing 13.28 Implementation of Authenticate for the Admin and User

We'll keep this implementation simple for the sake of illustration. Next, we'll add a login function that will call the authenticate method on the provided role. Listing 13.29 shows the definition of the function.

```
fn login<T: Authenticate>(role: T, username: &str, password: &str) -> bool {
    role.authenticate(username, password)
}
```

Listing 13.29 Definition of the login Function

The function takes in a role, user, and password and makes a call to the authenticate method. T is a generic in this case with a bound of Authenticate. If the authentication is successful, the function will return a true, and in case of failure, it will return a false.

Let's use the code in main, as shown in Listing 13.30.

```
fn main() {
    let admin = Admin;
    let'ser = User;

    let admin_login = login(admin, "admin", "adminpass");
    let'ser_login = login(user, "user", "userpass");

    if admin_login {
        println!("Admin login successful!");
    } else {
        println!("Admin login failed!");
    }
```

```
    if user_login {
        println!("User login successful!");
    } else {
        println!("User login failed!");
    }
}
```

Listing 13.30 Using the Login Authentication in main

We first created instances of Admin and User and then tried to login using valid credentials for both roles. Finally, the result of each login attempt is printed to the console.

This example demonstrates how you can use marker types to differentiate roles and behaviors within a simple login authentication system. Many other powerful use cases of unit structs could be described, but they are well beyond the scope of this book.

Compared to other zero-sized types, Unit structs are non-copy by default. In general, all structs are non-copy by default, meaning that they are moved through ownership changes when assigned to variables and are not copied. Consider the code shown in Listing 13.31.

```
struct ABC;
fn main() {
    let a = ();
    let b = a;
    let c = a;
    let x = ABC;
    let y = x;
    let z = x; // Error
}
```

Listing 13.31 Unit Type Copied While Unit Structs Are Moved Instead of Copied

The unit variable a, which has a unit type, can be copied many times. However, the unit struct variable x is not copied but rather moved during the assignment y = x to variable y. The compiler, therefore, now throws an error on line let z = x stating that "use of moved value: x." The ownership of variable x has already been moved to variable y on the second-to-last line and therefore later access to variable x is not valid.

This behavior is important in situations where you want to enforce strict ownership and prevent accidental data duplication. For instance, in the use case of being marker types, they may be used to enforce different behaviors based on their presence or absence. If they were copyable, it could lead to unexpected behavior and unintended sharing of traits or marker information. For instance, in our login authentication example, we'll want only a single Admin. If Admin is copyable, you could unexpectedly create two Admins, which may violate the requirement of having a single Admin of the system.

13.5.4 PhantomData

PhantomData is a marker struct of zero size. This struct helps in expressing the relationships and constraints between types without introducing any runtime overhead.

Consider a situation where we want to opt out of the auto marker traits of Send and Sync for some struct. Doing so might be useful in a situation where a developer intentionally wants to prevent a struct from being sent across threads or from being accessed concurrently from multiple threads. (We'll cover threads in detail in Chapter 14.) One approach to opting out of marker traits that we saw earlier in Section 13.3.1 was to use the external crate of negative_impl and then negatively implement the Send and Sync traits for the struct, as shown in Listing 13.32.

```
struct ABC;
use negative_impl::negative_impl;
#[negative_impl]
impl !Send for ABC {}

#[negative_impl]
impl !Sync for ABC {}
```

Listing 13.32 Opting Out of Default Send and Sync Traits for ABC

The approach shown in Listing 13.32 has a drawback, however: It depends on an external crate. Bringing external crates into scope is associated with extra overhead.

Let's explore another approach. We can opt out of a marker trait by adding another member field. A type is only Send and Sync if all of its members are also Send in Sync. If we introduce a field in the struct that is neither Send nor Sync, then the type will also be neither Send nor Sync.

Luckily, we have such a type, which is an Rc smart pointer. The Rc pointer is neither Send nor Sync. Let's revise the struct definition so that it contains a field that is an Rc pointer. The revised definition is as follows:

```
struct ABC {
    ensuring_no_send_sync: Rc<()>,
}
```

Why Is Rc Not Send or Sync?

Rc pointers are neither Send nor Sync, which ensures Rc can only be used within a single thread, as its internal reference count is not thread safe and could lead to data races if accessed or modified concurrently across threads.

This approach is not ideal because it increases the size of every struct instance and requires creating an Rc smart pointer each time a new instance is created. For instance, let's print the size of the struct, as shown in Listing 13.33.

```
fn main() {
    println!("{}", size_of::<ABC>());
}
```

Listing 13.33 Printing the Size of ABC

Its size is 8 bytes compared to zero bytes in the previous case.

You can use PhantomData to solve such issues. Instead of defining the field to be an Rc pointer, change its type to that of a PhantomData wrapping an Rc type. The revised definition of the struct is shown in Listing 13.34.

```
use std::marker::PhantomData;
struct ABC {
    ensuring_no_send_sync: PhantomData<Rc<()>>,
}
```

Listing 13.34 Revised Definition of the ABC Struct Using PhantomData

Since PhantomData is zero sized, the print statement in Listing 13.33 will print zero size. Additionally, the struct ABC will have the necessary property of not being Send nor Sync.

Adding a PhantomData field to a type informs the compiler that the type behaves as if it stores a value of a certain type, such as Rc in this case, even though it doesn't actually hold that value. This approach does not cause any execution time overhead. This zero-sized type can be used for compile-time type checking and optimizations that don't impact runtime performance.

13.6 Practice Exercises

In this section, we'll provide some practice questions to help reinforce your learning. The exercise questions cover almost all the concepts introduced in the chapter. Solutions to each exercise will follow in Section 13.7.

1. **Fixing the size calculation of unsized types**
   The following code snippet attempts to print the size of a raw array slice, which results in a compile-time error. Fix the issue and provide the correct way to handle the size of unsized types.

   ```
   use std::mem::size_of;
   fn main() {
       println!("[i32] size is: {}", size_of::<[i32]>()); // Error: Size cannot
   be known at compile time
   }
   ```

2. **Implementing correct handling of `?Sized` trait bound in functions**
   Fix the function `some_fn` so that it can correctly handle unsized types. Modify the function to ensure that it compiles and can `print` out both sized and unsized types.

```
fn some_fn<T: ?Sized + std::fmt::Debug>(val: T) {
    println!("{:?}", val)
}
fn main() {}
```

3. **Implementing generic structs with dynamically sized fields**
   The following code does not compile and contains an error. Change the `struct` definition only to fix the error based on the concepts covered in the chapter.

```
struct FlexibleStruct {
    fixed_part: u32,
    dynamic_part: str,
}
fn main() {
    let instance = FlexibleStruct {
        fixed_part: 42,
        dynamic_part: "Hello",  // Error
    };
}
```

4. **Simulating never type in custom functions**
   Complete the code by implementing the following two functions:
   - `function_never_succeeds()` should always return an error (but never succeed).
   - `function_never_fails()` should always succeed (but never return an error).

   Call both functions in `main` and `print` appropriate messages for each result.

```
enum CustomNeverType {}
fn function_never_succeeds() -> Result<CustomNeverType, i32> {
    // Task: Implement a function where success is unattainable
}
fn function_never_fails() -> Result<i32, CustomNeverType> {
    // Task: Implement a function where failure is impossible
}
fn main() {
    // Task: Call both functions and handle their results
}
```

5. **Implementing a function with `unit` type**
   Consider the following code. Implement the `perform_operation` function, which will perform the following:
   - Accept an integer as input.

- If the value is positive, print a success message and return `Ok(())`.
- If the value is negative or zero, return an error message.

In the `main` function, call `perform_operation` with both a positive and a negative value and handle the results by printing appropriate messages.

```
fn perform_operation(value: i32) -> Result<(), String> {
    // Task: Implement the operation where success returns unit value and
error returns a message
}
fn main() {
    let result = perform_operation(10);
    // Task: Handle the result and print appropriate messages
}
```

13.7 Solutions

This section provides the code solutions for the practice exercises in Section 13.6. The code is largely self-explanatory; however, we have included comments and additional explanations wherever necessary to enhance understanding.

1. **Fixing the size calculation of unsized types**

```
use std::mem::size_of;
fn main() {
    println!("[i32] size is: {}", size_of::<&[i32]>());
}
```

2. **Implementing correct handling of `?Sized` trait bound in functions**

```
fn some_fn<T: ?Sized + std::fmt::Debug>(val: &T) {
    println!("{:?}", val)
}
fn main() {}
```

3. **Implementing generic structs with dynamically sized fields**

```
struct FlexibleStruct<T: ?Sized> {
    fixed_part: u32,
    dynamic_part: T,
}
fn main() {
    let instance = FlexibleStruct {
        fixed_part: 42,
        dynamic_part: "Hello",
    };
}
```

4. **Simulating never type in custom functions**

```
enum CustomNeverType {}
fn function_never_succeeds() -> Result<CustomNeverType, i32> {
    Err(100)  // Return an arbitrary error code
}
fn function_never_fails() -> Result<i32, CustomNeverType> {
    Ok(200)  // Return a success value
}
fn main() {
    match function_never_succeeds() {
        Ok(_) => println!("This will never print because success is not
possible"),
        Err(e) => println!("Error occurred as expected: {}", e),
    }
    match function_never_fails() {
        Ok(value) => println!("Success as expected: {}", value),
        Err(_) => println!("This will never print because failure is not
possible"),
    }
}
```


5. **Implementing a function with unit type**

```
fn perform_operation(value: i32) -> Result<(), String> {
    if value > 0 {
        println!("Operation performed successfully with value: {}", value);
        Ok(()) // Return unit type on success
    } else {
        Err(String::from("Error: Value must be positive")) // Return error
message
    }
}
fn main() {
    let positive_result = perform_operation(10);
    match positive_result {
        Ok(_) => println!("Positive operation completed."),
        Err(e) => println!("{}", e),
    }
    let negative_result = perform_operation(-5);
    match negative_result {
        Ok(_) => println!("This will not print."),
        Err(e) => println!("{}", e),
    }
}
```

13.8 Summary

This chapter provided a comprehensive look at understanding size in Rust, focusing on how the sizes of types influences memory management and performance. We began by explaining the differences between sized and unsized types, highlighting how Rust handles each and the importance of understanding these concepts when designing efficient programs. This chapter then explored working with references to unsized types and how optionally sized traits and generic parameters offer flexibility. Additionally, we introduced unsized coercion, a feature allowing for seamless handling of unsized types. A significant portion of this chapter was dedicated to zero-sized types, including the `never` type, `unit` type, `unit` structs, and `PhantomData`, offering deep insights into how Rust's type system operates. This knowledge is crucial for making informed design choices regarding memory and performance. The chapter concludes with exercises to solidify these concepts.

In the next chapter, we'll explore concurrency, and you'll learn how to safely work with multiple threads and asynchronous programming in Rust.

Chapter 14
Concurrency

Concurrency brings the power of parallelism to your applications. In this chapter, we'll explore threading and asynchronous programming so you can unlock new levels of performance.

In this chapter, we'll explore concurrency in Rust, starting with the basics of threads and how to manage ownership in a multithreaded context. The chapter covers various methods of thread communication, such as messages passing through channels and sharing states. Synchronization techniques, including barriers and scoped threads, are discussed to ensure safe concurrent operations. The chapter also introduces thread parking, async/await for asynchronous programming, and Tokio tasks for high-performance applications. Practical examples like web scraping using threads illustrate these concepts in action.

14.1 Thread Basics

Let's start with some background concepts before diving into the implementation details. You'll learn how to create threads and the different approaches for executing them to completion in this section.

14.1.1 Fundamental Concepts

We start with *concurrency*, which refers to when sections of code run in parallel. Thus, if you have some sort code segments, concurrency will allow these segments to execute at the same time. In most current operating systems, an executed program's code is run in a process, and the operating system manages multiple processes at once. Within a program, you can also have independent parts that run simultaneously. The feature that runs these independent parts are called *threads*.

Threads have many use cases. For instance, one part of our program may wait for some input, thereby unnecessarily blocking the remaining parts of your code. We can make such programs efficient by creating threads, where the input/output (I/O)-related code may be assigned to one thread, and the remaining computation to the other. In this way, the program will not block the remainder of the code unnecessarily.

Parallelism is another closely related term to threads. However, fundamental differences exist between the two. Concurrency is about multiple tasks which start, run, and complete in overlapping time periods in no specific order. Parallelism is about multiple tasks that literally run at the same time on hardware with multiple computing resources, like multicore processors. From a programming perspective, we are interested in concurrency, and from the hardware perspective, we are interested in parallelism. The programming community, however, uses the two terms interchangeably.

With this background, let's look at creating threads in Rust.

14.1.2 Creating Threads

The threads module in the standard library provides the basic implementation of threads in Rust. Consider the code shown in Listing 14.1.

```
use std::thread;
fn main() {
    println!("This will be printed");
    println!("This will also be printed");
    println!("The concurrency will start after this line");

   thread::spawn(|| {
        println!("Hello 1 from the thread");
        println!("Hello 2 from the thread");
        println!("Hello 3 from the thread");
        println!("Hello 4 from the thread");
        println!("Hello 5 from the thread");
        println!("Hello 6 from the thread");
        println!("Hello 7 from the thread");
    });
    println!("Hello 1 from the main");
    println!("Hello 2 from the main");
}
```

Listing 14.1 Creating a Thread in main

First, we included the threads from the standard library, that is, `std::threads`. In `main`, we have a few `print` statements at the start. Next, we create a thread using the `thread::spawn` function. This function takes a closure as an input. The input to the closure in this case is empty, and the body of the `closure` basically contains the code inside the thread. Seven `print` statements exist in the thread body. At the end of the thread body, we have a couple of `print` statements in `main`.

The three `print` statements at the start of the program will run sequentially and not concurrently. The reason for sequential execution is that we are creating the threads after the first three lines, and therefore, concurrency in the program starts after those lines.

Each program by default has one thread, which is the `main` thread. When we call the `spawn` function on the fourth line in `main`, we've created another thread in addition to the `main` thread. After this point, the remaining code either belongs to the `main` thread or the newly created thread. Threads can execute in parallel; therefore, the `print` statements inside the thread and the two `print` statements at the end of the `main` will execute in parallel.

Two things to note about the execution, or scheduling order of threads, include the following:

- First, the execution is not deterministic. Therefore, each time you run your program, it may execute in a different order.
- Second, the scheduling order of threads is handled by the operating system. The operating system will divide the CPU time into small chunks and will run different threads in these chunks turn by turn.

By executing the program shown in Listing 14.1 a few times, you may note that the `print` lines before the thread `spawn` function are always printed first. These lines will always be printed first, because as mentioned earlier, the concurrency in the code starts after these lines. There is no unique order in which the remaining `print` statements are printed. Each time we execute, we may obtain a different result. This is because thread scheduling is being handled by the operating system, and there is no deterministic order in which threads are executed.

In all the executions, the `print` statements at the end of the `main` are always executed; however, during some executions, all the `print` statements in the thread may not execute. This problem may arise because the `main` thread will always go to completion before the termination of the program. Once the `main` thread completes, all spawned threads are shut down, whether or not they have finished running.

14.1.3 Thread Completion Using Sleep

To enable a spawned thread to always go to completion, two approaches can be used. In the first approach, you can use the thread `sleep` function. Listing 14.2 shows how to use the thread `sleep` function.

```
use std::thread;
use std::time::Duration;

fn main() {
    println!("This will be printed");
    ...
```

```
    thread::spawn(|| {
        println!("Hello 1 from the thread");
        ...
    });
    thread::sleep(Duration::from_millis(1));
    println!("Hello 1 from the main");
    println!("Hello 2 from the main");
}
```

Listing 14.2 Code from Listing 14.1 Updated to Use the thread::sleep Function

The `thread:sleep` function is defined in the standard library, and therefore, you'll first include it. The call to `thread::sleep` inside a certain thread forces that thread to stop its execution for a short duration, allowing a different thread to execute. The `thread::sleep` is called in the `main` thread; therefore, it will block the `main` thread for `1` millisecond, thereby giving a chance for the other thread to execute. If we execute the other thread, it is highly likely that that thread will go to completion. If, however, you only see output from the `main` thread, or don't see any overlap of output between `main` and the thread, try increasing the sleep time.

The call to `thread::sleep` only ensures that the calling thread is blocked for a certain amount of time. During that time, we expect the other threads to take turns and therefore go to completion, but completion is not guaranteed. Everything depends on the size of the other threads and how the operating system manages threads. This approach is appropriate in scenarios where you need a simple, non-critical synchronization mechanism for tasks like debugging, testing, or simulating delays, but it is not suitable for precise or production-grade thread management due to its inefficiency and lack of reliability in timing.

14.1.4 Thread Completion Using Join

Let's look at the second approach, which is using a `join` method. This method will guarantee that the spawned thread will go to completion before the end of `main`. Each call to `thread::spawn` returns a type of `JoinHandle`, which we can store in a variable and then call a `join` method on it. Consider the code shown in Listing 14.3.

```
...
fn main() {
    ...
    let t = thread::spawn(|| {
    });
t.join();
    println!("Hello 1 from the main");
    println!("Hello 2 from the main");
```

```
    }
}
```

Listing 14.3 Ensuring Thread Completion Using a Call to join Method

The variable `t` is now of type `JoinHandle`. A call to `join` on `t`, in other words, `t.join()`, will block the execution of the thread in which it is called (in this case, the execution of the `main` thread), until the thread for which it is called goes to completion (in this case, the thread `t`). After thread `t` completes, the remaining code in `main` will be executed. Running the program shown in Listing 14.3 will result in the following output:

```
This will be printed
This will also be printed
The concurrency will start after this line
Hello 1 from the thread
Hello 2 from the thread
Hello 3 from the thread
Hello 4 from the thread
Hello 5 from the thread
Hello 6 from the thread
Hello 7 from the thread
Hello 1 from the main
Hello 2 from the main
```

The code in the thread is executed first. Once the thread completes, the remaining code in `main` was executed. This scenario arises because the `main` was blocked due to the call to `join` method. In contrast, let's try moving the call to `join` to the end of `main`, as shown in Listing 14.4.

```
...
fn main() {
    ...
    let t = thread::spawn(|| {
    });
    println!("Hello 1 from the main");
    println!("Hello 2 from the main");
t.join();
   }
}
```

Listing 14.4 Calling join at the End of main

In this scenario, the `main` thread may yield control to the spawned thread, but it will wait for the thread to complete before exiting, thereby ensuring that the thread finishes its execution.

Using `join` for thread completion is the right choice in practice when you need a robust and precise way to block the `main` thread until a spawned thread finishes its execution, particularly in production scenarios where thread synchronization and proper resource cleanup are critical.

14.2 Ownership in Threads

In Rust, the concept of ownership plays a crucial role in managing data across multiple threads. As threads can potentially operate on shared data, understanding ownership ensures that data is accessed safely and concurrently, preventing issues like data races and ensuring memory safety. (For a refresher on the basics of ownership, refer to Chapter 4, Section 4.1.)

Consider the code shown in Listing 14.5.

```
use std::thread;
fn main() {
    let x = "some string".to_string();
    thread::spawn(|| {   // Error
    println!("{x}");
    });
}
```

Listing 14.5 Accessing a Variable Defined in main inside the Thread

In this code, we define a `String` variable and then try to access it inside the thread. The code does not compile and throws an error, "closure may outlive the current function, but it borrows `x`, which is owned by the current function." The `print` statement will be executed in a spawned thread. The closure is not updating the variable `x` so Rust will infer that, inside the closure, this variable will be used immutably using an immutable reference (see Chapter 9, Section 9.1.4, for more details).

As already highlighted, thread execution is non-deterministic and is controlled by the operating system. Rust therefore can't tell how long a spawned thread may live. The spawn thread may live longer than the `main` thread, which will lead to a problem because the closure shown in Listing 14.5 borrows `x` immutably. At the end of `main`, the variable `x` will go out of scope, and as a result, the reference inside the closure will start pointing to invalid memory. To fix this problem, move the variable `x` to inside the closure, as shown in Listing 14.6.

```
use std::thread;
fn main() {
    let x = "some string".to_string();
    thread::spawn(move || {
        println!("{x}");
```

```
    });
}
```

Listing 14.6 Fixing the Code from Listing 14.5 Using the move Keyword

In Rust, the `move` keyword in closures indicates that the closure should take ownership of the variables it captures from its environment. As shown in Listing 14.6, the closure defining the thread takes the ownership of the variable `x`.

Due to the ownership transfer, variable `x` is no longer accessible in `main`. For instance, adding the following line to `main` (shown in Listing 14.6) will generate an error:

```
println!("{x}"); // Error
```

This error will go away if `x` happens to be a primitive type. For instance, the code shown in Listing 14.7 will compile with no errors.

```
use std::thread;
fn main() {
    let x = 5;
    thread::spawn(move || {
        println!("{x}");
    });
    println!("{x}");
}
```

Listing 14.7 Moving a Primitive Variable inside the Thread without Generating an Error

The code compiles because the primitives are not moved but copied instead (see Chapter 4, Section 4.1.1). The `move` keyword in this case will make another copy of `x`, which resides inside the thread. The variable `x` is therefore accessible in `main`.

The `move` keyword may not be required if the closure is implementing the `FnOnce` trait. Recall from Chapter 9, Section 9.1.4, that a closure implements an `FnOnce` trait if it is taking ownership of variables from its environment. For instance, consider the code shown in Listing 14.8.

```
use std::thread;
fn main() {
    let x = "some string".to_string();
    thread::spawn(|| {
        let y = x;
        println!("{y}");
    });
}
```

Listing 14.8 The move Keyword Isn't Required if the Closure Is Implementing the FnOnce Trait

The closure takes the ownership of x inside the closure. The code compiles without the move keyword. In this case, the Rust compiler infers that the closure is implementing FnOnce trait since it is using the variables from its environment through a transfer of ownership. The transfer of ownership, which was performed with the move keyword, is now performed by the closure's implementation of the FnOnce trait.

Threads in Rust are isolated from each other through ownership to ensure that data races never occur. However, if the threads are isolated, how will they communicate? We'll explore this topic next.

14.3 Thread Communication

Threads need some mechanism to communicate for solving complex problems. The two strategies commonly used in this regard are known as *message passing* and *sharing states*. We'll discuss both strategies in the following sections.

14.3.1 Message Passing

Message passing is a powerful and flexible way to enable threads to communicate and share data safely in Rust. In this section, we'll explore how to create channels for message passing, manage the flow of data between sending and receiving threads, scale communication across multiple threads, handle blocking behavior, and use non-blocking techniques like try_receive for responsive and efficient thread management.

Creating a Channel

Rust achieves message passing using the concept of *channels*. Although a fairly simple concept, using an analogy of a river will help clarify the basics.

Like a river, if you put something like a boat in it, the boat will travel downstream to the end of the waterway. Let's say our river is a "channel" with two halves: the transmitter and the receiver. In our river analogy, the transmitter is the upstream location where you would place the boat, and the receiver is the downstream location where the boat will end up and will be received.

Listing 14.9 shows how you can create a channel using the mpsc module in Rust's standard library.

```
use std::sync::mpsc;
fn main() {
    let (tx, rx) = mpsc::channel::<String>();
}
```

Listing 14.9 Creating a Channel Using the mpsc Module

Rust provides implementation of channels in its mpsc module. Mpsc stands for "multiple producer, single consumer." As the name suggests, this module allows multiple sending ends for sending messages and one receiving end that consumes those values. The function channel creates a new channel, which returns a tuple with two elements called Sender and Receiver. (Sender is also sometimes referred to as "transmitter.") The abbreviations tx and rx are traditionally used for referring to the transmitter and receiver, respectively.

Now, one part of our code will call methods on the transmitter, passing in the data we want to send, and another part of our code is listening to the receiver for arriving messages. The channel is said to be closed if either the transmitter or the receiver half is dropped. Note that the channel function works with a generic type T, which must be mentioned during the call to the function using the turbo fish syntax (see Chapter 9, Section 9.4). In this case, we'll be sending and receiving integer values; therefore, we'll mention String.

Sending and Receiving Threads

Let's create a thread that will send a message to the main thread, as shown in Listing 14.10.


```
use std::{sync::mpsc, thread};
fn main() {
    let (tx, rx) = mpsc::channel::<String>();
    thread::spawn(move || {
        let val = "Hi from thread".to_string();
        println!("Sending Value: {val}");
        tx.send(val).unwrap();
    });
}
```

Listing 14.10 Creating a Thread to Send a Message to the main Thread

This code first created the value we intend to send. The send method on the transmitter tx will send the value and returns a Result type. If the receiver has already been dropped and there's nowhere to send a value, the send operation will return an Err and will panic. Notice the move keyword in the closure passed to the thread::spawn. The spawned thread must take the ownership of the transmitter tx so that the thread can send messages through the channel, as ensured using the move keyword.

Let's now look at the receiver end. The recv method on the receiver allows for the receiving of values. The code shown in Listing 14.11 illustrates how this method can be used in main to receive the value that is sent out by the thread.

```
...
fn main() {
    ...
    thread::spawn(|| {
        ...
    });
    let received_val = rx.recv().unwrap();
    println!("Received: {received_val}")
}
```

Listing 14.11 Receiving Values in main Using the recv Method

This `recv` method blocks the thread that calls it and waits until the value is sent down the `channel`. This function also returns a `Result`: either return a value that is being sent out or an `Err`. The latter case arises when the transmitter closes before even sending out a useful value. This problem can occur if the transmitting thread panics, encounters an error, or terminates before sending a value through the channel. If we execute the preceding code, notice the following output:

```
Sending Value: Hi from thread
Received: Hi from thread
```

Once a value is sent out by the thread, that value is no longer available, and its ownership no longer resides with the thread. For instance, adding the following line to the end of the code in the thread body will generate an error:

```
println!("Val is: {val}"); // Error
```

However, no issues will arise with stack-allocated data like primitives, which are not moved but rather copied.

Adding Multiple Threads

Multiple threads can send messages to a receiving thread. Let's add a loop to the code shown earlier in Listing 14.11 so that it now creates 10 threads, as shown in Listing 14.12.

```
use std::{sync::mpsc, thread};
fn main() {
    let (tx, rx) = mpsc::channel::<String>();
    for i in 0..10 {
        thread::spawn(move || { // Error
            let val = "Hi from thread".to_string();
            println!("Sending Value: {val}");
            tx.send(val).unwrap();
        });
    }
```

```
    let received_val = rx.recv().unwrap();
    println!("Received: {received_val}");
}
```

Listing 14.12 Creating Multiple Threads Where Each Thread Sends a Message to the main Thread

Although similar to the code shown in Listing 14.11, we included the code of the thread inside a `for` loop. This placement creates multiple threads where each thread sends out a message to the `main` thread. However, an error arises, "use of moved value: `tx`."

Let's examine this problem: We are creating 10 threads. The first thread gets a chance to execute and will take the ownership of the sender `tx` due to the `move` keyword. The `tx` is therefore no longer available for other threads. To ensure that the other threads can send messages using the `tx`, we must pass a `clone` of `tx` to each thread. The updated code is shown in Listing 14.13.

```
use std::{sync::mpsc, thread};
fn main() {
    let (tx, rx) = mpsc::channel::<String>();
    for i in 0..10 {
        let tx_clone = tx.clone();
        thread::spawn(move || {
            let val = "Hi from thread".to_string();
            println!("Sending Value: {val}");
            tx_clone.send(val).unwrap();
        });
    }
    let received_val = rx.recv().unwrap();
    println!("Received: {received_val}");
}
```

Listing 14.13 Passing in a Clone of tx to Each Individual Thread

During each iteration, a `clone` of `tx` is created that is next passed into the individual threads, which consumes it. Note that the `clone` of `tx` is now used inside each thread to send the message. Note that, unlike the transmitter, we cannot `clone` the receiver. The `channels` only allow for multiple senders and a single receiver.

Instead of an arbitrary string, let's update the code to send the value of looping variable `i` instead. The updated code is shown in Listing 14.14.

```
use std::{sync::mpsc, thread};
fn main() {
    let (tx, rx) = mpsc::channel(); // Type mentioning not needed now
    for i in 0..10 {
        let tx_clone = tx.clone();
```

```
        thread::spawn(move || {
            println!("Sending Value: {i}");
            tx_clone.send(i).unwrap();
        });
    }
    let received_val = rx.recv().unwrap();
    println!("Received: {received_val}");
}
```

Listing 14.14 Instead of Arbitrary Value, We Now Send the Looping Variable i

We do not need to mention the type using the turbo fish syntax in the call to channel. The Rust compiler can infer the type based on the call to send. Although multiple threads send out messages, executing the program will generate an output that may seem a bit strange:

```
Sending Value: 0
Sending Value: 1
Sending Value: 2
Sending Value: 3
Received: 0
Sending Value: 4
thread 'Sending Value: <unnamed>5' panicked at
```

Each time you execute the program, you may notice a slightly different output. A common pattern across multiple outputs is that, although different threads send out the values, only a single value is received. This problem arises because the call to receive only receives a single value. Additionally, the output may also include a panic message, which indicates that not all the threads were able to send out their messages. This issue may occur if the main thread finishes early, in which case the program will terminate, ultimately leading to the channel ending abruptly.

Let's add some code for receiving one more message, as shown in Listing 14.15.

```
use std::{sync::mpsc, thread};
fn main() {
    ...
    let received_val = rx.recv().unwrap();
    println!("Received: {received_val}");

    let received_val = rx.recv().unwrap();
    println!("Received: {received_val}");
}
```

Listing 14.15 Adding Code for Receiving One More Message

Executing this code will produce a result similar to the following output:

```
Sending Value: 2
Sending Value: 0
Sending Value: 1
Sending Value: 3
Sending Value: 4
Received: 2
Received: 0
Sending Value: 6
thread 'Sending Value: 7
<unnamed>Sending Value: 8
' panicked at thread 'src/main.rs<unnamed>:
```

If you execute this program multiple times, notice how the value that is sent out first is always received first. This order occurs because the `channels` works like a queue that follows *first in, first out (FIFO)* order. The sender or transmitter may push many items in the queue. However, they will always be received in the same order in which they are being sent. In the preceding output, the value `2` is sent out first and is therefore received first. In the same way, the second value that is sent out is the value `0` and is therefore received next.

Receiving messages in this manner works for single messages; however, to properly receive all messages, you must treat the receiver like an iterator. The code shown in Listing 14.16 illustrates how you can receive all the messages.

```
use std::{sync::mpsc, thread};
fn main() {
    let (tx, rx) = mpsc::channel();
    for i in 0..10 {
        let tx_clone = tx.clone();
        thread::spawn(move || {
            println!("Sending Value: {i}");
            tx_clone.send(i).unwrap();
        });
    }
    for message in rx {
        println!("Received: {message}");
    }
}
```

Listing 14.16 Receiving All the Messages

In the last `for` loop, the program iterates over the receiver (`rx`), which is receiving values sent by the spawned threads. Each iteration waits for a message from the channel, and once a message is received, the program prints the value. The loop will continue until all values have been received and the channel is closed. After receiving each message, the `main` thread is blocked until new messages become available. This iteration will only end when the `channel` closes, which only happens when all the transmitters are dropped.

Now, when you execute the code shown in Listing 14.16, notice how all the values being sent are also received. However, the program does not terminate and is kept in a running state. To understand why this happened, consider the code shown in Listing 14.16 again. The receiver end of the `channel` will stop listening for new messages when all the transmitters are dropped. During each iteration, we are creating a `clone` of the transmitter that is next moved to the thread. When the thread finishes, the `clone` of transmitter is dropped. However, when all the threads complete, the original transmitter still remains because this transmitter is never consumed by any thread. To ensure that the original transmitter is also dropped, call the `drop` method on the transmitter after the thread, as shown in Listing 14.17.

```
use std::{sync::mpsc, thread};
fn main() {
    let (tx, rx) = mpsc::channel();
    for i in 0..10 {
        ...
    }
        drop(tx);
    ...
}
```

Listing 14.17 Calling drop on tx to Ensure That No Transmitters Remain

This step ensures that the program finishes successfully.

Blocking Threads

Let's now discuss the blocking nature of the `recv` method in more detail. Consider the code shown earlier in Listing 14.10 and Listing 14.11 as well as the code shown in Listing 14.18.

```
use std::{sync::mpsc, thread};
fn main() {
    let (tx, rx) = mpsc::channel();
    thread::spawn(move || {
        let val = "some_val".to_string();
        println!("Sending Value: {val}");
        tx.send(val).unwrap();
    });
```

```
    let received_val = rx.recv().unwrap();
}
```

Listing 14.18 Thread Sending a Message to the main Thread Using Channels

In this code, we are creating a thread where we are sending out a value to `main`. The `main` receives the value using the `recv` method. The call to `recv` is *blocking*, which means that the thread that calls it is blocked until a message for which it is waiting is received. To clearly see this scenario, let the thread go to sleep for `3` seconds, as shown in Listing 14.19.

```
use std::{sync::mpsc, thread, time::Duration};
fn main() {
    let (tx, rx) = mpsc::channel();
    thread::spawn(move || {
        let x = "some_value".to_string();
        println!("Sending value {x}");
        thread::sleep(Duration::from_secs(3));
        tx.send(x).unwrap();
    });
    let received_val = rx.recv().unwrap();
    println!("I will not excute until the value is recieved");
}
```

Listing 14.19 Causing the Thread to Sleep for 3 Seconds before Sending a Value to the main Thread

The value will now be only sent after the sleep time has passed. In `main`, the call to `recv` will block the `main` thread until the value is received. The `print` statement at the end of the `main` will therefore not execute until the value is received, which will only happen after 3 seconds. Confirm this works as expected by executing the code.

Try Receive

The blocking nature of the `recv` method is not efficient in situations where the time for sending the value may be substantial, and any remaining code that does not depend on the value may be unnecessarily blocked. Rust provides an alternative to `recv` called `try_recieve`, which is non-blocking. This method returns a `Result`. If a message is available, an `Ok` is returned; otherwise, an `Err` is returned. Use this method inside a loop that calls `try_recv` after some intervals, handles a message if one is available, and otherwise does other work for a little while, until checking again. The code shown in Listing 14.20 illustrates the usage of the `try_recv` method.

```
use std::{sync::mpsc, thread, time::Duration};
fn main() {
    ...
```

```
    let mut received_status = false;
    while received_status != true {
        match rx.try_recv() {
            Ok(received_value) => {
                println!("Received value is {:?}", received_value);
                received_status = true;
            }
            Err(_) => println!("I am doing some other stuff"),
        }
    }
}
```

Listing 14.20 Illustration of Using the try_recv Method

The variable `received_status` is defined to check whether the value is received or not and is used to terminate the `while` loop. Once a value is received, the variable will be set to `true`. Inside the loop, we are matching on the `try_recv()`. If a value is available, we'll print it and set the `received_status` to `true`. Otherwise, we'll "do some other stuff."

Caution!

You must carefully inspect the code that you want to execute while waiting for the message to be received, in our case, the code associated with the arm corresponding to the `Err` variant. This code must not depend on the messages from the thread or, in general, on the results produced by the thread. If it depends on messages produced by the thread and no message is received, the program would end up running code that expects data that isn't there, which could cause problems or lead to the program getting stuck.

14.3.2 Sharing States

Message passing is a one-way data flow in which a sender thread passes a message to a receiving thread. The ownership is transferred from the sending thread to the receiving thread. With *shared state* concurrency, some piece of data residing inside the memory can be accessed by multiple threads, and access is controlled by a locking mechanism. This capability is made possible with a special type called `Mutex`. You'll learn how this type works in the following sections.

Acquiring a Lock on Mutex

The `Mutex` type is defined in the Rust standard library and stands for mutual exclusion. Let's start with an example definition, shown in Listing 14.21.

```
use std::sync::Mutex;
fn main() {
    let m = Mutex::new(5);
    {
        let mut num = m.lock().unwrap();
        *num = 10;
    }
}
```

Listing 14.21 Defining a Variable of Type Mutex and Acquiring a Lock on It

The new constructor function creates an instance of the type Mutex. The data wrapped by a Mutex can only be accessed by a single thread at any given time. To gain access to the data, a locking mechanism is used. In particular, when a thread wants to gain access to the data behind a Mutex, it will acquire a lock on the Mutex. Once a lock is acquired, no other thread can access that data. The lock will be released once the thread is done with the data, allowing other threads to acquire the lock.

As shown in Listing 14.21, a lock is acquired inside a code block using the lock method. The call to lock method will block the current thread, until it is able to acquire the lock. If multiple threads are trying to call the lock, then only the first thread that makes a call will be given access, and the remaining threads will be blocked until the lock is released. The call to lock returns a result. If there's already a thread that has acquired a lock and if that thread panics, then the call to lock will return an Err.

Mutexes have a reputation for being hard to manage. Each time you acquire a lock, you need to make sure you unlock it explicitly so that the data is available for other parts of your code. Fortunately, Rust's type system and ownership rules guarantee that you can't get locking and unlocking wrong.

The lock will be automatically released once the variable num shown in Listing 14.21 is dropped or goes out of scope. For instance, after the code block, we can acquire the lock again and print it, as shown in Listing 14.22.

```
use std::{sync::Mutex, thread};
fn main() {
    let m = Mutex::new(5);
    {
        let mut num = m.lock().unwrap();
        *num = 10;
    }
    let lock_m = m.lock().unwrap();
    println!("m is: {:?}", *lock_m);
}
```

Listing 14.22 Acquiring the lock Again after the Code Block

Any further attempt to lock_m will block the current thread. For instance, the code shown in Listing 14.23 will not go to completion due to the blocking of main by the lock method.

```
use std::{sync::Mutex, thread};
fn main() {
    let m = Mutex::new(5);
    {
        let mut num = m.lock().unwrap();
        *num = 10;
    }

    let lock_m = m.lock().unwrap();
    println!("m is: {:?}", *lock_m);

    let lock_m1 = m.lock().unwrap();
    println!("This code is blocked");

}
```

Listing 14.23 Calling the Lock in the Presence of an Existing Lock, Resulting in a Blocking State

When you execute the code, note that the print line at the end of main does not print, and the program does not go to completion. This error arises because the main thread is blocked.

To ensure that the main thread goes to completion, we'll call a drop before acquiring the second lock, as shown in Listing 14.24.

```
use std::{sync::Mutex, thread};
fn main() {
    ...
    let lock_m = m.lock().unwrap();
    println!("m is: {:?}", *lock_m);
    drop(lock_m);
    let lock_m1 = m.lock().unwrap();
    println!("This code is blocked");
}
```

Listing 14.24 Calling drop on lock_m Releases the lock and Allows main to Go to Completion

A call to drop will release the lock. The lock_m1 will be released at the end of the main. When main completes, all the locks are released. Then, the program is allowed to go to completion.

Using Mutex to Share Data between Threads

Consider a File struct that needs to be accessed by multiple threads. Let's start with a basic definition of the File struct:

```
struct File {
    text: Vec<String>,
}
```

For simplicity, assume that the File contains a single field of text representing some textual data. Next, we would like to pass an instance of the File between different threads. To ensure that a single thread has access to the instance of the File at any given time, we'll wrap the instance by a Mutex as in the following example:

```
use std::sync::Mutex;
fn main() {
    let file = Mutex::new(File { text: vec![] });
}
```

This function will ensure that the text is updated by a single thread at any given time.

14

Now, let's say we want to create a few threads, where each thread will add content to the text of the File, as shown in Listing 14.25.

```
use std::{sync::Mutex, thread};
fn main() {
    let file = Mutex::new(File { text: vec![] });
    let mut thread_vec = vec![];
    for i in 0..10 {
        let handle = thread::spawn(move || { // Error
            let mut file_lock = file.lock().unwrap();
            file_lock.text.push(format!("Hello from Thread {i}"));
        });
        thread_vec.push(handle);
    }
}
```

Listing 14.25 Creating Multiple Threads for Updating the Contents of the File

We first created a vector of threads (i.e., thread_vec), which stores the thread handles. Next, during each iteration of the loop, we created one thread and stored it in a variable of handle. To update the file inside a thread, we acquired a lock and then added some text to the file. Finally, we added the thread to the thread_vec.

The code logic is reasonable, but an error arises, "use of a moved value: file, value moved into closure here, in previous iteration of loop." In the first iteration of the loop, the variable file, which is defined in main, will be moved into the thread; therefore, that variable cannot be used again by something else. Moreover, we don't want to clone the file variable because we want each thread to use the same file. In this case, what we want is shared ownership of the file instance.

As described earlier in Chapter 10, Section 10.2.2, the solution was to use the Rc (reference counting) smart pointer. Let's update the code shown earlier in Listing 14.25 to use the Rc smart pointer, as shown in Listing 14.26.

```
use std::{rc::Rc, sync::Mutex, thread};
fn main() {
    let file = Rc::new(Mutex::new(File { text: vec![] }));
    let mut thread_vec = vec![];
    for i in 0..10 {
        let file = Rc::clone(&file);
        let handle = thread::spawn(move || { // Error
            let mut file_lock = file.lock().unwrap();
            file_lock.text.push(format!("Hello from Thread {i}"));
        });
        thread_vec.push(handle);
    }
}
```

Listing 14.26 Wrapping the Mutex by an Rc Pointer to Allow Shared Ownership

We first wrapped the Mutex with an Rc pointer to allow shared ownership. During each iteration, we increment the owners by calling a clone on the file. Notice that the file variable inside the loop is shadowing the file variable in the main.

We expect this code to work, but another error is thrown, "Rc cannot be sent between threads safely," further elaborating, "the trait Send is not implemented for Rc." The Rc smart pointer is not thread safe, meaning that the pointer cannot be sent between threads in a safe way. To enable thread-safe shared ownership, you'll need to use the Arc smart pointer. Let's update the code to use the Arc smart pointer instead of the Rc smart pointer, as shown in Listing 14.27.

```
use std::{sync::{Arc, Mutex}, thread};
fn main() {
    let file = Arc::new(Mutex::new(File { text: vec![] })); // updated
    let mut thread_vec = vec![];
    for i in 0..10 {
        let file = Arc::clone(&file); // updated
        ...
    }
}
```

Listing 14.27 Replacing the Rc Smart Pointer with the Arc Smart Pointer

The code now compiles successfully.

Arc versus Rc

Arc stands for *atomic reference counting*, which means the reference count of the shared value is updated in a thread-safe manner using atomic operations, ensuring that multiple threads can modify the count without causing data races or inconsistencies.

Recall from Chapter 10 that the Rc manages a reference count for keeping track of the owners, adding to the count for each call to clone and subtracting from the count when each clone or an owner is dropped. However, the pointer doesn't use any concurrency primitives to ensure that changes to the count can't be interrupted by another thread. Therefore, the count always has a correct value.

The Arc pointer, meanwhile, ensures that reference counts are updated in a consistent manner between the threads. The rest of the functionality is more or less the same.

To ensure that all threads go to completion, call join on all the threads. Finally, to display the contents of the file, we'll acquire a lock on the file and then iterate through its contents. The code corresponding to these tasks is shown in Listing 14.28.


```
use std::{sync::{Arc, Mutex}, thread};
fn main() {
    let file = Arc::new(Mutex::new(File { text: vec![] }));
    let mut thread_vec = vec![];
    for i in 0..10 {
            ...
    }
    for handle in thread_vec {
        handle.join().unwrap();
    }
    let file_lock = file.lock().unwrap();
    for t in &file_lock.text {
        println!("{t}");
    }
}
```

Listing 14.28 Calling join on All the Threads and Displaying the Contents of the File

This code will produce the following output:

```
Hello from Thread 0
Hello from Thread 2
Hello from Thread 1
Hello from Thread 3
Hello from Thread 4
Hello from Thread 5
```

```
Hello from Thread 6
Hello from Thread 7
Hello from Thread 8
Hello from Thread 9
```

All the threads have successfully added the text to the variable `file`. The order is, however, arbitrary because threads are executed in a non-deterministic order.

> **Mutexes and Interior Mutability**
>
> Let's inspect the variable `file` shown in Listing 14.27 and Listing 14.28. Notice how this variable is not declared as mutable. However, its value did mutate in the loop. This mutation is possible because `Mutexes` uses interior mutability just like the `RefCell` smart pointer (refer to Chapter 10, Section 10.2.3).

14.4 Synchronization through Barriers

Barriers enable multiple threads to synchronize the beginning of some computation. It represents a point in the code where the execution of calling threads is paused, until all threads have reached that specific point in the code. Let's look at an example to understand the process and then walk through the process of implementing barriers.

14.4.1 Motivating Example for Barriers

In this scenario, we have some computationally expensive tasks. To efficiently complete these tasks, we can divide the tasks among the threads. However, due to dependencies among the tasks, the individual tasks must be processed in sequential order. Thus, a certain task, such as `Task 2` processing, can only commence once `Task 1` is complete. To simulate this scenario, consider the code shown in Listing 14.29.

```
use std::{sync::{Arc, Barrier, Mutex}, thread};
fn main() {
    let mut threads_vec = Vec::new();
    let task = Arc::new(Mutex::new(vec![]));

    for i in 0..5 {
        let task = task.clone();
        let handle = thread::spawn(move || {
            // Tasks 1
            task.lock()
                .unwrap()
                .push(format!("Thread {i}, Completed its part on Task 1"));
```

```
            // Task 2
            task.lock()
                .unwrap()
                .push(format!("Thread {i}, Completed its part on Task 2"));
        });
        threads_vec.push(handle);
    }

    for handle in threads_vec {
        handle.join().unwrap();
    }

    let task_lock = &*task.lock().unwrap();
    for contents in task_lock {
        println!("{contents}");
    }
}
```

Listing 14.29 Simulating a Computationally Expensive Task Divided among the Threads

First, we created a thread_vec for storing multiple threads that are working on completing the task. The variable task represents a computationally expensive task. Since the task must be shared among the threads, we've wrapped it with an Arc and Mutex. The task in this case is simply a vector where each thread that completes its assigned work will push a string to the vector to indicate that it has completed its part. The code next creates five threads inside a loop. Each thread will first try to gain access to the task and then push some string to it. This step simulates some computation that has been assigned to the thread as part of completing the task. Once a thread completes its assigned work on Task 1, it can work on the part of Task 2 assigned to it.

Not, we call the join method on all the threads. Finally, we print the task vector. The syntax &*task.lock().unwrap(); might be a bit confusing. The deref (*) on the result of calling the lock provides the actual contents, in this case, the vector that we are not allowed to move. (We cannot move the string vector out of the task.) We can read the vector, but we cannot take it out and assign it to something else (i.e., change its ownership). The reference at the start provides a reference to the inner contents that we can use to read in the data. Executing the code shown in Listing 14.29 results in an output similar to the following:

```
Thread 1, Completed its part on Task 1
Thread 1, Completed its part on Task 2
Thread 0, Completed its part on Task 1
Thread 0, Completed its part on Task 2
Thread 2, Completed its part on Task 1
Thread 2, Completed its part on Task 2
```

```
Thread 3, Completed its part on Task 1
Thread 3, Completed its part on Task 2
Thread 4, Completed its part on Task 1
Thread 4, Completed its part on Task 2
```

The threads completed the tasks but not in sequential order. Recall that we want the threads to only start working on `Task 2`, after work on `Task 1` is completed by all the threads. To achieve this functionality, we turn to the topic of barriers next.

14.4.2 Synchronizing Threads Using Barriers

Let's modify the code shown in Listing 14.29 by using barriers, as shown in Listing 14.30.

```
use std::{sync::{Arc, Barrier, Mutex}, thread};
fn main() {
    let mut threads_vec = Vec::new();
    let task = Arc::new(Mutex::new(vec![]));
    let barrier = Arc::new(Barrier::new(5)); // added
    for i in 0..5 {
        let task = task.clone();
        let barrier = barrier.clone(); // added
        let handle = thread::spawn(move || {
            // Tasks 1
            task.lock()
                .unwrap()
                .push(format!("Thread {i}, Completed its part on Task 1"));
            barrier.wait(); // added
                            // Task 2
            task.lock()
                .unwrap()
                .push(format!("Thread {i}, Completed its part on Task 2"));
        });
        threads_vec.push(handle);
    }
    ...
}
```

Listing 14.30 Updated Code Based on Barriers

A `barrier` is created by calling the `new` constructor function. The input to the constructor indicates the number of threads that must reach the `barrier` before the other threads can proceed. The `barrier` will be used in multiple threads, and therefore, it has been wrapped around by an `Arc` pointer. During each iteration, we pass a `clone` of the `barrier` to each thread to avoid ownership issues.

A call to the `wait` method on the `barrier` creates a barrier point in the code. This step will block the calling thread until all the threads reach the `barrier` point. As shown in Listing 14.30, all the threads will be blocked before the commencement of `Task 2`. If you execute the code, you'll see an output similar to the following:

```
Thread 0, Completed its part on Task 1
Thread 1, Completed its part on Task 1
Thread 2, Completed its part on Task 1
Thread 3, Completed its part on Task 1
Thread 4, Completed its part on Task 1
Thread 4, Completed its part on Task 2
Thread 3, Completed its part on Task 2
Thread 1, Completed its part on Task 2
Thread 2, Completed its part on Task 2
Thread 0, Completed its part on Task 2
```

The tasks are now completed in sequential order. The threads were only allowed to work on `Task 2` after `Task 1` was completed by all the threads.

Multiple `barrier` points may be created inside the code. For instance, you may have another `task` (i.e., `Task 3` in the code) and need to synchronize the threads before proceeding to `Task 3` from `Task 2`. This sequence can be created by adding one more barrier point in the code shown in Listing 14.30. The updated code is shown in Listing 14.31.

```
...
fn main() {
    ...
    for i in 0..5 {
        ...
            // Task 2
            task.lock()
                .unwrap()
                .push(format!("Thread {i}, Completed its part on Task 2"));
            barrier.wait();
            // Task 3 // added
            task.lock()
                .unwrap()
                .push(format!("Thread {i}, Completed its part on Task 3"));
        });
        threads_vec.push(handle);
    }
    ...
}
```

Listing 14.31 One More Barrier Point Added to the Code from Listing 14.30

14.5 Scoped Threads

In concurrent programming, controlling thread lifetimes and their interactions with data is critical for safety. *Scoped threads* in Rust offer a way to spawn threads that are guaranteed to terminate before a given scope ends, ensuring no dangling references. This feature allows for safer and more efficient use of data shared between threads.

Consider the code shown in Listing 14.32.

```
use std::thread;
fn main() {
    let mut vec = vec![1, 2, 3];
    thread::spawn(|| {
        println!("{:?}", vec);
    });
}
```

Listing 14.32 A Simple Thread Trying to Print a Vector Defined in main

In this case, we have a vector defined in `main`, which is used inside a thread. As expected, this code throws an error, "closure may outlive the current function, but it borrows `vec`, which is owned by the current function." The thread is not updating the vector, so Rust will infer that it will be used as an immutable borrow. The problem in this case is that Rust can't tell how long the spawned thread may live. The spawned thread may live longer than the `main` thread, which will lead to a problem because the closure passed to the `thread::spawn` borrows `x` immutably. Previously, we fixed this issue using the `move` keyword. However, the `move` will move the ownership of the `vec` to inside the closure, thereby preventing further use of the vector in subsequent code.

Scoped threads are a new feature introduced in Rust version 1.63.0. They enhance the ability of threads to borrow local variables more effectively. In particular, they provide clearer control over the lifetime of borrowed variables. Listing 14.33 shows a revised version of the code shown in Listing 14.32, now using scoped threads.

```
use std::thread;
fn main() {
    let mut vec = vec![1, 2, 3];
    thread::scope(|some_scope| {
        some_scope.spawn(|| {
   println!("Thread inside scope);
            println!("{:?}", vec);
        });
    });
}
```

Listing 14.33 Revising the Code from Listing 14.32 Using Scoped Threads

A call to the thread::scope creates a new thread scope. The input to the function is a closure. You can use the input to the closure (i.e., some_scope) to spawn threads inside the thread::scope by calling spawn on it. In contrast to the code shown earlier in Listing 14.32, we are now accessing the vector inside the scope and no errors arise. All threads spawned within the scope will be automatically joined before the scope finishes. All the threads that reside inside a thread scope remain in that scope and will complete its execution in that scope. If we add some code to main after the scope, the added lines will only be executed when the scope ends. For instance, consider the code shown in Listing 14.34.

```
fn main() {
    let mut vec = vec![1, 2, 3];
    thread::scope(|some_scope| {
        some_scope.spawn(|| {
   println!("Thread inside scope);
            println!("{:?}", vec);
        });
    });

    println!("The scope finished");
    vec.push(5);
    println!("vec: {:?}", vec);
}
```

Listing 14.34 Accessing the Vector after the Scope Finishes Doesn't Lead to an Error

At the end of the scope, the thread is guaranteed to have finished its execution; therefore, no references to vec exist, and we can therefore use it.

Threads within the same scope will execute in parallel and will thus have all the usual borrowing issues. For instance, consider the code shown in Listing 14.35, which basically adds one more thread to the scope.

```
use std::thread;
fn main() {
    let mut vec = vec![1, 2, 3];
    thread::scope(|some_scope| {
        some_scope.spawn(|| {
   println!("Thread inside scope");
            println!("{:?}", vec);
        });

        some_scope.spawn(|| { // Error
            println!("Another Thread inside scope");
            vec.push(4);
```

```
            println!("vec: {:?}", vec);
        });
    });
    println!("The scope finished");
    vec.push(5);
    println!("vec: {:?}", vec);
}
```

Listing 14.35 Two Threads within the Same Scope

The second added thread to the scope now modifies the vector. In this case, we get an error, "cannot borrow vec as mutable because it is also borrowed as immutable mutable." This error arises because the borrowing rules are being violated within the scope. Inside the scope, the first thread may live longer than the second thread, which will violate the borrowing rule (i.e., immutable and mutable should not coexist). Let's use an immutable reference in the code shown in Listing 14.35 by commenting out or removing the line `vec.push(4);` in the body of the second thread, as shown in Listing 14.36.

```
use std::thread;
fn main() {
    let mut vec = vec![1, 2, 3];
    thread::scope(|some_scope| {
        ...
        some_scope.spawn(|| {
            println!("Another Thread inside scope");
            // vec.push(4); // updated line
            println!("vec: {:?}", vec);
        });
    });
...
}
```

Listing 14.36 Updated Code from Listing 14.35 Removing the Mutable Access to vec in the Second Scope

Note that the push method on vectors accesses the calling vector through a mutable reference. The thread scopes simplifies the borrowing, and you only need to pay attention to the borrowing rules inside the scope.

Finally, you may be wondering whether you can explicitly call `join` on all the threads, instead of including it in a thread scope to achieve the same behavior. For instance, consider the code shown in Listing 14.37.

```
use std::thread;
fn main() {
    let mut vec = vec![1, 2, 3];
    let handle_1 = thread::spawn(|| { // Error
    println!("Thread 1");
        println!("{:?}", vec);
    });

    let handle_2 = thread::spawn(|| { // Error
        println!("Thread 2 ");
        println!("vec: {:?}", vec);
    });
    handle_1.join();
    handle_2.join();
    println!("The scope finished");
    vec.push(5); // Error
    println!("vec: {:?}", vec);
}
```

Listing 14.37 Issue with Code Implementing the Behavior of Scoped Threads Using Simple Threads

The two threads are accessing the vector immutably because they are only printing its contents. However, Rust's ownership rules require that any variable accessed within a thread must be owned by that thread. Thus, the vector must be *moved* into each thread's body, transferring ownership. Since only one owner is allowed, attempting to move the same vector into both threads results in a compilation error. Additionally, even though the threads complete after `handle_2.join`, Rust cannot guarantee that the threads will complete at compile time, so it prevents any mutation of the vector after the threads start, thus ensuring memory safety.

This uncertainty about thread completion further highlights the need for scoped threads.

14.6 Thread Parking

Thread parking is a mechanism used in Rust to temporarily suspend the execution of a thread without consuming CPU resources. With thread parking, you can allow a thread to be efficiently paused and resumed later when needed, which is quite useful for managing concurrency. Let's understand the need for and the workings of thread parking through an example.

14.6.1 Motivating Example for Thread Parking

Let's say we have two threads named thread_1 and thread_2. We want thread_1 to do some work and then read some shared data, which is being updated by thread_2. In contrast, thread_2 should also do some work and then update the shared data. In this scenario, thread_1 should only read the shared data after that data has been updated by thread_2. The code shown in Listing 14.38 provides an implementation for this scenario.

```
use std::{sync::{Arc, Mutex}, thread, time::Duration};
fn main() {
    let data = Arc::new(Mutex::new(5));
    let data_clone = data.clone();
    let thread_1 = thread::spawn(move || {
        println!("Thread 1: I am doing some work");
        println!("Thread 1: I am doing some more work");
        println!("Thread 1: Printing the updated data");
        println!("Thread 1: Data: {:?}", *data.lock().unwrap());
    });

    let thread_2 = thread::spawn(move || {
        println!("Thread 2: I am working on updating the data");
        thread::sleep(Duration::from_secs(1));
        *data_clone.lock().unwrap() = 10;
        println!("Thread 2: Data updated completed");
    });
    thread_2.join();
    thread_1.join();
}
```

Listing 14.38 Simulating thread_1 Doing Some Work and Then Displaying Updated Data Provided by thread_2

We first created a shared data source. Since data will be shared among the threads, it is a Mutex wrapped around by an Arc pointer. In this program, thread_1 first engages in some work, simulated by the print statements in this case. After completing the work, thread_1 will print the value of shared data. In contrast, thread_2 works on updating the shared data and afterwards updates the shared data. The call to sleep simulates the behavior that the thread_2 is involved in some underlying computation, which takes time to update the data. Note that the variable data is consumed by thread_1 (the reason for using the move was discussed in the earlier section), so therefore we'll send in a clone of the data to thread_2. Finally, we'll call join on each of the two threads.

Executing the code shown in Listing 14.38 should result in the following output:

```
Thread 1: I am doing some work
Thread 1: I am doing some more work
Thread 2: I am working on updating the data
Thread 1: Printing the updated data
Thread 1: Data: 5
Thread 2: Data updated completed
```

Unfortunately, this code does not achieve the desired functionality. The data is being printed by thread_1 before the data can be updated by thread_2. The thread_1 was supposed to only print the data after the data has been updated by thread_2.

We have several different solutions to solve this problem. One possible solution is to make sure that thread_1 goes to sleep for some time before printing, thereby giving thread_2 some time to update the data. This approach is simulated in the code shown in Listing 14.39 by updating the code shown in Listing 14.38.

```
...
fn main() {
    ...
    let thread_1 = thread::spawn(move || {
        println!("Thread 1: I am doing some work");
        println!("Thread 1: I am doing some more work");
        thread::sleep(Duration::from_secs(2)); // added
        println!("Thread 1: Printing the updated data");
        ...
    });
    ...
}
```

Listing 14.39 Calling the Sleep Function on thread_1 to Achieve the Desired Functionality

Executing this code will produce the following output:

```
Thread 1: I am doing some work
Thread 1: I am doing some more work
Thread 2: I am working on updating the data
Thread 2: Data updated completed
Thread 1: Printing the updated data
Thread 1: Data: 10
```

thread_1 displays the correct updated value in this case and therefore achieves the desired functionality. However, its execution is not ideal because we must know in advance for how long thread_1 must sleep. If we make it sleep for more time, we'll be wasting CPU time. On the other hand, if we only make it sleep for some time that is less than the update time, we'll be printing the wrong data.

14.6.2 Temporarily Blocking a Thread

The thread::park function provides an efficient solution. When a thread calls thread::park, that thread will be blocked until it's unparked by some other thread. Recalling the code shown earlier in Listing 14.38, we'll block thread_1 until the data is updated. Listing 14.40 shows the updated code from Listing 14.38 now using the thread::park function.

```
use std::{
    sync::{Arc, Mutex},
    thread,
    time::Duration,
};
fn main() {
    let data = Arc::new(Mutex::new(5));
    let data_clone = data.clone();
    let thread_1 = thread::spawn(move || {
        println!("Thread 1: I am doing some work");
        println!("Thread 1: I am doing some more work");
        println!("Thread 1: Parked"); // updated
        thread::park();               // updated
        println!("Thread 1: Printing the updated data");
        println!("Thread 1: Data: {:?}", *data.lock().unwrap());
    });

    let thread_2 = thread::spawn(move || {
        println!("Thread 2: I am working on updating the data");
        thread::sleep(Duration::from_secs(1));
        *data_clone.lock().unwrap() = 10;
        println!("Thread 2: Data updated completed");
    });
    thread_2.join();
    thread_1.thread().unpark(); // updated
    thread_1.join();
}
```

Listing 14.40 Using thread::park to Make Sure thread_1 Reads Correct Data

The call to thread::park by thread_1 will block thread_1. Therefore, thread_1 will only resume its execution when some other thread calls the unpark method on it. When thread_2 completes, which happens on the line thread_2.join, we are sure that the data is updated. Therefore, we can call the unpark on thread_1 to resume the execution of thread_1. The call to the join method on thread_1 ensures that thread_1 goes to completion.

Executing the code shown in Listing 14.40 produces the correct output now:

```
Thread 2: I am working on updating the data
Thread 1: I am doing some work
Thread 1: I am doing some more work
Thread 1: Parked
Thread 2: Data updated completed
Thread 1: Printing the updated data
Thread 1: Data: 10
```

14.6.3 Park Timeout Function

Similar to thread parking, you can use the `thread::park_timeout` function to block the calling thread for a specified time interval. The thread will remain blocked until the specified time expires or an `unpark` call is made during that time.

Consider the code shown in Listing 14.41, which uses the `thread::park_timout` function instead of the `thread::park` function.


```
...
fn main() {
    ...
    let thread_1 = thread::spawn(move || {
        println!("Thread 1: I am doing some work");
        println!("Thread 1: I am doing some more work");
        println!("Thread 1: Parked");
        thread::park_timeout(Duration::from_secs(4));
        ...
    });

    let thread_2 = thread::spawn(move || {
    ...
    });
    thread_2.join();
    thread_1.thread().unpark();
    thread_1.join();
}
```

Listing 14.41 Using thread::park_timeout instead of thread::park

When you execute this code, notice how it achieved the desired functionality. Moreover, `thread_1` was supposed to wait for 4 seconds, but it completed immediately. This immediate execution occurred because, before the time mentioned in the call to `thread::park_timeout` has expired, a call to `thread::unpark` was made in the `main` thread. If you comment out the call to thread `unpark` in `main` and execute this code again, notice now how `thread_1` was blocked for some time, even though `thread_2` completed.

Thread Sleep versus Park Timeout

An important distinction to make clear is the difference between `thread::sleep` and `thread::park_timeout` function. `thread::sleep` unconditionally blocks the calling thread. However, `thread::park_timeout` conditionally blocks the calling thread based on the `unpark` function.

14.7 Async Await

Typical code in Rust is *synchronous*, which executes line by line in sequential order without yielding control back to the runtime. However, Rust provides the ability to write *asynchronous* code as well so you can write function closures and blocks that can pause execution and yield control back to the runtime. In this way, you can allow other code to make progress and then pick back up from where they left off. The pauses to wait for are generally for some input or output to occur. Due to this rather yielding nature, this approach is also sometimes called *cooperative scheduling*, as it gives control back to the executor or runtime, thereby cooperatively allowing other code to make progress and execute.

Cooperative Scheduling

Unlike threads, `async` code uses cooperative scheduling instead of preemptive scheduling. If we have two threads, the operating system can switch between the two threads at any given time. However, in `async` code, we as developers tell the runtime when a block of `async` code is ready to yield, so that other `async` code can run on the same thread. For instance, in the code of the `tasks` later in Listing 14.49, there are two points where we are yielding execution by calling `await`. This grants developers more control, but it also increases their responsibility. In particular, we must ensure that our `async/await` code is efficient. For example, it's important to avoid placing CPU-intensive operations within an `async` function.

Rust uses `async` and `await` syntax to write asynchronous code. With the help of these two constructs, you can basically write asynchronous code that looks like synchronous code. Let's walk through some code examples for creating async functions and using a special `tokio` runtime.

14.7.1 Creating Async Functions

Using the `async` keyword makes a function asynchronous. For instance, consider the example code shown in Listing 14.42.

```
async fn printing() {
    println!("I am async function");
}

fn main() {
    printing();
}
```

Listing 14.42 Defining an Async Function and Calling It in main

The async keyword at the start of the printing function signature changes the behavior of the function code, in terms of how it will execute. If we execute this code, perhaps you'll be surprised that there is no output from the function.

The async function returns something that implements the Future trait. If we assign the call to printing to some variable let x = printing(); and inspect the type of the variable x, notice how it has a type of impl Future<Output = ()>. This type means something that implements the Future trait, with an associated type of Output. In this case, we are not returning anything, so Output is a unit value.

The Future trait is fairly complex, but at an abstract level, it has a poll method that can either return a ready state when a return value is available or a pending state indicating that the value is currently not available. The executor will poll it after some interval to check again the status of the Future. The executor will stop polling the Future trait once a value becomes available.

14.7.2 Deriving Futures to Completion with the Await Method

The Future trait behaves like a promise, promising you that the function will generate some value in the future. We do not know exactly when; all we know is that it will provide a value at some point. Futures can only start to execute and work on generating the value when we call the await method on it. For instance, consider the code shown in Listing 14.43.

```
async fn printing() {
    println!("I am async function");
}

fn main() {
    let x = printing().await; // Error
}
```

Listing 14.43 Calling await on a Future

The variable x is now resolved into the returning value from the function (a unit value in this case). The call to await derive the future to completion.

> **Note**
>
> Rust futures are lazy, meaning that they won't do anything unless driven to completion. The await drives the futures to completion.

Unfortunately, the code throws an error, "await is only allowed inside async functions and blocks." Let's make the main function asynchronous, as shown in Listing 14.44.

```
async fn printing() {
    println!("I am async function");
}

async fn main() { // Error
    let x = printing().await;
}
```

Listing 14.44 Making the main async

This change leads to another error, "main function is not allowed to be async." This error occurred because async returns a future, which we just described as lazy. Therefore, the future is not executed until we use an await call on main, as shown in Listing 14.45.

```
async fn printing() {
    println!("I am async function");
}

async fn main() {
    let x = printing().await;
}.await // Error
```

Listing 14.45 Calling await on main

However, this change leads to another error. So now, how do we fix this problem?

The solution is provided by what is known as the *executor* or *runtime*. The executor or runtime program's task is to take the topmost future, like the main function and manually poll them for completion, especially if some futures nested within other futures. The executor is also responsible for running multiple futures in parallel, to make the concurrency possible between multiple futures.

14.7.3 Tokio

The standard Rust library does not provide an asycn runtime. Therefore, to run our async code, we must use a community built async runtime. The most popular one is called tokio. We'll work with this runtime in the following sections, using it to execute tasks on both single threads and multiple threads as well as using the sleep function.

Tokio Runtime

First, let's add the tokio runtime as a dependency by adding the following code in the dependencies section of the *Cargo.toml* file:

```
tokio = {version = "1.17", features = ["full"]}
```

Now, in main, let's annotate the main function with #[tokio::main]. The code shown earlier in Listing 14.45 has been updated into the code shown in Listing 14.46.

```
async fn printing() {
    println!("I am async function");
}
#[tokio::main]
async fn main() {
    let x = printing().await;
}
```

Listing 14.46 Using Tokio to Run Our async Code

#[tokio::main] is an attribute macro that allows our main function to be async and specifies that our async code will be executed by the tokio runtime. In other words, this attribute macro derives our main to go to completion. When you execute this code, it will generate the output from the printing function. The conclusion is that the call to await derives the future to completion, but the future must be used inside an async block.

As pointed out earlier, futures are lazy, which means that they won't do anything until we call the await method on them. For instance, consider the code shown in Listing 14.47.

```
async fn printing() {
    println!("I am async function");
}
#[tokio::main]
async fn main() {
    let x = printing();
    println!("The future has not been polled yet");
    x.await;
}
```

Listing 14.47 Futures Are Lazy and Don't Do Anything Until We Call await on Them

In this code, the print statement will be executed first, and then, the future corresponding to the async function will get a chance to execute. This approach can be super helpful in situations where you need to delay some costly operation in the code that may take considerable time.

Hopefully, now you see how being lazy provides some benefits. The first benefit of futures being lazy is that they are zero-cost abstractions. In other words, you won't incur any runtime cost unless you actually use the future. Another benefit of futures being lazy is that they're easy to cancel. In order to cancel a particular future, simply call the drop method on the future, the remaining code will be barred from polling it, as shown in Listing 14.48.

```
async fn printing() {
    println!("I am async function");
}
#[tokio::main]
async fn main() {
    let x = printing();
    println!("The future has not been polled yet");
        drop(x);
    // x cannot be polled now in the remaining code
}
```

Listing 14.48 Canceling a Future

Tokio Tasks

In the previous section, we haven't really taken any advantage of our async code because everything was running serially. To make our async code run concurrently, we'll use tokio tasks. A *task* is a lightweight, non-blocking unit of execution. Tasks allow top-level futures to execute concurrently.

In main, let's spawn a few tasks, which will execute the printing function, as shown in Listing 14.49.

```
async fn printing(i: i32) {
    println!("Task {i}");
}
#[tokio::main]
async fn main() {
    let mut handles = vec![];
    for i in 0..3 {
        let handle = tokio::spawn(async move {
            println!("Task {i}, printing, first time");
            printing(i).await;
            println!("Task {i}, printing, second time");
            printing(i).await;
```

```
            println!("Task {i}, completed");
        });
        handles.push(handle);
    }
    for handle in handles {
        handle.await.unwrap();
    }
    println!("All Tasks are now completed");
}
```

Listing 14.49 Using tasks for Running async Code Concurrently

In this code, we have an async printing function that just prints the variable passed in. In main, we first created an empty vector for storing *handles* (or identifiers) to tasks. Next, we spawn multiple tasks inside a loop. The tokio::spawn function creates a new task. Inside each task, we call the printing function with the iterating variable representing the task number. This call to printing will therefore print the task number which calls it. The tokio::spawn function takes a future as an argument and returns a join handle. At the end of each task, we add the task handle to the handles vector.

Just like threads, the move keyword is used to take ownership of the variables from the environment. In this case, we are taking ownership of the variable i, which is passed to the printing function. At the end of main, we loop through the tasks and call await on them. Calling await on handles returns a Result, which may be an Err if the task panics.

Notice that the syntax for tasks is similar to the syntax for spawning a thread. This similarity is purposeful so you can easily switch from using threads to using tasks. When you execute the program, you should see an output similar to the following:

```
Task 0, printing, first time
Task 0
Task 1, printing, first time
Task 1
Task 1, printing, second time
Task 1
Task 1, completed
Task 2, printing, first time
Task 2
Task 2, printing, second time
Task 2
Task 2, completed
Task 0, printing, second time
Task 0
Task 0, completed
All Tasks are now completed
```

The tasks are executed concurrently. Like threads, the order of execution is not deterministic. In other words, when you execute again, you may get different results.

Executing Tasks on a Single Thread

By default, tokio uses the thread pool to execute tasks on multiple threads. You can, however, force the tokio runtime to run all the tasks on a single thread by changing the *flavor* to the current thread, as shown in Listing 14.50.

```
async fn printing(i: i32) {
    println!("Task {i}");
}
#[tokio::main(flavor = "current_thread")]
async fn main() {
    ...
}
```

Listing 14.50 Changing the Flavor to current_thread

Now, when you execute the code shown in Listing 14.50, notice how all the tasks execute sequentially. We may want the tasks to execute sequentially in situation when the tasks depend on each other's results.

Recall that the code belonging to a single thread executes sequentially. Since the code inside the tasks always make progress, that is, the two Futures corresponding to the lines of

```
printing(i).await;
```

inside the tasks are immediately resolved, therefore the tasks are driven to completion in sequential order. If for some reason, the call to printing waits for some I/O in some task, then the other tasks will be given a chance since tasks are non-blocking.

Asynchronous I/O Using the Tokio Sleep

Let's simulate one such scenario by considering asynchronous I/O using the tokio sleep function. The code shown in Listing 14.51 illustrates how you can use the function.

```
use std::time::Duration;
use tokio::time::sleep;
async fn printing(i: i32) {
    sleep(Duration::from_secs(1)).await;
    println!("Task {i}");
}

#[tokio::main(flavor = "current_thread")]
```

```
async fn main() {
...
}
```

Listing 14.51 Using the tokio sleep Function to Simulate Asynchronous I/O

The tokio sleep function is provided by the tokio::time module. The sleep function needs the desired duration for the sleep time as an input, which is provided by the Duration::from_secs, which itself is provided by the std::Time::Duration. You can call the sleep function inside the printing function. The sleep function in tokio is similar to the thread sleep function except that it will stop the current future from executing for a given duration, instead of stopping an entire thread. Running the code shown in Listing 14.51 will produce output similar to the following:

```
Task 0, printing, first time
Task 1, printing, first time
Task 2, printing, first time
Task 0
Task 0, printing, second time
Task 1
Task 1, printing, second time
Task 2
Task 2, printing, second time
Task 0
Task 0, completed
Task 1
Task 1, completed
Task 2
Task 2, completed
All Tasks are now completed
```

The tasks have executed concurrently. First, task 0 is executed. After printing the line in main for task 0, the Future on the line after the first print inside the task cannot be resolved immediately. Consider the following lines of code:

```
println!("Task {i}, printing, first time");
printing(i).await; // this line
```

This error arises because the printing function needs to sleep for 1 second. At this time, task 0 yields the control back to the scheduler to run some other task. The next task is then picked up, which also prints the first line and then goes to sleep for one second. In summary, when a task cannot make further progress, it will yield the control back so that other tasks may be executed.

The Futures are scheduled in this case on a single thread; however, these tasks do not block the other tasks on the same thread.

You might be thinking that the same non-blocking behavior of `tasks` can be achieved if we removed the `flavor = current_thread`. This approach is generally true since the `tasks` will be executed on possibly different threads. However, in many applications, the number of created `tasks` may grow in size substantially, compared to the number of available threads. Thus, multiple `tasks` will be executed on the same thread. In that scenario, we probably don't want to block a thread simply because some other `task` on the same thread is waiting for I/O.

14.8 Web Scraping Using Threads

In this section, we'll illustrate a use case for threads. We hope this use case will provide some intuitive wisdom regarding the advantages of concurrent execution of code using threads over sequential execution.

Our example involves *web scraping*, which refers to the extraction of data from a webpage. The goal in this example is to extract textual information from webpages using threads and also using conventional sequential code and then compare the time required by each of these coding approaches.

Consider the code shown in Listing 14.52.

```
use std::sync::{mpsc, Arc, Mutex};
use std::thread;
use std::time::{Duration, Instant};
use ureq::{Agent, AgentBuilder};
fn main() -> Result<(), ureq::Error> {
    let webpages = vec![
        "https://gist.github.com/recluze/1d2989c7e345c8c3c542",
        "https://gist.github.com/recluze/a98aa1804884ca3b3ad3",
        "https://gist.github.com/recluze/5051735efe3fc189b90d",
        "https://gist.github.com/recluze/460157afc6a7492555bb",
        "https://gist.github.com/recluze/5051735efe3fc189b90d",
        "https://gist.github.com/recluze/c9bc4130af995c36176d",
        "https://gist.github.com/recluze/1d2989c7e345c8c3c542",
        "https://gist.github.com/recluze/a98aa1804884ca3b3ad3",
        "https://gist.github.com/recluze/5051735efe3fc189b90d",
        "https://gist.github.com/recluze/460157afc6a7492555bb",
        "https://gist.github.com/recluze/5051735efe3fc189b90d",
        "https://gist.github.com/recluze/c9bc4130af995c36176d",
        "https://gist.github.com/recluze/1d2989c7e345c8c3c542",
        "https://gist.github.com/recluze/a98aa1804884ca3b3ad3",
        "https://gist.github.com/recluze/5051735efe3fc189b90d",
        "https://gist.github.com/recluze/460157afc6a7492555bb",
        "https://gist.github.com/recluze/5051735efe3fc189b90d",
        "https://gist.github.com/recluze/c9bc4130af995c36176d",
```

```
    ];

    let agent = ureq::AgentBuilder::new().build();
    let now = Instant::now();

    for web_page in &webpages {
        let web_body = agent.get(web_page).call()?.into_string()?;
    }
    println!("Time taken wihtout Threads: {:.2?}", now.elapsed());
}
```

Listing 14.52 Reading the Content of Some Webpages Using Sequential Code and Measuring Time

In this code, first, we include the relevant modules. The `ureq` is a module from crates.io and serves as a simple safe `HTTP` client. To properly include `ureq`, include the following line in the *Cargo.toml* file under the dependencies:

```
ureq = "2.5.0"
```


In `main`, we'll use the question mark operator (`?`), which returns a `Result`. Therefore, `main` will return a `Result`. Next, we have a vector containing some GitHub webpages. We want to read all the textual information from these webpages and then return this data in a `String` variable. To grab the textual information from the webpages, first we send out HTTP `GET` requests by creating an agent for keeping the state between the requests. The agent is created using the `AgentBuilder::new` function to basically accumulate options. Various options include sending out requests such as timeouts, timeouts for read, timeouts for write, proxy, delay-related info, and others. In this case, we are building a simple agent with all the possible default options.

Next, we'll compute the time before reading the webpages. The `Instant::now` function from `std::time::instant` provides a sort of current system time. Then, we loop through all the `webpages` and read in all the textual information from the individual `webpages`. The call to `get` method with an input of `web_page` will send out a request for reading the `web url`. The call method on the request will fetch the contents from the `webpage` if the request is successful. Further, the caller is blocked until the job is done. The call method returns a `Result`, and therefore, we used a question mark at the end to check whether it returns successfully. Finally, we convert the received information into a string using the `into_string` function.

Finally, outside the loop, we print the time that has been taken by the code inside a `print` statement. The `elapsed` function returns the amount of time elapsed from the time when the variable `now` was created. In other words, it simply measures the amount of time taken by the code to grab the contents of all the `webpages`. The placeholder of `.2` means that the fraction part should only contain a couple of digits.

When the loop starts, the first webpage contents will be fetched. The contents of the second webpage can only be fetched after the contents of the first webpage are completed. Thus, while fetching the contents of the first page, the code will be in blocking state until all the contents are being fetched. Once all the contents are fetched, the code will be allowed to proceed and further fetch the remaining pages. This order is due to the sequential nature of the code.

Let's now add the code for the same process using threads. Listing 14.53 shows the updated code.

```
use std::sync::{mpsc, Arc, Mutex};
use std::thread;
use std::time::{Duration, Instant};
use ureq::{Agent, AgentBuilder};
fn main() -> Result<(), ureq::Error> {
    ...
    let now = Instant::now();
    let agent = Arc::new(agent);
    let mut handles: Vec<thread::JoinHandle<Result<(), ureq::Error>>> =
Vec::new();
    for web_page in webpages {
        let agent_thread = agent.clone();
        let t = thread::spawn(move || {
            let web_body = agent_thread.get(web_page).call()?.into_string()?;
            Ok(())
        });
        handles.push(t);
    }
    for handle in handles {
        handle.join().unwrap();
    }
    println!("Time taken using Threads: {:.2?}", now.elapsed());
    Ok(())
}
```

Listing 14.53 Reading the Contents of Webpages Using Threads

As usual, we first call the now function to start counting the time. Then, we have an agent that will be used inside multiple threads for sending out requests to read webpages. Due to its usage in multiple threads, we wrapped it inside the Arc smart pointer. Next, we have a vector of threads. Then, we iterate as many times as there are webpages to read, and during each iteration, we create one thread that will be responsible for reading out one webpage. Each thread will be using the agent; therefore, we make a clone of the agent inside each thread. We next push each thread identifier to the thread vector and call joins on each thread to ensure completion of all the threads. At the end, we compute the time and print its value.

From the execution of the code, you may notice that the code using threads takes substantially less time when compared to the code that executes sequentially. This difference arises because, in the case of threads, when one thread was busy grabbing the contents from the webpages, the other threads were not blocked and could perform their jobs in parallel.

14.9 Practice Exercises

In this section, we'll provide some practice questions to help reinforce your learning. The exercise questions cover almost all the concepts introduced in the chapter. Solutions to each exercise will follow in Section 14.10.

1. **Completing multithreaded execution with print statements**
   Complete the following code. Your task is to spawn multiple threads, each printing the message "Hi from Thread" to the console. Inside the loop, implement the code to create a thread, add it to the vector thread_vec, and print the message from each spawned thread. Finally, ensure all threads run to completion by using join to wait for each thread.

```
use std::thread;
fn main() {
    let mut thread_vec = vec![];
    for i in 0..10 {
        thread_vec.push(
        // insert code here
        // Spawn a thread
        // include a simple print statement with a msg of "Hi from Thread"
        );
    }

    // The code below will make sure that all the threads go to completion
    for i in thread_vec {
        i.join();
    }
}
```

2. **Parallel summation using multiple threads**
   Complete the following code by creating two additional threads, each calculating the summation of a specific range of numbers. The first thread, handle_1, computes the summation from 0 to 1000. Your task is to add code to spawn a second thread that sums the range from 1001 to 2000, and a third thread that sums from 2001 to 3000.

Finally, ensure that all threads complete their execution and their results are summed up correctly to produce the final summation result.

```
use std::thread;
fn main() {
    let handle_1 = thread::spawn(|| {
        let mut sum = 0;
        let range = 0..=1_000;
        for num in range {
            sum += num;
        }
        sum
    });
// Note: The thread spawn returns a joinhandle type. If there is anything
// returned from

// closure inside the thread, it will be inside the joinhandle type. In this
// case, it will be Joinhandle<i32>.

// You can access the returned i32 value by calling .unwrap() on join.

// Todo!: Insert a code for creating another thread which will compute the
// summation from 1001 - 2000

// Todo!: Insert a code for creating another thread which will compute the
// summation from 2001 - 3000
    let mut sum = 0;

// Todo!: Insert code to make sure that the summation is computed correctly.
// Summation will be computed correctly, if all the threads go to completion.

    println!("Final Summation Result {sum}");
}
```

3. **Resolving ownership issues in threaded code**
   Fix the following code so it compiles successfully. The code attempts to push a new element into the vector v from within a spawned thread. Currently, the compiler flags an error due to ownership and borrowing issues, with v being used in a separate thread. Modify the code to handle the ownership of v in such a way that allows concurrent access by the spawned thread.

```
use std::thread;
fn main() {
    let mut v = vec!["Nouman".to_string()];
    let handle = thread::spawn(|| {
```

```
        v.push("Azam".to_string());
    });
}
```

4. **Correcting ownership and lifetime issues in threads**
   Fix the following code so it compiles successfully. The code spawns a thread that tries to access the vector `v` and the variable `x`. However, `v` is moved into the thread, making it inaccessible in the `main` thread, leading to a compilation error. Modify the code to ensure both `v` and `x` are accessible where needed, while still allowing the thread to print them.

```
use std::thread;
fn main() {
    let v = vec![1, 2, 3];
    let x = 5;
    let handle = thread::spawn(move || {
        println!("Here's a vector: {:?}", v);
        println!("Here's a variable : {:?}", x);
    });
    println!("The variable x is still alive {}", x);
    println!("The variable v is not alive {}", v);    // something wrong here
    handle.join();
}
```

5. **Implementing data sending via channels in threads**
   Complete the following code to send integer values from multiple threads to the `main` thread using Rust's multi-producer, single-consumer (`mpsc`) channels. Implement the missing code in the `thread_fn` to send the integer `d` through the channel. Additionally, in the `main` function, add code to call `thread_fn` with values from `0` to `4`, ensuring each call uses a `clone` of the `tx` sender. Ensure that the `main` thread correctly receives and prints the sent values.

```
use std::sync::mpsc;
use std::thread;
fn thread_fn(d: i32, tx: mpsc::Sender<i32>) {
    thread::spawn(move || {
        println!("{} send!", d);
        // Add code for sending d
    });
}
fn main() {
    let (tx, rx) = mpsc::channel();
    for i in 0..5 {
        // Add code for calling the function with value i and a clone of tx
    }
    drop(tx);
```

14

```
    for recieving_val in rx {
        println!("{} received!", recieving_val);
    }
}
```

6. **Implementing concurrent task execution with message passing**
   The provided code attempts to utilize a multithreading approach with Rust's message passing to receive a value while performing other tasks. However, the `try_recv()` method does not allow the program to execute the "I am doing some other stuff" line since the main thread is blocked until the spawned thread sends the value. Your task is to modify the code so that it can successfully perform other operations concurrently while still waiting for messages from the spawned thread. Follow these instructions:
   - Ensure that the `main` thread can perform other operations while waiting to receive a message from the `spawned` thread.
   - Utilize an appropriate way to manage the thread's execution to achieve this goal.

```
use std::sync::mpsc;
use std::thread;
use std::time::Duration;
fn main() {
    let (tx, rx) = mpsc::channel();
    let t = thread::spawn(move || {
        let x = "some_value".to_string();
        println!("Sending value {x}");
        tx.send(x).unwrap();
    });
    t.join(); // Something wrong here
    let mut received_status = false;
    while received_status != true {
        match rx.try_recv() {
            Ok(received_value) => {
                println!("Received value is: {received_value}");
                received_status = true;
            }
            Err(_) => println!("I am doing some other stuff"), /* This line
                       never executes. Make appropriate changes in the code,
                       so that it executes. */
        }
    }
}
```

7. **Running asynchronous tasks concurrently with `tokio`**
   In this code, two asynchronous functions, `fn1` and `fn2`, execute sequentially. However, we now need to run both tasks concurrently using spawned tasks. Your goal is

to modify main to call fn1 and fn2 inside separate spawned tasks so that both functions can execute concurrently, thus reducing the total runtime. Follow these instructions:

- Adjust the main function to spawn separate tasks for fn1 and fn2.
- Ensure that both tasks can start and complete concurrently without blocking each other.

```
use tokio::time::{sleep, Duration};
async fn fn1() {
    println!("Task 1 started!");
    sleep(Duration::from_secs(3)).await;
    println!("Task 1 completed!");
}
async fn fn2() {
    println!("Task 2 started!");
    sleep(Duration::from_secs(2)).await;
    println!("Task 2 completed!");
}
#[tokio::main]
async fn main() {
    fn1().await;
    fn2().await;
}
```


14.10 Solutions

This section provides code solutions for the practice exercises in Section 14.9. The code is largely self-explanatory; however, we have included comments and additional explanations wherever necessary to enhance understanding.

1. **Completing multithreaded execution with print statements**

```
use std::thread;
fn main() {
    let mut thread_vec = vec![];
    for i in 0..10 {
        thread_vec.push(thread::spawn(|| {
            println!("Hi from the thread");
        }));
    }
    // The code below will make sure that all the threads go to completion
    for i in thread_vec {
```

```
        i.join();
    }
}
```

2. **Parallel summation using multiple threads**

```
use std::thread;
fn main() {
    let handle_1 = thread::spawn(|| {
        let mut sum = 0;
        let range = 0..=1_000;
        for num in range {
            sum += num;
        }
        sum
    });
    let handle_2 = thread::spawn(|| {
        let mut sum = 0;
        let range = 1_001..=2_000;
        for num in range {
            sum += num;
        }
        sum
    });
    let handle_3 = thread::spawn(|| {
        let mut sum = 0;
        let range = 2_001..=3_000;
        for num in range {
            sum += num;
        }
        sum
    });
    let mut sum = 0;
    sum += handle_1.join().unwrap();
    sum += handle_2.join().unwrap();
    sum += handle_3.join().unwrap();
    println!("Final Summation Result {sum}");
}
```

3. **Resolving ownership issues in threaded code**

```
use std::thread;
fn main() {
    let mut v = vec!["Nouman".to_string()];
    let handle = thread::spawn(move || {
```

```
        v.push("Azam".to_string());
    });
}
```

4. **Correcting ownership and lifetime issues in threads**

```
use std::thread;
fn main() {
    let v = vec![1, 2, 3];
    let x = 5;
    let handle = thread::spawn(move || {
        println!("Here's a vector: {:?}", v);
        println!("Here's a variable : {:?}", x);
    });
    println!("The variable x is still alive {}", x); // Note: primitives are
                                                      // not moved but copied
                                                      // so no issues here
    / println!("The variable v is not alive {}", v); // Note: Heap allocated
                                                      // data is moved so no
                                                      // longer usable
    handle.join();
}
```

5. **Implementing data sending via channels in threads**

```
use std::sync::mpsc;
use std::thread;
fn thread_fn(d: i32, tx: mpsc::Sender<i32>) {
    thread::spawn(move || {
        println!("{} send!", d);
        tx.send(d).unwrap();
    });
}
fn main() {
    let (tx, rx) = mpsc::channel();
    for i in 0..5 {
        thread_fn(i, tx.clone());
    }
    drop(tx);
    for recieving_val in rx {
        println!("{} recieved!", recieving_val);
    }
}
```

6. **Implementing concurrent task execution with message passing**

```
use std::sync::mpsc;
use std::thread;
use std::time::Duration;
fn main() {
    let (tx, rx) = mpsc::channel();
    let t = thread::spawn(move || {
        let x = "some_value".to_string();
        println!("Sending value {x}");
        tx.send(x).unwrap();
    });
    let mut received_status = false;
    while received_status != true {
        match rx.try_recv() {
            Ok(received_value) => {
                println!("Received value is: {received_value}");
                received_status = true;
            }
            Err(_) => println!("I am doing some other stuff"),
        }
    }
    t.join();  // do not be surprised if the code does alot of other stuff
               // seems like Rust Playground is not good at scheduling the
               // threads
}
```

7. **Running asynchronous tasks concurrently with tokio**

```
use tokio::time::{sleep, Duration};
async fn fn1() {
    println!("Task 1 started!");
    sleep(Duration::from_secs(3)).await;
    println!("Task 1 completed!");
}
async fn fn2() {
    println!("Task 2 started!");
    sleep(Duration::from_secs(2)).await;
    println!("Task 2 completed!");
}
#[tokio::main]
async fn main() {
    let mut handles = vec![];
    let handle_1 = tokio::spawn(async move {
        fn1().await;
    });
```

```
        handles.push(handle_1);
        let handle_2 = tokio::spawn(async move {
            fn2().await;
        });
        handles.push(handle_2);
        for handle in handles {
            handle.await.unwrap();
        }
    }
```

14.11 Summary

This chapter focused on threads and concurrency in Rust, providing a thorough understanding of thread management and communication. We began with the fundamentals of thread basics, laying the groundwork for concurrent programming. Our discussion then shifted to ownership in threads, emphasizing how Rust's ownership model applies in multithreaded contexts.

Thread communication was explored through two main approaches: message passing through channels, which enables safe communication between threads, and sharing states, which covers shared state management techniques. This chapter also introduced synchronization through barriers to coordinate thread execution, followed by an overview of scoped threads to ensure proper thread management within specific contexts.

Additionally, we delved into thread parking, explored the concept of pausing threads, and introduced `async await` to handle asynchronous programming efficiently. `tokio` tasks were discussed as a framework for building asynchronous applications, culminating in practical applications such as web scraping using threads.

In the next chapter, we'll introduce the Rust macros, and you'll learn how to write your own macros for extending the language features.

Chapter 15
Macros

Macros empower you to extend the language itself. In this chapter, we'll discover how to leverage macros for more concise and expressive code.

This chapter introduces Rust's powerful macro system, beginning with the basics of macro creation and usage. You'll learn how to capture types and repeat patterns, thus enabling code generation and even *metaprogramming*, which is the practice of writing code that can generate or manipulate other code. We'll walk through practical examples of using macros to simplify and enhance Rust code, demonstrating their flexibility and power. By mastering macros, you can write more concise and expressive code, leveraging Rust's metaprogramming capabilities.

15.1 Macro Basics

In Rust, macros offer a powerful way to write code that can generate other code, enabling flexibility and efficiency in programming. By diving into macros, you'll learn how to engage in metaprogramming in Rust using declarative macros—the kind most commonly encountered and utilized in the Rust community. As we unravel their syntax, rules, and usage, you'll see how they enable cleaner, more concise code.

15.1.1 Basic Syntax

The basic syntax of declarative macros is quite similar to a Rust `match` expression. Consider the basic syntax of macros as shown in Listing 15.1.

```
macro_rules! macro_name {
//    |--- Match rules
    (...) => { ... };
    (...) => { ... };
    (...) => { ... };
}
```

Listing 15.1 The Basic Syntax of Macros

The syntax starts with `macro_rules!`, which is also the called *macro declaration*, followed by a name for the macro and then the body of the macro. Inside the body of the macro, we have match rules. Each macro must have at least one rule and may contain many rules. The rules have a similar syntax to that of a match statement. The left side is a matching pattern, and the right side indicates the code substitution that should be made when a pattern is matched. We'll examine both the matching pattern and the code substitution in Section 15.1.2. Each rule ends with a semicolon. A semicolon is optional for the last rule.

The code shown in Listing 15.2 defines a simple macro.

```
macro_rules! our_macro {
    () => {
        1 + 1
    };
}
fn main() {
    our_macro!();
}
```

Listing 15.2 Defining a Simple Macro

The name of the macro in this case is our_macro, and it contains a simple rule. The pattern is empty, and for the code substitution part, we have 1+1. In the main function, we can invoke the macro by writing its name, similar to that of a function call. The macro will execute but some warning messages may arise, which you can just ignore for now. our_macro has no associated pattern, and therefore, its invocation in main will be substituted by its code, which is 1+1. In other words, the calling of the macro does nothing but substitute the code of 1+1 into the main program. We can confirm this substitution by enclosing the invocation inside a print statement, as in the following example:

```
fn main() {
    println!("{}", our_macro!());
}
```

When executed, the result is an output of 2.

Macros can be invoked using any type of brackets. For instance, all the invocations shown in Listing 15.3 are valid.

```
fn main() {
    our_macro!();
    our_macro![];
    our_macro! {};
}
```

Listing 15.3 Any Type of Brackets Can Be Used for Invoking a Macro

The same is also true for the left sides and right sides of the rules themselves. For instance, consider the code shown in Listing 15.4.

```
macro_rules! our_macro {
    [] => {   // you can use (), {} or [] for left side of the rule
        1 + 1
    };

    () => [
        1 + 1
    ]; // you can use (), {} or [] for right side of the rule
}
```

Listing 15.4 Any Type of Brackets Can Be Used for Left Side and Right Side of the Rule

The general convention, however, is to use parentheses for the left side of the rule and curly brackets for the right side of the rule. We'll stick to this convention in our remaining examples.

Before we move further, note that macro invocation is not strictly the same as a function call. To see this distinction clearly, recall that a function returns some expression that has no semicolon at the end of the function. However, let's add a semicolon to the end statement of the expression `1+1`, which is the last statement in the macro body, as shown in Listing 15.5.

```
macro_rules! our_macro {
    () => {
        1 + 1;
    };
}
```

Listing 15.5 Updated Definition of our_macro

The code in `main` will still produce a value of `2` when executed.

15.1.2 Matching Pattern in the Rule

The left side of the rule may contain any type of matching expression for containing anything that can be parsed and matched. Or, should we say, almost anything. Let's modify the macro shown earlier in Listing 15.3 and add another rule with some random pattern, as shown in Listing 15.6.

```
macro_rules! our_macro {
    () => {
        1 + 1;
    };
```

```
    (something@_@) => {
        println!("You found nonsense here")
    };
}
fn main() {
    our_macro!(something@_@);
}
```

Listing 15.6 Updated Macro Definition with One More Rule

The pattern in the added rule does not make any sense but is something that can be matched. The right side of the rule is the substitution code, which must be valid Rust code. This is because it is something that the macro will be expanded to. In this case, when the rule matches, the invocation of the macro will be replaced with the substitution part which is the `print` statement. Executing the code in will `print` statement inside the second rule.

The summary so far is that each rule inside a macro consists of two main parts: the left side, which holds the *matching pattern*, and the right side, which defines the Rust code to be expanded. The key point to note here is that the left side, representing the matching pattern, can include nearly any syntactically correct expression, meaning anything that can be parsed by the compiler. On the other hand, the right side, also known as the *expansion* or *body* of the rule, must contain valid Rust code written with correct syntax.

15.1.3 Captures

The patterns we've just seen don't make any sense. More useful patterns can be constructed by making use of captures, which are variables from the surrounding scope that the macro can refer to and use within its body. Captures allow a macro to include dynamic values or data from outside the macro's pattern, making the macro more flexible and powerful. Captures have the following syntax:

```
$name: (expression or type or identifier)
```

The `$` in `$name` denotes a capture variable within a macro pattern. In Rust, an *expression* is any piece of code that produces a value. Therefore, expressions could be function calls, arithmetic operations, even whole blocks of code. Let's walk through an example of expression. Consider the code shown in Listing 15.7.

```
macro_rules! our_macro {
    ...

    ($e1:expr, $e2:expr) => {
        $e1 + $e2
    };
```

```
}
fn main() {
    println!("{}", our_macro!(2, 2));
}
```

Listing 15.7 Updated our_macro Definition with One More Rule Added

The left side of the rule will match any two expressions. The rule, when matched, will be expanded to the addition of the two expressions. In this case, the result is a value of 4. Recall that an expression can be anything that produces a value; therefore, the following invocation is also valid:

```
println!("{}", our_macro!(2 + 2 + (2 * 2), 2));
```

In this case, the output is a value of 10.

In general, you can match on any number of expressions in the left side of the rule. Consider another rule in a macro, as shown in Listing 15.8.

```
macro_rules! our_macro {
    ...
    ($a:expr, $b:expr, $c:expr) => {
        $a * ($b + $c)
    };
}
fn main() {
    println!("{}", our_macro!(5, 6, 3));
}
```

Listing 15.8 Updated Definition of our_macro with an Added Rule

We'll come back to the topic of captures in Section 15.2.

15.1.4 Strict Matching

The left side of the macro needs to strictly match in the invocation. For instance, if we change the commas in the last rule shown in Listing 15.8 to that of semicolons, then the invocation in `main` will throw an error, as shown in Listing 15.9.

```
macro_rules! our_macro {
    ...
    ($a:expr, $b:expr; $c:expr) => {
        $a * ($b + $c)
    };
}
```

15

```
fn main() {
    println!("{}", our_macro!(5, 6, 3)); // Error
}
```

Listing 15.9 Changed Definition of the Left Side of the Rule with a Semicolon instead of Comma

This errors arises because invocation does not match any rule. The error can be fixed by changing the comma in the invocation to a semicolon, as shown in Listing 15.10.

```
macro_rules! our_macro {
    ...
    ($a:expr, $b:expr; $c:expr) => {
        $a * ($b + $c)
    };
}
fn main() {
    println!("{}", our_macro!(5, 6; 3)); // fixed by changing comma to
                                         // semicolon
}
```

Listing 15.10 Error in Listing 15.7 Fixed by Replacing a Comma with Semicolon

The essential point to note is that you must provide something in the invocation itself that would match at least one rule.

15.1.5 Macro Expansion

Macro expansion in Rust is the process by which a macro's code is transformed into valid Rust code before compile time. This step is necessary because macros allow you to write more concise and reusable code, and when the macro is invoked, it expands into the actual code that the compiler will process. Expanding the macro code ensures that the logic defined by the macro is applied in the correct context, allowing for flexible and efficient code generation.

The `cargo expand` command provides particularly useful insights into the macro's expansion. It basically displays the expanded code. To enable the command, we need to install `cargo-expand` using the following command:

```
c:\> cargo install cargo-expand
```

This command works on the nightly version. To install the nightly version, the following command is used:

```
c:\> rustup install nightly
```

If nightly is already installed, then you'll you enable it using the following command:

```
c:\> rustup override set nightly
```

After installing switching to the nightly version, everything will probably be fine; however, if for any reason the commands in the remaining of the same section are not working, then you may consider running the following commands:

```
c:\> rustup component add rustfmt
C:\> rustup component add rustfmt --toolchain nightly
```

Now let's see the usage of the command by considering the following single invocation of the macro in `main`:

```
fn main() {
    our_macro!();
}
```

The following command will expand the code in `main`:

```
c:\> cargo expand
```


Running the command will produce the output shown in Listing 15.11.

```
#![feature(prelude_import)]
#[prelude_import]
use std::prelude::rust_2021::*;
#[macro_use]
extern crate std;
fn main() {
    1 + 1;
}
```

Listing 15.11 Expansion of the Macro Invocation Using the Cargo Expand

The expansion shows us the underlying code that the macro generates, allowing us to see what the compiler will work with after the macro has been expanded. The prelude and the standard library crate are included by default by almost all the Rust programs. In `main`, we only have a single statement which corresponds to the expansion of the macro invocation. Note that there are also some extensions that expand the code such as Rust Macro Expand.

It is important to highlight that we have been using macros from the very beginning of the book. The `print` statement is also a macro. Considering the following code in `main`:

```
fn main() {
    println!("Hello to macros world");
}
```

Running the cargo expand for the preceding code will generate the output shown in Listing 15.12.

```
#![feature(prelude_import)]
#[prelude_import]
use std::prelude::rust_2021::*;
#[macro_use]
extern crate std;
fn main() {
    {
        ::std::io::_print(format_args!("Hello to macros world\n"));
    };
}
```

Listing 15.12 Expansion of the Simple print Statement

This shows us the long syntax for the print statement.

> **Macros for Reducing Complexity**
> Macros provide strong motivation for simplifying code. Without macros, you would need to write separate code for each type, resulting in extensive duplication and unnecessary complexity. While generics can sometimes reduce this redundancy, in some scenarios, macros are especially effective in removing unwanted complexity, ultimately making your code more readable and compact. Take the print! macro, for example. If we had to rewrite the full underlying code each time we wanted to print something, our code would quickly become unwieldy. Macros thus help by extracting repetitive details, allowing the core logic to be clearer and more manageable.

15.2 Capturing Types

Captures were introduced briefly in the previous section. In this section, we explore it in greater detail. We'll start with a special form of capture, called type capture, and then we'll move on to identifier capture.

15.2.1 Type Capture

A type capture refers to a macro pattern that matches and captures a type, allowing the macro to use it dynamically within its expanded code. It is denoted by `$name:ty` where `$name` is the variable capturing the type and `ty` indicates that the capture is specifically for types. Let's go through a user input taking example.

Recall from Chapter 3 that handling user input in Rust requires writing several lines of code, which can feel cumbersome for users accustomed to simpler syntax in other languages. To streamline this process, we can define a macro that abstracts away the complexity, making user input more accessible and straightforward. Let's create a macro to simplify this task.

Consider the implementation of the input macro in Listing 15.13.

```
macro_rules! input {
    ($t: ty) => {
        let mut n = String::new(); // Error
        std::io::stdin()
            .read_line(&mut n)
            .expect("failed to read input");

        let n: $t = n.trim().parse().expect("invalid input");
        n
    };
}
```

Listing 15.13 Partially Implemented Macro for Taking User Input

You may recall from Chapter 3, Section 3.3.3, that the last line in the code is used for converting the user input to a desired data type. The desired type will be passed to this macro when it is invoked and will be captured by a special form of capture (i.e., ty) designed for grabbing or matching on types in Rust. The ty is a type capture, which can be any Rust data type. In the last line of the code, we used the type passed in to set the type of variable n. Thus, the variable n can be any type that has been matched by the type capture, and it will be provided at the time of macro invocation. At the end of the macro, we return the desired input by writing n without a semicolon. The lack of a semicolon at the end of the variable n indicates that it is the return value of the block, as Rust implicitly returns the last expression in a block when there is no semicolon. In this way, the macro can evaluate to the value of n, making it the result of the macro invocation.

Let's invoke this now from the main with the input of i32, as shown in Listing 15.14.

```
fn main() {
    println!("Please enter a floating point number");
    let some_input_0 = input!(f32);
}
```

Listing 15.14 Invoking the input Macro in main

You may note that the compiler is not happy and giving us errors in the definition of the macro. Let's try to fix the errors.

When the compiler tries to expand the code in main, it will substitute the code of the macro. When this substitution is made, the opening and closing curly brackets of the right side of the macro rule are not part of the expansion. This means that only the lines shown in Listing 15.15 will be substituted in `main`.

```
let some_input_0 =
    let mut n = String::new();
    std::io::stdin()
        .read_line(&mut n)
        .expect("failed to read input");

   let n: $t = n.trim().parse().expect("invalid input");
    n
```

Listing 15.15 Lines from the Macro That Are Substituted in main after Invocation

The results from multiple lines of code cannot be assigned to a variable using the `let` keyword. In this case, the lines are all assigned to the variable `some_input_0`, which is not valid Rust syntax. Remember that the right side of the rule must have valid Rust syntax which is not the case here. To make it valid Rust code, we'll enclose the code using curly brackets. The valid Rust code is given in Listing 15.16.

```
let some_input_0 = {
        let mut n = String::new();
        std::io::stdin()
            .read_line(&mut n)
            .expect("failed to read input");

        let n: f32 = n.trim().parse().expect("invalid input");
        n
    };
```

Listing 15.16 Valid Rust Code

The code lines from Listing 15.15 are now enclosed in curly brackets.

To fix the code in the macro rule, we'll enclose the right side of the rule in an additional curly bracket. The updated code is shown in Listing 15.17.

```
macro_rules! input {
    ($t: ty) => {{
        let mut n = String::new();
        std::io::stdin()
            .read_line(&mut n)
            .expect("failed to read input");

        let n: $t = n.trim().parse().expect("invalid input");
        n
```

```
    }};
}
```

Listing 15.17 Updated Code of the input Macro

This fixes the code in main in Listing 15.14. The additional brackets now treat the code in the expansion part as a code block which returns the variable n.

Let's consider one more example of the type capture. Consider an add_as macro, which will add numbers in different types. The macro is defined in Listing 15.18.

```
macro_rules! add_as {
    ($a: expr, $b: expr, $typ: ty) => {
        $a as $typ + $b as $typ
    }
}
```

Listing 15.18 Definition of the add_as Macro for Adding Numbers in Different Types

The macro has three captures consisting of two expressions and one type. The expressions passed in will be added together based on the specified type passed in to the macro. Listing 15.19 shows the invocation of the macro from main with a sample input.

```
fn main() {
    println!("{}", add_as!(15,2.3,f32));
    println!("{}", add_as!(15,2.3,f32));
}
```

Listing 15.19 Invocations of the Macro add_as in main

The two expressions in the macro invocation will match the values of 15 and 2.3 respectively, while the type will match to that of f32 in case of the first invocation. The second invocation will consider the type to be an i32 and will do the computation in i32 type.

15.2.2 Identifiers Capture

Identifiers in a program are names associated with elements such as variables and functions, and these names allow you to reference and identify specific parts of the code. Let's first look at some examples that highlight the need for identifiers in macros.

Consider the macro and its invocation in main, as shown in Listing 15.20.

```
macro_rules! some_macro {
    () => {
        let mut x = 4;
    };
}
```

```
fn main() {
    some_macro!();
    x = x+1; // Error
}
```

Listing 15.20 A Simple Macro and Its Invocation in main

The left side of the rule is empty, meaning that it will match if nothing is passed to the macro. The right side simply expands to a variable initialized from a value of 4. In main, we invoke the macro and then add the statement of x = x+1 after the invocation. This process should work fine because the macro expands to let mut x = 4, and then after that step, we are adding 1 to it. Let's look at the expansion of the code using the cargo expand command, as shown in Listing 15.21.

```
#![feature(prelude_import)]
#[prelude_import]
use std::prelude::rust_2021::*;
#[macro_use]
extern crate std;
fn main() {
    let mut x = 4;
    x = x + 1;
}
```

Listing 15.21 Expansion of the Code from Listing 15.20

From this expansion, note that you do have valid Rust code in main. However, the compiler is not happy, saying that it "cannot find x in this scope."

This error arises in this case because the identifiers (in this case, the variable x), which reside inside the macro scope or body, cannot be directly accessed or referenced outside of it. Attempting to do so would mean *crossing boundaries*, as it would involve using an identifier from the macro's internal scope in the surrounding code, which is not allowed due to Rust's strict scoping rules. For this reason, macros are also sometimes described as *hygienic*. A hygienic macro in the context of Rust means that the identifiers declared within a macro are distinct and do not unintentionally conflict with or interfere with identifiers in the surrounding code. This goal guarantees that variables or other elements defined inside the macro remain isolated, so that you can prevent naming clashes and unexpected behavior.

Let's change the code shown earlier in Listing 15.20 to the code shown in Listing 15.22.

```
macro_rules! some_macro {
    () => {
        x = x + 1; // Error
    };
```

```
}
fn main() {
    let mut x =4;
    some_macro!();
}
```

Listing 15.22 Updated Code from Listing 15.20: The Variable Now Defined in main Still Throws an Error

In this case, the variable x is now explicitly defined in main via the let keyword (which was not the case earlier in Listing 15.20). Since the identifiers in the main scope are distinct from those in the macro scope, the variable x will not be transferred to the macro scope, which leads to an error in the code of macro. In this scenario, identifiers come into play because they allow you to cross boundaries or scopes that might otherwise not be possible.

To correct the code shown in Listing 15.22, we'll use an identifier in the left side of the rule. Listing 15.23 shows the correct definition of the macro using identifiers.

```
macro_rules! some_macro {
    ($var: ident) => {
        $var = $var + 1; // Error
    };
}
fn main() {
    let mut x = 4;
    some_macro!(x);
}
```

Listing 15.23 Updated Definition of the Macro Using Identifiers

The syntax for an identifier capture is $var: ident, where $var is a capture placeholder that matches and captures an identifier (such as a variable name) in the code where the macro is invoked. The ident part specifies that the macro is expecting an identifier, meaning it will match variable names, function names, or any valid Rust identifier. As shown in Listing 15.23, the left side of the rule is using an identifier of $var. In the expansion part, we use that identifier again instead of simple x. Finally, in the invocation, we pass in the variable x. The macro now knows that $var is now an identifier, which is basically the name x, which will be expanded to x in the expansion part.

Now that you understand the need for identifier capture, let's go over an example of identifiers. Consider a macro that will create a new function, as shown in Listing 15.24.

```
macro_rules! create_function {
    ($func_name:ident, $input: ident, $type_input: ty, $type_output: ty) => {
        fn $func_name($input: $type_input) -> $type_output {
```

```
            println!(
                "You called {:?}() with the input of {:?}",
                stringify!($func_name),
                stringify!($input)
            );
            $input
        }
    };
}

create_function!(f1, x, i32, i32);
fn main() {
    println!("Function f1 should returns: {}", f1(15));
}
```

Listing 15.24 Macro for Creating a New Function

The macro will create a function with one input and one output. The left side of the rule contains four types of captures. The first capture is an identifier, which will be the function name. The second capture is the input variable name; therefore, it is also an identifier. The third capture is the type of the input and is therefore a type capture. The fourth capture will be the output data type and therefore is a type capture. In the expansion part, we used the captures to define a function using the syntax of a function signature. Inside the function, we are printing the function name and the input and returning the input value passed in. The `stringify` macro converts the input passed into its respective `String`. We are invoking this macro outside the `main` to create a new function `f1`. In `main`, we call the function `f1` to check whether it has been created.

Macros and Ownership

Notice that macros do not take ownership of anything. You need to keep an eye on the expansion only. Variables retain their ownership as long as we do not change their ownership in the expansion of the code.

15.3 Repeating Patterns

Repetition in code often invites opportunities for refinement, and macros are powerful tools for such cases. By capturing and reusing patterns, macros simplify code structure, making it both more compact and easier to manage.

Let's consider defining a macro for concatenating strings that can handle any number of input strings. If there are no inputs, the macro will produce an empty string. If there's only one input string, it will simply return that string. For multiple inputs, the macro will create a new string that concatenates all the input strings. Let's begin by defining this macro. Listing 15.25 shows the definition of the macro rule for the first case, when the macro is invoked with no arguments.

```
macro_rules! string_concat {
    () => {
        String::new();
    };
}
```

Listing 15.25 Definition of the string_concat Macro for Concatenating Strings

Let's add one more rule for cases when we have only a single input. The updated definition of the macro is shown in Listing 15.26.

```
macro_rules! string_concat {
    () => {
        String::new();
    };

    ($some_str: expr) => {{
        let mut temp_str = String::new();
        temp_str.push_str($some_str);
        temp_str
    }};
}
```

Listing 15.26 Updated string_concat Macro with an Added Rule for Handling Input String

In the second rule, we first create an empty `String` and then add the contents of the input `String` captured by the expression to it. Finally, we return the created `String`. Note that the additional curly brackets are added to properly return a value, as explained in the previous section.

Next, we'll add one more rule to handle cases when two input strings are available to concatenate. Listing 15.27 shows the updated definition with the added third rule.

```
macro_rules! string_concat {
    () => {
        String::new();
    };

    ($some_str: expr) => {{
        let mut temp_str = String::new();
        temp_str.push_str($some_str);
        temp_str
    }};
```

```
    ($some_s1: expr, $some_s2:expr) => {{
        let mut temp_str = String::new();
        temp_str.push_str($some_s1);
        temp_str.push_str($some_s2);
        temp_str
    }};
}
```

Listing 15.27 Updated string_concat Macro with an Added Rule for Handling Two Input Strings

The code in the third rule is similar to the code in the second rule. First, we created an empty `String` and then added the two strings to it and finally returned the created string.

This approach works correctly but is not efficient. Consider a case where we have three, four, or five or more input strings. You would need to keep on adding rules for each case. To properly handle such cases, the Rust compiler provides repeated macro arguments. The repeating arguments can handle variable numbers of inputs. This flexibility allows the macro to match and process multiple inputs flexibly, making it especially useful for tasks like concatenating multiple strings or generating similar code structures for various inputs.

The repeated sequences are mentioned using the syntax `$((...),*)`. The repeated arguments are inside the inner parentheses. For instance, `$(($a: expr),*)` means that we want to repeat the expression multiple times. After the inner parentheses, we need to mention a delimiter, which is typically a comma (,). The delimiter is next followed by a repetition operator which can be a (+), a star (*), or a question mark (?). Plus means 1 or more times of the repeated sequence which is given by the code inside the inner parenthesis, a star will mean 0 or more times and a question mark will mean 0 or 1 times.

Consider the revised version of the `string_concat` in Listing 15.28 based on the repeating arguments.

```
macro_rules! string_concat {
    ($($some_str:expr),*) => {{
            let mut temp_str = String::new();
            $(temp_str.push_str($some_str);)*
            temp_str
    }};
}
```

Listing 15.28 Updated string_concat Macro Based on Repeating Arguments

The left side of the rule will match something that is an expression and that repeats 0 or more times and will be separated by a comma. In the expansion part of the rule, we declare an empty vector. The next line `$(temp_str.push_str($some_str);)*` means that, for as many times as we've matched the expression in the left side, we want to want to push that many strings into the `temp_str`. The syntax of `$(...)*` will repeat as the code inside the parentheses as many times as we have matches in the left side of the rule. Finally, we return the `temp_str` back.

This code will work for zero, one, or two and more strings. For instance, consider the code shown in Listing 15.29.

```
fn main() {
    let str_null = string_concat!();
    let str_single = string_concat!("First");
    let str_double = string_concat!("First", "Second");
}
```

Listing 15.29 Calling the Macro string_concat in main

This now revised definition of the macro shown earlier in Listing 15.28 eliminates the need for defining separate rules for different numbers of inputs.

The code in shown Listing 15.30 shows the expansion of the code in `main`.


```
fn main() {
    let str_null = String::new();
    let str_single = {
        let mut temp_str = String::new();
        temp_str.push_str("First");
        temp_str
    };
    let str_double = {
        let mut temp_str = String::new();
        temp_str.push_str("First");
        temp_str.push_str("Second");
        temp_str
    };
}
```

Listing 15.30 Expansion of the Code from Listing 15.29

For the first invocation of the macro, we have a single line expansion. For the second invocation and third invocation, however, we have the correct expansion of the macro through its repeating arguments.

A few points to note with regard to delimiters: First, a space is the default delimiter. For instance, the code shown in Listing 15.31 will compile.

```
macro_rules! string_concat {
    ($($some_str:expr)*) => {{ // no delimiter => default space delimiter
            let mut temp_str = String::new();
            $(temp_str.push_str($some_str);)*
            temp_str
        }
    };
}
fn main() {
    let str_null = string_concat!();
    let str_single = string_concat!("First");
    let str_double = string_concat!("First" "Second"); // Space is used to
                                                        // separate values
}
```

Listing 15.31 Default Delimiter Is a Space

Other allowed delimiters are ; and =>. You must pay special attention when adding a delimiter. For instance, if you put the delimiter inside the inner parentheses in the syntax for left side of the rule (e.g., `($(...,)*)`, the comma will be part of the repeated sequence and will be mandatory after each and every repeated sequence. In other words, the comma is no longer treated as a delimiter but is instead part of the repeated sequence itself. The delimiter in this case is absent, and therefore, the program will use the default delimiter, which is a space. For instance, consider the code shown in Listing 15.32.

```
macro_rules! string_concat {
    ($($some_str:expr,)*) => {{
            let mut temp_str = String::new();
            $(temp_str.push_str($some_str);)*
            temp_str
        }
        };
}

fn main() {
    let str_null = string_concat!();
    let str_single = string_concat!("First"); // Error
    let str_double = string_concat!("First" "Second"); // Error
}
```

Listing 15.32 Delimiter inside the Repeated Pattern

Note how the compiler is not happy with the second and third invocations. These errors arise because the sequence that we want to match is not followed by the mandatory comma (,). To correct the code, update the code in `main` by adding the mandatory comma after each provided input, as shown in Listing 15.33.

```
fn main() {
    let str_null = string_concat!();
    let str_single = string_concat!("First",);
    let str_double = string_concat!("First", "Second",);
}
```

Listing 15.33 Fixing the Code from Listing 15.32 by Adding the Mandatory Comma after Each Input

15.4 Practice Exercises

In this section, we'll provide some practice questions to help reinforce your learning. The exercise questions cover almost all the concepts introduced in the chapter. Solutions to each exercise will follow in Section 15.5.

1. **Expanding a macro for generating multiple functions with return expressions**
   Consider the following code. Show the expansion part of this code, especially for the invocation to macro.

```
macro_rules! make_functions {
    ($($func_name:ident: $return_type:ty => $return_expr:expr),+) => {
        $(
            fn $func_name() -> $return_type {
                $return_expr
            }
        )+
    };
}
make_functions!(foo: i32 => 42, bar: String => "hello world".to_owned());
fn main() {
    let result1 = foo();
    let result2 = bar();
    println!("foo result: {}", result1);
    println!("bar result: {}", result2);
}
```

2. **Expanding a macro for struct creation with custom fields**
   Let's say you want to create a macro called `make_struct` that will create a new `struct` containing some fields. The input to the macro is the name of the `struct` and the name of the fields along with their types. The skeleton of the macro along with its left sides of the rules are given.

   You're required to write the code for the expansion (i.e., the right side of the rule).

```
macro_rules! make_struct {
    ($name:ident {$($field:ident: $ty:ty),* }) => {
       // Your code here
        }
    };
}
```

3. **Implementing a custom vector initialization macro**
   Design a custom macro called vec_mac! to initialize a Vec and populate it with elements. Only a skeleton of the macro is provided. Your task is to complete its implementation.

```
macro_rules! vec_mac {
        ($(),*) => {{ // This line needs to be completed
            let mut some_vec = Vec::new();
            // needs a line
            some_vec
        }
    };
}
```

4. **Expanding a macro for summing multiple expressions**
   Consider the following code. Write the expanded version of the code that can be viewed using the cargo expand utility.

```
macro_rules! sum_macro {
    ($($x:expr),*) => {
        {
            let mut sum = 0;
            $(sum += $x;)*
            sum
        }
    };
}
fn main() {
    let result = sum_macro!(1, 2, 3, 4, 5);
}
```

15.5 Solutions

This section provides the code solutions for the practice exercises in Section 15.4. The code is largely self-explanatory; however, we have included comments and additional explanations wherever necessary to enhance understanding.

1. **Expanding a macro for generating multiple functions with return expressions**

```
fn foo() -> i32 {
    42
}
fn bar() -> String {
    "hello world".to_owned()
}
fn main() {
    let result1 = foo();
    let result2 = bar();
}
```

2. **Expanding a macro for struct creation with custom fields**

```
macro_rules! make_struct {
    ($name:ident {$($field:ident: $ty:ty),*}) => {
        struct $name {
            $($field: $ty),*
        }
    };
}

// Sample usage
make_struct!(MyStruct {
    field1: i32,
    field2: String
});

fn main(){}
```

3. **Implementing a custom vector initialization macro**

```
macro_rules! vec_mac {
        ( $($element: expr), *) => {{
            let mut some_vec = Vec::new();
            $(some_vec.push($element);)*
            some_vec
        }
    };
}
```

4. **Expanding a macro for summing multiple expressions**

```
fn main() {
    let result = {
        let mut sum = 0;
        sum += 1;
```

```
            sum += 2;
            sum += 3;
            sum += 4;
            sum += 5;
            sum
        };
    }
```

15.6 Summary

This chapter introduced macros in Rust, focusing on their fundamental aspects and practical applications. We began with the basics, explaining what macros are, how they differ from functions, and their role in enhancing code efficiency. We then explored capturing types, demonstrating how macros can capture and manipulate types dynamically to create flexible and reusable code structures.

Next, we examined the concept of repeating patterns, showcasing how macros can streamline repetitive tasks by enabling concise code generation. This feature is particularly useful for reducing boilerplate code and improving maintainability. This chapter concludes with exercise questions and solutions to help solidify understanding and encourage hands-on practice with Rust macros.

Next, we'll cover web programming, including the basics of setting up a web server to handle HTTP requests and responses.

Chapter 16
Web Programming

The web is a vast frontier for programmers. This chapter will introduce you to the basics of web programming with Rust, where your ideas can come to life online.

This chapter covers the fundamentals of web programming in Rust, including setting up a basic web server and handling HTTP requests and responses. You'll learn how to create and manage multiple requests using threads, ensuring high performance and scalability. The chapter provides practical examples of building web applications and services, highlighting Rust's strengths in web development. By understanding these concepts, you'll be equipped to create robust and efficient web applications using Rust.

16.1 Creating a Server

A web server delivers content for a website to a client that requests it. Typically, the client is a web browser. The client will make a request, and the web server will respond back to the request by providing the correct requested information.

We'll cover the fundamentals of implementing servers in the following sections, but first, let's go over a few key concepts.

16.1.1 Server Basics

Communication between a client and server relies on several protocols, with the two primary ones being the Hypertext Transfer Protocol (HTTP) and the Transmission Control Protocol (TCP). Both are *request-response protocols*, meaning a client initiates requests, and a server listens and responds. The structure of these requests and their responses is defined by the protocols themselves.

TCP operates at a lower level, detailing how information is transmitted between servers without specifying the content of the information. While we won't focus on TCP's lower-level mechanics in this book, TCP is a reliable mechanism for data transfer between two endpoints, even over an unreliable physical medium.

HTTP builds on top of TCP by defining the content of requests and responses. While technically HTTP can work with other protocols, it most commonly sends its data over

TCP. In this chapter, we'll work directly with the raw bytes of TCP and HTTP requests and responses.

Additionally, a few important terms to understand include the following:

- **IP address**
  This value is a unique identifier formatted to identify a device on the internet or on a local network.
- **Port number**
  A port number identifies the specific process to which an incoming network message should be forwarded on the given IP address. When a program is running, it becomes a process, and its port number specifies the exact process on a system.

IP Address versus Port Number

To understand IP addresses and port numbers, think of a house and its residents. The IP address is like the house address, identifying a specific location. The port number is like a resident in the house, identifying the specific person a message should reach.

- **Socket**
  A socket is one endpoint of a two-way communication link between two network processes. It consists of an IP address and a port number, allowing TCP to identify the exact process at a specific IP to which data should be delivered. In summary, a socket is simply a combination of an IP address and a port number.

16.1.2 Implementing a Server

In the standard library, the `std::net` module provides the basic implementation for creating a web server using the `TcpListener` and `TcpStream` types. Consider the code shown in Listing 16.1.

```
use std::net::{TcpListener, TcpStream};
fn main() {
    let listener = TcpListener::bind("127.0.0.1:8000").unwrap();
    let stream = listener.accept();
    println!("The {:?}", stream.as_ref().unwrap().0);
}
```

Listing 16.1 Implementation of a Simple Server

The `bind` function in Rust's `TcpListener` module creates a new TCP listener, binding it to a specified IP address and port. In this example, `TcpListener::bind("127.0.0.1:8000")` binds the `listener` to the IP address `127.0.0.1` on port `8000` of a localhost, setting up the server to listen for incoming TCP connections on this address. If the binding is successful, it returns a `TcpListener` instance; otherwise, `unwrap` is called, which will panic if an error arises. The idea of localhost is to access the network services that are running on the same machine via the *loopback* network interface. This bypasses any local *network interface hardware*. The 8000 is the port number in this case. The variable `listener` is now ready to accept any new incoming TCP connections.

The `accept` method on the `listener` accepts a new connection from client. This method returns a `Result`. If the connection is successful, a tuple is returned in an `Ok` variant containing an instance of `TcpStream` type and the socket address. `TcpStream` in Rust represents an active TCP connection between a client and a server, allowing bidirectional communication. This type contains a file descriptor for the underlying network socket, the local and remote socket addresses as well as methods for sending and receiving data. The second part of the tuple returned in the `Ok` variant of `listener.accept()`, shown in Listing 16.1, contains an instance of `SocketAddr`, which represents the remote client's IP address and port number. This address identifies the client that has established a connection with the server. The `print` statement displays the information corresponding to the `stream`. Remember that the value inside the `Ok` variant is a tuple, and we can use the index notation to access the elements within this tuple.

If you execute the code shown in Listing 16.1, you may not see any output in the terminal. The prompt will not be released because the server is running in the background. To communicate with the server, open up any web browser and try to connect to the localhost server on port number 8000 by entering "http://127.0.0.1:8000/" in the address bar and then pressing `Enter`. This step will display the following output in the terminal:

```
The TcpStream { addr: 127.0.0.1:8000, peer: 127.0.0.1:52780, socket: 216 }
```

Let's break down this output:

- `addr: 127.0.0.1:8000` indicates the address and port where the server (`listener`) is bound.
- `peer: 127.0.0.1:52780` shows the client's IP address and port. The client has the same IP address as that of the server because both the client and server are currently running on the same machine. The port number of the client, however, is different, which makes sense because the server and the client are different entities residing on the same system. In particular, the client is the web browser in our case, and the server is the program that is currently in execution. The specific port is assigned by the operating system and may vary with each connection.
- `socket: 216` is a system-assigned identifier for the TCP socket, unique to this connection within the application.

16.1.3 Handling Multiple Connections

The current server only accepts a single connection. Once a connection is requested, the server accepts the request and then terminates with no response. Let's extend our server to listen to multiple connections by adding a loop to the code shown earlier in Listing 16.1. The updated code is shown in Listing 16.2.

```
use std::net::{TcpListener, TcpStream};
fn main() {
    let listener = TcpListener::bind("127.0.0.1:8000").unwrap();
    for i in 0..10 {
        match listener.accept() {
            Ok((socket, addr)) => println!("The client info: {:?}", addr),
            Err(e) => println!("Couldn't get client: {:?}", e),
        }
    }
}
```

Listing 16.2 Server for Handling Multiple Connections

Inside a `for` loop, we are matching on the `listener.accept`. As mentioned earlier, a call to `accept` returns a `Result` therefore we have added a couple of arms. When the connection is successful, we want to print the client information, which is stored in the second item of the tuple inside the `Ok` variant. The second arm occurs when the connection is not successful.

Now, if we execute the code shown in Listing 16.2 and try to connect through the browser, the client information is immediately displayed on the terminal. Note that, in this case, the server will terminate only after handling `10` connections. Also, when you attempt a single connection, the `Ok` arm's `print` statement executes multiple times, which suggests that more than one connection is being established and managed. This repetition occurs because no response was received from the server, thus prompting retries in the hope of obtaining a reply from the server. We'll look at the response from a server in more detail shortly. Each time you attempt to reconnect, multiple retries follow in expectation that the server will eventually respond. When the number of attempts exceeds 10, the server will be shut down.

To keep listening for incoming connections indefinitely, rather than stopping after a fixed number of connections, you can use the `incoming` function. Consider the code shown in Listing 16.3.

```
use std::net::{TcpListener, TcpStream};
fn main() {
    let listener = TcpListener::bind("127.0.0.1:8000").unwrap();
    for stream in listener.incoming() {
        println!("The stream {:?}", stream.unwrap());
    }
}
```

Listing 16.3 Using the incoming Function to Listen Indefinitely for Incoming Connections

The incoming method returns an iterator over the connections received on the listener. The type of the stream in this case is a Result containing either a TcpStream or error.

16.1.4 Adding a Connection Handling Function

To properly handle a request and then send a meaningful reply back to the client based on the contents of the TcpStream, let's now add a function called handle_connection. This function will essentially accept a TcpStream and display the contents of the HTTP request from the client. The updated code in the main function, along with the definition of the handle_connection function is shown in Listing 16.4.

```
use std::io::{BufRead, BufReader};
use std::net::{TcpListener, TcpStream};
fn main() {
    let listener = TcpListener::bind("127.0.0.1:8000").unwrap();
    for stream in listener.incoming() {
        let stream = stream.unwrap();
        handle_connection(stream);
    }
}

fn handle_connection(mut stream: TcpStream) {
    let buf_reader = BufReader::new(&mut stream);
    let http_request = buf_reader
        .lines()
        .map(|result| result.unwrap())
        .take_while(|lines| !lines.is_empty())
        .collect::<Vec<String>>();

    println!("Request: {:#?}", http_request);
}
```

Listing 16.4 Adding a handle_connection Function to the Code from Listing 16.3

The handle_connection function takes a mutable variable called stream, which reads the data from the TcpStream passed in. The BufReader::new(&mut stream) wraps the TcpStream in a buffered reader type, which helps improve the efficiency of reading data from the stream. This type is defined in the BufReader module.

BufReader Improves Efficiency

BufReader::new(&mut stream) wraps the TcpStream in a buffered reader, which improves efficiency by reducing the number of system calls made when reading data. Without buffering, each read operation would directly interact with the underlying operating system, potentially reading small chunks of data at a time, which is slow and inefficient. Instead, BufReader reads larger chunks of data at once and stores them in an internal buffer, allowing subsequent reads to access the buffered data without making additional system calls. This approach reduces overhead and improves performance, especially when processing multiple lines of an HTTP request.

The next part of the code extracts the lines contained in the requested TcpStream. The lines method on buf_reader creates an iterator that reads the TcpStream line by line. In particular, the lines method reads data from the BufReader until it encounters a newline character (\n). Each line is returned as a Result<String, std::io::Error>. Next, the map is applied to each line returned by the lines. More specifically, the map(|result| result.unwrap()) calls .unwrap() on each Result, extracting the string if it's Ok, or panicking if it encounters an error. The take_while(|lines| !lines.is_empty()) function keeps reading lines until it encounters an empty line, which often signifies the end of an HTTP request header. It stops the iterator at the first empty line, ignoring any lines after that. Finally, the collect function converts the result into a vector of strings. At the end of the function, we display the contents of the http_request.

When you execute the code shown in Listing 16.4 and try to send a request to server, you should see the response in the terminal shown in Listing 16.5.

```
Request: [
    "GET / HTTP/1.1",
    "Host: 127.0.0.1:8000",
    "Connection: keep-alive",
    "Cache-Control: max-age=0",
    "sec-ch-ua: \"Chromium\";v=\"130\", \"Google Chrome\";v=\"130\", \"Not?A_
Brand\";v=\"99\"",
    "sec-ch-ua-mobile: ?0",
    "sec-ch-ua-platform: \"Windows\"",
    "Upgrade-Insecure-Requests: 1",
    "User-Agent: Mozilla/5.0 (Windows NT 10.0; Win64; x64) AppleWebKit/537.36
(KHTML, like Gecko) Chrome/130.0.0.0 Safari/537.36",
    "Accept: text/html,application/xhtml+xml,application/xml;q=0.9,image/
avif,image/webp,image/apng,*/*;q=0.8,application/signed-exchange;v=b3;q=0.7",
    "Sec-Fetch-Site: none",
    "Sec-Fetch-Mode: navigate",
    "Sec-Fetch-User: ?1",
    "Sec-Fetch-Dest: document",
    "Accept-Encoding: gzip, deflate, br, zstd",
    "Accept-Language: en-US,en;q=0.9,ur;q=0.8",
]
```

Listing 16.5 Captured HTTP Request Headers from an Incoming Connection

On some browsers, you may notice multiple displays for the request in the terminal. As mentioned previously, this duplication occurs because some browsers will try to reattempt a connection with the server since we have not included any response code from the server. In some cases, you may also notice repeated empty requests, which may occur if a connection is made without sending data. Alternatively, data might be incomplete, and the resulting `http_request` vector could be empty, thus leading to an empty print output.

Let's now inspect the request. The first line `GET / HTTP/1.1` is called the request line and it contains information such as the HTTP method used the unique resource identifier (URI), which is similar to the URL and the HTTP version number. In this case, we have a `GET` request on the root path indicated by the backslash (\), which corresponds to local host. Even though it's not visible here, the request line ends with a carriage return (`\r`) and line feed sequence (`\n`) which is basically something that separates the request line from the rest of the request data underneath the request line. We have various request headers and request bodies. Since this is a `GET` request, we don't have a body.

16.2 Making Responses

16

Now, let's learn how to write out a response to a request from a client. The response from the server to a connection request from a client must be a certain format. Let's first look at the syntax and then dive into how to make a valid response and how to return different responses.

16.2.1 Response Syntax

Listing 16.6 shows the syntax of a response.

```
HTTP-Version Status-Code Reason-Phrase CRLF
headers CRLF
message-body

ex: HTTP/1.1 200 OK\r\n\r\n
```

Listing 16.6 Response Syntax

First, we have the status line, which consists of the HTTP version, the status code, the reason phrase, and a carriage return line feed sequence. The status codes indicate whether a specific HTTP request has been successfully completed or not. These codes are issued by a server in response to a client's request made to the server. The `Reason-Phrase` is intended as a short description of the `Status-Code` for human users. Finally, `CRLF` refers to the special character elements carriage return (`\r`) and line feed (`\n`). These elements are embedded in HTTP headers to signify an end of line marker. Next, we list the headers followed by a carriage return line feed sequence. The header field passes additional context and metadata about the request or response. For example, a request message can use headers to indicate its preferred media formats. Finally, we have the message body. The message body contains the data bytes transmitted in an HTTP transaction message.

The last line in Listing 16.6 shows an example response using the syntax. First, we specify the HTTP version followed by the status code 200, which is the standard status code for success. Next, we mention the reason phrase, which is OK, and finally two carriage return line feed sequences. The first one is for the status line, and the second carriage return sequence is for the header, which in this case is empty.

Let's use this example response in the code shown in Listing 16.7.

```
use std::io::Write;
...
fn handle_connection(mut stream: TcpStream) {
    let buf_reader = BufReader::new(&mut stream);
    let http_request = buf_reader
        .lines()
        .map(|result| result.unwrap())
        .take_while(|lines| !lines.is_empty())
        .collect::<Vec<String>>();
    println!("Request: {:#?}", http_request);

    let response = "HTTP/1.1 200 OK\r\n\r\n";
    stream.write(response.as_bytes()).unwrap();
    stream.flush().unwrap();
}
```

Listing 16.7 Updated Definition of handle_connection

We've added the last three lines of code to the definition of `handle_connection` shown earlier in Listing 16.4. First, we store the response in a variable and next write it to the stream. The `write` function (defined in `use std::io::Write`) sends an HTTP response back to the client over the `TcpStream`. The `flush` function forces any data buffered in the `TcpStream` to be sent immediately to the client. When you call `stream.flush().unwrap();`, `flush` ensures that the response data ("`HTTP/1.1 200 OK\r\n\r\n`") is transmitted without delay. Without calling `flush`, the data might remain in the buffer temporarily, especially if the buffer isn't full, which could delay the client's receipt of the response.

Now, when you access localhost from the web browser, notice how, instead of an error page or a page not working message, you see a blank page. The server is responding but with an empty message body—a good sign because we are at least providing a response to the client's request.

16.2.2 Responding with Valid HTML

To respond with some valid HTML, let's create a new file called *index.html* at the root of our project. The updated directory structure will look as follows:

```
web_programming/
├── Cargo.toml
├── index.html
├── src/
│   └── main.rs
└── target/
```

In the *index.html* file, let's place some valid HTML, as shown in Listing 16.8.

```
<!DOCTYPE html>
<html lang="en">
    <head>
        <meta charset="utf-8">
        <title>Hello!</title>
    </head>
    <body>
        <h1>Hello!</h1>
        <p> Hi from web programming in Rust </p>
    </body>
</html>
```

Listing 16.8 Valid HTML in the index.html File at the Root Level

The HTML has content in the body that will be visible on the webpage. We can now return the HTML file in response to the request. Consider the code shown in Listing 16.9.

```
use std::fs;
...
fn handle_connection(mut stream: TcpStream) {
    let buf_reader = BufReader::new(&mut stream);
    let http_request = buf_reader
        .lines()
        .map(|result| result.unwrap())
        .take_while(|lines| !lines.is_empty())
        .collect::<Vec<String>>();
    println!("Request: {:#?}", http_request);

    let status_line = "HTTP/1.1 200 OK \r\n";
    let contents = fs::read_to_string("index.html").unwrap();
    let length = contents.len();
```

16

```
    let responce = format!(
        "{} Contents-Length: {}\r\n\r\n{}",
        status_line, length, contents
    );
    stream.write_all(responce.as_bytes()).unwrap();
    stream.flush().unwrap();
}
```

Listing 16.9 Responding to Client with a Valid HTML

In this case, we have broken down the response into three pieces: status_line, the message body or contents, and the length of the body. The status_line sets the HTTP status line to indicate a successful (200 OK) response. The variable contents read in the contents contained in the index.html into String. The function read_to_string is used for this purpose, which is defined in the std::fs module. Next, we compute the length of the contents, and then using the format! macro, we concatenate the response in the proper format to make up a valid response. The last two lines are same as the code shown earlier in Listing 16.7, which first writes to the stream and then ensures that it immediately reaches its intended destination.

Now, when you execute the program and try to access the localhost again, the response is immediately available in the web browser, as shown in Figure 16.1.

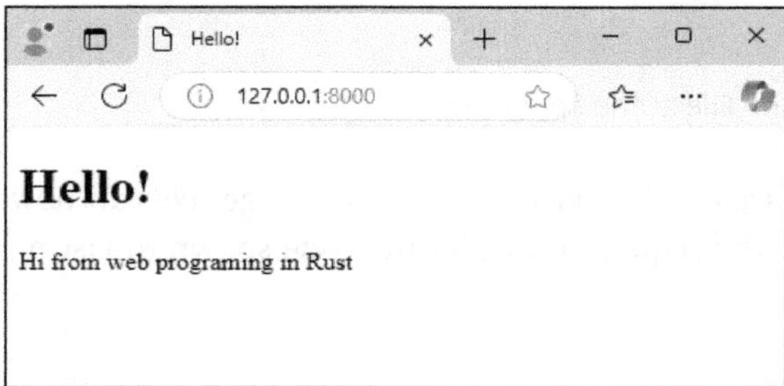

Figure 16.1 Server Response in the Web Browser

The response shown in Figure 16.1 will appear for any type of request. For instance, if you try to access something like http://127.0.0.1:8000/page1 or http://127.0.0.1:8000/page2 in your web browser, the server will return the same HTML response. Of course, this behavior is not desirable.

16.2.3 Returning Different Responses

We need some mechanism that will validate the request and then respond back with an appropriate HTML response. To simulate this scenario, let's add a few more pages to the project root, as shown in Listing 16.10, Listing 16.11, and Listing 16.12. These pages represent different responses a web server might serve based on the requested URL.

Listing 16.10 shows the HTML structure for *page1.html*, a simple webpage displaying a greeting message. This page serves as a basic example of an HTML document that a web server can deliver to a client.

```
<!DOCTYPE html>
<html lang="en">
    <head>
        <meta charset="utf-8">
        <title>Hello!</title>
    </head>
    <body>
        <h1>Hello!</h1>
        <p> Hi from page 1 </p>
    </body>
</html>
```

Listing 16.10 HTML Code for page1.html

Similarly, Listing 16.11 contains the HTML for *page2.html*, which is another page with a slightly modified message.

```
<!DOCTYPE html>.
<html lang="en">
    <head>
        <meta charset="utf-8">
        <title>Hello!</title>
    </head>
    <body>
        <h1>Hello!</h1>
        <p> Hi from page2 </p>
    </body>
</html>
```

Listing 16.11 HTML Code for page2.html

By adding another page, you can simulate the navigation between different resources so that we can test how the server serves multiple files.

Finally, Listing 16.12 defines a *404.html* page, which serves as an error page when a requested resource is not found. This page is a crucial part of any web server, providing a user-friendly response when an invalid URL is entered.

```
<!DOCTYPE html>
<html lang="en">
    <head>
        <meta charset="utf-8">
        <title>Not Found</title>
```

```
    </head>
    <body>
        <h1>This page does not exist</h1>
        <p> Sorry </p>
    </body>
</html>
```

Listing 16.12 HTML Code for 404.html

The *404.html* page helps improve the user experience by clearly indicating when a requested resource is unavailable, thus preventing generic error messages from appearing.

The project directory structure after adding the HTML files will look as follows:

```
web_programming/
├── Cargo.toml
├── index.html
├── page1.html
├── page2.html
├── 404.html
├── src/
│   └── main.rs
└── target/
```

Now, consider the revised definition of the `handle_connection` function, as shown in Listing 16.13.

```
fn handle_connection(mut stream: TcpStream) {
    let buf_reader = BufReader::new(&mut stream);
    let mut request_line = buf_reader.lines().next();
    let (status_line, file_name) = match request_line.unwrap().unwrap().as_
str() {
        "GET / HTTP/1.1" => (Some("HTTP/1.1 200 OK\r\n"), Some("index.html")),
        "GET /page1 HTTP/1.1" => (Some("HTTP/1.1 200 OK\r\n"),
Some("page1.html")),
        "GET /page2 HTTP/1.1" => (Some("HTTP/1.1 200 OK\r\n"),
Some("page2.html")),
        _ => (Some("HTTP/1.1 404 NOT FOUND\r\n"), Some("404.html")),
    };

    let contents = fs::read_to_string(file_name.unwrap()).unwrap();
    let responce = format!(
        "{} Content-Length: {}\r\n\r\n{}",
        status_line.unwrap(),
```

```
        contents.len(),
        contents
    );

    stream.write_all(responce.as_bytes()).unwrap();
    stream.flush().unwrap();
}
```

Listing 16.13 Revised handle_connection Function with Different Responses

The line `let mut request_line = buf_reader.lines().next();` reads the first line from the incoming HTTP request, which is typically the request line (e.g., "`GET / HTTP/1.1`"). The `next` method retrieves the first item from the iterator, which is wrapped in a `Result` and `Option` to handle potential errors and end-of-stream cases. Next, we match on the `request_line`. If the request is made for a valid resource, then we return the resource (which in this case is an HTML file) along with the status line.

The first arm will handle the request for the root. The slash (/) after the `GET` represents the request for the root. The second arm is the request for `page1`, the third arm for `page2` and the last arm for any other invalid request. In summary, the match will return a tuple containing the `status_line` and the `file_name` based on the type of request.

Finally, let's make a response and write it to the stream as in our earlier examples. When you execute the code and try to access http://127.0.0.1:8000/page1, the server will respond with the contents of `page1.html` instead of `index.html`. In the same way, when you try to access http://127.0.0.1:8000/page2, the server will respond with the contents of `page2.html`. The server will respond with the `404.html` page in the event of any invalid resource request.

If you keep this server running for an extended period, eventually it may crash due to an `unwrap` call on a `None` value. Let's explore why this problem may arise.

Earlier, we noted that when the server does not provide a valid response, the client keeps sending requests, some of which may be empty. This behavior can also occur if the server responds with minimal or no information, thus causing the browser to send repeated empty requests, sometimes arbitrarily, if it receives no response for a while. This behavior may vary across browsers. When an empty request is received and stored in `request_line`, calling `unwrap` on it will lead to a panic.

Note

The communication between the client and server can occur through various HTTP methods such as `POST`, `PUT`, `DELETE`, `PATCH`, and others. In this section, we focused on the `GET` method to illustrate how basic communication operates.

16.3 Multithreaded Server

In this section, we'll explore a simple example of multithreaded servers, which can handle multiple requests at any one time. For a refresher on multithreading, see Chapter 14.

The server code that we wrote in the previous section was based on a single thread, which is the main thread. The main thread is responsible for handling all incoming requests to server. A problem with a single threaded server is that it cannot handle multiple requests simultaneously. For example, if two requests arrive around the same time, the server will process the first request it receives and won't handle the second one until it has finished processing the first. If, for any reason, the first request takes a while to process, the second request must wait until the server completes the first one. Naturally, this situation could lead to poor user experience.

To illustrate multithreaded servers more clearly in code, consider the modified handle_connection function definition shown in Listing 16.14.

```
use std::time::Duration;
use std::thread;
...
fn handle_connection(mut stream: TcpStream) {
    let buf_reader = BufReader::new(&mut stream);
    let mut request_line = buf_reader.lines().next();
    let (status_line, file_name) = match request_line.unwrap().unwrap().as_
str() {
        "GET / HTTP/1.1" => (Some("HTTP/1.1 200 OK\r\n"), Some("index.html")),
        "GET /page1 HTTP/1.1" => {
            thread::sleep(Duration::from_secs(10)); // added a delay
            (Some("HTTP/1.1 200 OK\r\n"), Some("page1.html"))
        }
        "GET /page2 HTTP/1.1" => (Some("HTTP/1.1 200 OK\r\n"),
Some("page2.html")),
        _ => (Some("HTTP/1.1 404 NOT FOUND\r\n"), Some("404.html")),
    };

    let contents = fs::read_to_string(file_name.unwrap()).unwrap();
    let responce = format!(
        "{} Content-Length: {}\r\n\r\n{}",
        status_line.unwrap(),
        contents.len(),
        contents
    );
```

```
    stream.write_all(responce.as_bytes()).unwrap();
    stream.flush().unwrap();
}
```

Listing 16.14 Simulating Slow Handling of Request for page1

This code is almost the same as the code shown earlier in Listing 16.13, with an added line of code to add a 10-second delay in the arm corresponding to request for page1. If we open two browsers and first make a request to page1, then immediately make another request to the index page, the index page request will be delayed until the request for page1 completes. The request to index page is being blocked unnecessarily in this case, and threads allow us to avoid such blocking behavior.

The essential idea behind using threads is to make a new thread corresponding to each request. Each created thread will then call the handle_connection function to properly handle the request without blocking any other requests. Consider the code shown in Listing 16.15.

```
use std::{
    fs,
    io::{BufRead, BufReader, Write},
    net::{TcpListener, TcpStream},
    sync::{Arc, Mutex},
    thread,
    time::Duration,
};

fn main() {
    let listener = TcpListener::bind("127.0.0.1:8000").unwrap();
    let mut active_requests = Arc::new(Mutex::new(0));
    for stream in listener.incoming() {
        let active_requests = Arc::clone(&active_requests);
        thread::spawn(move || {
            {
                let mut connection = active_requests.lock().unwrap();
                *connection += 1;
                if *connection >= 3 {
                    thread::sleep(Duration::from_secs(2));
                }
            }
            handle_connection(stream.unwrap());
            let mut connection = active_requests.lock().unwrap();
            *connection -= 1;
```

```
        });
    }
}
```

Listing 16.15 Multithreaded Server with Each Request Handled in a Separate Thread

Servers typically allow some predetermined number of connections to be handled at a time. This feature ensures a smooth user experience and also ensures that system resources are not being exhausted. To keep track of the number of active requests, we defined the variable `active_requests`. This variable will be updated by the threads, and therefore, we wrapped it in an `Arc` pointer and `Mutex`. Each created thread corresponding to a new request will increment this variable. Similarly, the variable will be decremented when the request is handled. Next, we loop through the requests.

During each loop, we create a `clone` of the `active_request` and pass it to a newly created thread. The created thread will take care of the incoming request. First, we update the `active_requests` by acquiring a `lock`. Next, we check whether the total number connections exceeds `3` or not. In this case, we assume that our system resources allow us to handle three requests at a time. In this case, we can just wait for a few seconds in the hope that during this time, some other request may have be completed. More sophisticated and meaningful logic can be added instead of just waiting, however, including routing the request to some other HTML page, letting the user know that the server is currently busy or working on completing the request.

Next, we call the `handle_connection` method. When this call finishes, the request has been handled, and therefore, the `active_request` variable is decremented.

Notice how the incrementing of `active_request` and the associated check using the `if` statement are performed inside a code block, but the decrementing of the `active_request` is performed without a code block. This apparent inconsistency relates to the blocking nature specific to `Mutex`, as explained in Chapter 14, Section 14.3. To summarize, a `Mutex` will block the current thread if it is unable to acquire a lock. A lock will remain with a thread as long as the variable `connection` remains in scope. If we do not use a code block for incrementing, then the lock on the `Mutex` will remain in the entire thread scope. Consider a request that will take more time to handle, such as the request to `page1` shown earlier in Listing 16.14. After acquiring the `lock`, the call to `handle_connection` will unnecessarily keep the `Mutex` locked, and therefore, no other thread will be able to acquire the `lock`, leaving other threads in a blocked state. The whole purpose of using threads in this case will fail due to blocking nature of the `lock`. To ensure that the threads are not blocked due to locks, we'll reduce the scope of the lock by including a code block. The `lock` is released as soon as the code block ends, and therefore, the slow resolution of call to `handle_connection` does not block other threads.

The decrementing of `active_request` is not followed by some other code; therefore, we do not include it in a code block. Our server is now ready to take on multiple connections. If you run the code and open up two separate browser windows, notice how making a request to page1 followed by a request to page2 or some other resource does not block the handling of other requests.

16.4 Practice Exercises

In this section, we'll provide some practice questions to help reinforce your learning. The exercise questions cover almost all the concepts introduced in the chapter. Solutions to each exercise will follow in Section 16.5.

1. **Reading and parsing HTTP request headers**
   Implement a server that listens for incoming connections on 127.0.0.1:9000. The server should accept client connections and read the HTTP request line by line. Extract and display the method (e.g., GET) and the HTTP version (e.g., HTTP/1.1). After parsing and displaying the request, send a simple acknowledgment back to the client (e.g., "Request received").
2. **Multithreaded HTTP server with response handling queue**
   Write a simple multithreaded HTTP server in Rust that listens for incoming TCP connections on port 8000. The server should accept incoming connections and limit the number of simultaneous active connections to 5. If more than 5 requests come in, the server should queue the excess requests and handle them later. Each incoming connection should receive a basic HTTP response: "HTTP/1.1 200 OK\r\nContent-Type: text/plain\r\n\r\nRequest Received".

   You're must ensure that the server does not block or fail when under heavy load and instead efficiently queues and processes requests as slots become available.

16.5 Solutions

This section provides the code solutions for the practice exercises in Section 16.4.

1. **Reading and parsing HTTP request headers**

```
use std::io::{BufRead, BufReader, Write};
use std::net::{TcpListener, TcpStream};

fn main() {
    let listener = TcpListener::bind("127.0.0.1:9000").unwrap();
    println!("Server listening on 127.0.0.1:9000...");

    for stream in listener.incoming() {
        match stream {
            Ok(mut stream) => {
                handle_connection(stream);
```

```
                }
                Err(e) => {
                    println!("Error accepting connection: {}", e);
                }
            }
        }
    }

    fn handle_connection(mut stream: TcpStream) {
        let buf_reader = BufReader::new(&mut stream);

        let request_lines: Vec<String> = buf_reader
            .lines()
            .map(|result| result.unwrap())
            .take_while(|line| !line.is_empty())
            .collect();

        let first_line = &request_lines[0];
        let parts: Vec<&str> = first_line.split_whitespace().collect();
        let method = parts[0];
        let protocol = parts[2];
        println!("Method: {}", method);
        println!("Protocol: {}", protocol);

        let response = "HTTP/1.1 200 OK\r\nContent-Type: text/plain\r\n\r
    \nRequest received";
        stream.write(response.as_bytes()).unwrap();
    }
```

2. **Multithreaded HTTP server with response handling queue**

```
    use std::{
        fs,
        io::{BufRead, BufReader, Write},
        net::{TcpListener, TcpStream},
        sync::{Arc, Mutex},
        thread,
        time::Duration,
    };

    fn handle_connection(mut stream: TcpStream) {
        let buf_reader = BufReader::new(&mut stream);

        let request_lines: Vec<String> = buf_reader
            .lines()
```

```
        .map(|result| result.unwrap())
        .take_while(|line| !line.is_empty())
        .collect();

    let response = "HTTP/1.1 200 OK\r\nContent-Type: text/plain\r\n\r
\nRequest Received";
    stream.write_all(response.as_bytes()).unwrap();
    stream.flush().unwrap();
}

fn main() {
    let listener = TcpListener::bind("127.0.0.1:8000").unwrap();
    let active_requests = Arc::new(Mutex::new(0));
    let request_queue = Arc::new(Mutex::new(Vec::new()));

    for stream in listener.incoming() {
        let stream = stream.unwrap();
        let active_requests = Arc::clone(&active_requests);
        let request_queue = Arc::clone(&request_queue);

        thread::spawn(move || {
            {
                let mut connection = active_requests.lock().unwrap();
                *connection += 1;
                if *connection > 5 {
                    let mut queue = request_queue.lock().unwrap();
                    queue.push(stream);
                    return;
                }
            }
            handle_connection(stream);
            {
                let mut connection = active_requests.lock().unwrap();
                *connection -= 1;
            }
            let mut queue = request_queue.lock().unwrap();
            if let Some(next_stream) = queue.pop() {
                handle_connection(next_stream);
            }
        });
    }
}
```

16.6 Summary

This chapter explored the exciting realm of web programming with Rust, serving as an introduction to building online applications and services. We started with web programming basics, providing an overview of the essential components involved in web development using Rust. We then guided you through the process of setting up a basic web server and managing HTTP requests and responses, thus laying a solid foundation for web applications.

Next, we covered techniques for handling multiple requests using threads, emphasizing Rust's ability to maintain high performance and scalability in concurrent environments. Practical examples illustrated how to build effective web applications, showcasing Rust's strengths in the context of web development. By the end of this chapter, you should have the knowledge and skills necessary to create robust and efficient web applications with Rust.

In the next chapter, we'll explore text processing techniques, efficient file handling and directory management, and tools that help you read, write, and organize data seamlessly.

Chapter 17
Text Processing, File Handling, and Directory Management

Understanding and working with data requires attention to how your data is structured and organized. In this chapter, you'll learn how to manage files and process text, equipping you with tools for handling information effectively.

This chapter explores text processing, file handling, and directory management in Rust. You'll learn about basic file operations, such as reading and writing files, as well as directory- and path-related functions. The chapter introduces regular expressions (regexes) for text processing, covering repetition quantifiers and capturing groups. String literals and their manipulation are also discussed, providing the tools needed to handle various file and text operations efficiently. This knowledge is essential for developing applications that interact with the file system and process text data.

17.1 Basic File Handling

Managing files is a fundamental aspect of many applications, allowing them to store, retrieve, and manipulate data persistently. In this section, we explore how Rust facilitates file handling, from basic operations to advanced functionalities.

17.1.1 Creating a File and Adding Content

In this section, you'll learn how to create a file and add content to it, laying the foundation for basic file operations in Rust. Consider the code shown in Listing 17.1.

```
use std::fs::*;
use std::io::{BufRead, BufReader, Read, Write};
use std::path::Path;

fn basic_file_handling() -> std::io::Result<()> {
    let path_loc = r"D:\my_text.txt";
    let path = Path::new(path_loc);
    let mut file = File::create(path)?;
```

```
    file.write(b"let's put this in the file")?;
    file.write("let's put this in the file".as_bytes())?;

    Ok(())
}
fn main() {
    basic_file_handling();
}
```

Listing 17.1 Creating a Text File and Adding Some Text

Notice how we've imported quite a few modules:

- **`std::fs` module**
  This module provides a comprehensive set of tools for working with the file system, such as creating, reading, writing, and deleting files and directories. This module serves as the foundation for handling file-related operations in Rust.
- **`std::io` module**
  The `BufRead`, `BufReader`, `Read`, and `Write` traits within this module facilitate efficient file input/output (I/O) operations. These traits enable buffered reading and writing, which are crucial for managing large files or streaming data.
- **`std::path::Path` module**
  The `Path` struct allows you to work with file system paths in a platform-independent manner. This module is integral to file handling as it helps represent, manipulate, and validate file and directory paths efficiently.

Following these modules, we have the `basic_file_handling` function. The output of the function (i.e., `std::io::Result`) is a specialized `Result` type for I/O operations. This type is commonly used for representing the output of any I/O-related operation that may produce an error. The next three lines are used to create a file at a specified location. First, we define a variable to store the file's intended location. The `r` before the string indicates that the string is a raw string literal in Rust. This ensures that escape sequences within the string are not processed, meaning backslashes (\) are treated literally rather than being interpreted as escape characters. The `path::new` converts the raw string `path_loc`, which represents the file's location, into a `Path` type using `Path::new`. The `Path` struct provides an abstraction over file system paths, thus enabling platform-independent path manipulation and file operations. The call to `File::create` creates a new file at the specified path. If the file already exists, its contents will be cleared. The `?` operator propagates any errors encountered during file creation, ensuring that error handling is seamless and idiomatic.

The call to write on the variable file writes something to the file. The first call to write will write the text `"let's put this in the file"` to the file. The `b` before the string makes it a byte string (a sequence of raw bytes), which is the format expected by the write method. The next line also writes the same text to the file. However, instead of using a byte string, it converts the regular string (&str) into bytes using the as_bytes method, so the string can be written to the file. Both lines achieve the same result but use slightly different syntaxes to provide the bytes for writing. The first method (byte string) is more direct, while the second method (using as_bytes()) provides more flexibility when working with variables or strings that need to be converted to bytes before writing. Including both shows different ways to handle data, but in most cases, the byte string literal is the simpler and preferred option.

If no errors arise, we return an Ok. Executing the code shown in Listing 17.1 will create a new file in the specified location with the contents provided in the call to write function.

17.1.2 Appending a File

You can use the append function when you want to edit a file while preserving its original contents. Consider the code shown in Listing 17.2.

```
...
fn basic_file_handling() -> std::io::Result<()> {
    let path_loc = r"D:\my_text.txt";
    let path = Path::new(path_loc);
    let mut file = OpenOptions::new().append(true).open(path)?;
    file.write("\n www.includehelp.com\n".as_bytes())?;
    Ok(())
}
fn main() {
    basic_file_handling();
}
```

Listing 17.2 Appending an Existing File

The first two lines are the same as our earlier example shown earlier in Listing 17.1. In the third line, the new constructor first creates a new instance of the OpenOptions struct, which configures how a file should be opened (e.g., for reading, writing, appending, etc.). By default, OpenOptions opens the file in read-only mode. The append method with the true parameter modifies the OpenOptions instance to specify that data should be appended to the file, rather than overwriting the file. Finally, the open method attempts to open the file at the given path. This method uses the previously set options (in this case, append mode) to determine how the file should be opened. At the end, we use the write method to add text to the file. When you execute the code, note that the contents added previously to the file are not cleared and instead new content is added.

17.1.3 Storing the Results

Frequently, you'll encounter situations where you need to store the result of your program in a file, which are contained in some variables. The code shown in Listing 17.3 illustrates how to store this result.

```
...
fn basic_file_handling() -> std::io::Result<()> {
    let path_loc = r"D:\my_text.txt";
    let path = Path::new(path_loc);
       let mut file = OpenOptions::new().append(true).open(path)?;

    // Storing string in a variable
    let str1 = "some text";
    file.write(str1.as_bytes())?;

    // Storing data in a vector
    let some_vec = vec![1,2,3,4,5,6];
    let str_from_vec = some_vec
    .into_iter()
    .map(|a| format!("{} ", a.to_string()))
    .collect::<String>();
    file.write(str_from_vec.as_bytes())?;

    // Storing data contained in multiple variables
    let (name, age) = ("Joseph", 40);
    let formatted_str = format!("I am {} and my name is {}", name, age);
    file.write(formatted_str.as_bytes())?;
    Ok(())
}
```

Listing 17.3 Storing Variables in a File

The first three lines open the file as earlier. Then, three examples illustrate different situations when storing data in a file. The first case is when you have data in a string format. The line `file.write(str1.as_bytes())?` stores the string value inside the variable `str1`.

The next case is when you have some computational results that are stored in a vector. In this case, we iterate through all the values and map them to a string followed by a space (which serves as a delimiter in this case). Finally, we collect all the values as a single long string and write this string to the file.

The final case is when you need to store data from separate sources. The `format!` macro can combine all pieces into one string, which we can write to a file. The essential idea is to concatenate all the information in one string and then write it as bytes to the file.

17.1.4 Reading from a File

Now that you know how to write to a file, let's look at reading from a file in detail. Reading data from a file is a fundamental operation that allows your programs to retrieve and process stored information.

Consider the code shown in Listing 17.4.

```
...
fn basic_file_handling() -> std::io::Result<()> {
    let path_loc = r"D:\my_text.txt";
    let path = Path::new(path_loc);
    let mut file = File::open(path)?;
    let mut contents = String::new();
    file.read_to_string(&mut contents)?;
    println!("The file contains {:?}", contents);
    Ok(())
}
```

Listing 17.4 Reading from a File

The `read_to_string` method reads the file's content and appends this content to the `contents` string. The `&mut` keyword indicates that `contents` is being passed as a mutable reference, meaning the string will be updated with the data read from the file. The execution of this code will result in a terminal output that is not well formatted. Additionally, the output contains many new line characters or `\n` sequences.

17

To properly format the output and make the response nicer, you can read the file line by line and display the contents, as shown in Listing 17.5.

```
fn basic_file_handling() -> std::io::Result<()> {
    let path_loc = r"D:\my_text.txt";
    let path = Path::new(path_loc);
    let mut file = File::open(path)?;
    let file_buffer = BufReader::new(file);
    for lines in file_buffer.lines() {
        println!("{:?}", lines?);
    }
    Ok(())
}
```

Listing 17.5 Properly Formatting the Output of the Text Read from a File

The call to constructor on `BufReader` converts the `file` to a `BufReader` types, which provides efficient buffered reading. The `BufReader` reduces the number of I/O operations by reading larger chunks of data into memory at once and then processing it line by line. The call to `lines` method on `buffer` returns an iterator that yields each line from the file as a Result<String, std::io::Error>, allowing you to process the lines individually. This function will produce an output that is a bit nicer and more closely resembles what the file actually contains.

17.2 Path- and Directory-Related Functions

Handling paths and directories is an essential aspect of many applications, especially when working with file systems. Rust's standard library provides powerful tools to interact with paths, retrieve directory information, and manipulate files. In this section, we'll dive into the core functions that allow you to manage and traverse directories, ensuring smooth file operations in your programs.

17.2.1 Working with Paths

We'll start with some path-related functions. In practice, path-related functions are essential for working with file systems, as they allow you to navigate directories, check for file existence, and manipulate file paths dynamically. Consider the code shown in Listing 17.6.

```
use std::path::{Path, PathBuf};
fn main() {
    let path = Path::new(r"D:\Rust\Examples\my_file.txt");
    println!("Folder containing the file: {:?}", path.parent().unwrap());
}
```

Listing 17.6 Retrieving the Folder of a File

The new constructor in the path creates a new path, representing a path to a specific file on the system. Calling the parent method on a path returns an Option<Path>, which represents the directory containing the file. In this case, path.parent will return Some(Path::new("D:\\Rust\\Examples")), which is the directory containing the file my_file.txt.

Next, let's review some methods for displaying the file name and the extension of the file, as shown in Listing 17.7.

```
use std::path::{Path, PathBuf};
fn main() {
    ...
    println!("Name of the file is {:?}", path.file_stem().unwrap());
    println!("Extension of the file is {:?}", path.extension().unwrap());
}
```

Listing 17.7 Method for Displaying File Name and Extension of a File

The call to file_stem returns the file name without its extension. In this case, the call will return Some("my_file"), which is the name of the file without its .txt extension. The call to the extension method returns the extension of the file.

Sometimes, when working with directories or file paths, you have paths that are not fixed or not known at compile time, for instance, paths based on user input, configuration files, or external data. You're therefore required to build paths step by step. Listing 17.8 shows two approaches for progressively creating paths.

```
use std::path::{Path, PathBuf};
fn main() {
    // approach 1
    let mut path = PathBuf::from(r"D:\");
    path.push(r"Rust");
    path.push(r"Examples");
    path.push(r"my_file");
    path.set_extension("txt");
    println!("The path is {:?}", path);

    // approach 2
    let path = [r"D:\", r"Rust", r"Examples", r"my_file.txt"]
        .iter()
        .collect::<PathBuf>();
    println!("The path is {:?}", path);
}
```

Listing 17.8 Progressively Creating Paths

In the first approach, we initialize a mutable variable of type PathBuf. A growable, mutable type in Rust, PathBuf is used to construct and manipulate file system paths dynamically. You can incrementally add path components using the push method and then use the set_extension method to specify file extension. This approach is ideal for building paths flexibly, without manually handling platform-specific separators, ultimately making the code more robust and portable across different operating systems.

In the second approach, an array of path components is iterated over, and the collect method combines them into a single PathBuf. This concise, functional approach is ideal for constructing paths when all components are already known.

Executing any of the two approaches will result in the following output:

```
The path is "D:\\Rust\\Examples\\my_file.txt"
```

Escaping Backslashes

Although a couple of backslashes exist in the path, the path is readable in Windows. Backslashes in file paths are not usually a problem in Windows, which uses this character as the standard directory separator on the system. Most Windows programs understand backslashes in paths and handle them appropriately. However, when writing code, especially cross-platform code, developers should be cautious about escaping backslashes, since a single backslash is also used as an escape character in many programming languages, including Rust. Confusion may arise if the character is not properly escaped (i.e., using \\ to represent a single backslash in strings).

Most programmers familiar with Windows are accustomed to this behavior, but we want to clarify the need for escaping backslashes when they appear in paths in code, as they might cause issues in certain contexts. Additionally, when working in cross-platform scenarios, using libraries like `std::path::Path` can help ensure that path separators are handled correctly across different operating systems.

17.2.2 Working with Directories

The existence of a directory and files can be checked using the `is_dir` and `is_file` methods, as shown in Listing 17.9.

```
use std::path::{Path, PathBuf};
fn main() {
    let path = Path::new(r"D:\Rust_learning");
    println!("Is the path a directory {:?}", path.is_dir());

    let path = Path::new(r"D:\my_text.txt");
    println!("Does the file exists: {:?}", path.is_file());
}
```

Listing 17.9 Checking the Existence of Directory and Files

The code demonstrates how to check whether a given path corresponds to a directory or a file using Rust's Path type. The `is_dir` method returns `true` if the specified path exists and is a directory, while the `is_file` method checks if the path exists and is a regular file. These methods help in verifying the existence and type of a file system entry before performing operations on it.

The code shown in Listing 17.10 displays metadata information, such as the file type, size, permissions, and timestamps.

```
use std::path::Path;
fn main() {
    let path = Path::new(r"D:\my_text.txt");
    let data = path.metadata().unwrap();
    println!("type {:?}", data.file_type());
    println!("length {:?}", data.len());
    println!("Permissions {:?}", data.permissions());
```

```
    println!("Modified {:?}", data.modified());
    println!("Created {:?}", data.created());

}
```

Listing 17.10 Displaying the Metadata Corresponding to a File

The call to `metadata` returns a `Result`. Specific methods, such as `file_type`, `len`, `permissions`, `modified`, and `created`, are then used to extract individual pieces of information about the file.

The code shown in Listing 17.11 displays all the files in the directory given by a path.

```
use std::path::Path;
fn main() {
    let path = Path::new(r"D:\");
    for files in path.read_dir().expect("read_dir call failed") {
        println!("{:?}", files.unwrap().path());
    }
}
```

Listing 17.11 Displaying All Files in a Directory

In this example, we first create a path. The call to `read_dir` returns an iterator of `Result<DirEntry>` for the directory's contents. Each `DirEntry` represents a file or subdirectory within `D:\`. The `expect` ensures that the program panics with the specified error message if the `read_dir` operation fails, such as when the directory doesn't exist or lacks proper permissions.

17

Executing the program in Listing 17.11 displays the complete paths of all the files and subdirectories within the `D:\`.

The `env` and the `fs` modules are two important modules for working with directories. Listing 17.12 shows some useful methods from these modules.

```
use std::env;
use std::fs;
//use std::path::Path;
fn main() {
    let mut curr_path = env::current_dir().expect("can't access current
directory");
    println!("{:?}", curr_path);

    println!("Create a new directory: {:?}", fs::create_dir(r"D:\rust1"));
    println!(
        "Create a new directory and sub directories: {:?}",
        fs::create_dir_all(r"D:\rust1\level1\level2")
```

```
    );

    println!("Remove a specific directory: {:?}", fs::remove_dir(r"D:\rust1\
level1\level2"));
    println!("Remove a specific directory when it is not empty {:?}",
fs::remove_dir(r"D:\rust1"));
    println!("Remove everything from a directory {:?}", fs::remove_dir_
all(r"D:\rust1"));
    println!("Remving a file {:?}", fs::remove_file(r"D:\my_text.txt"));
    println!("Renaming a file {:?}", fs::rename(r"D\prev.txt", r"D:\new.txt"));
    println!("Copying contents from one file to another: {:?}", fs::copy(r"D:\
new1.txt", r"D:\new2.txt"));
}
```

Listing 17.12 Useful Methods in the fs and env Modules

The env module in Rust provides functionality for interacting with the environment, such as retrieving the current working directory using `env::current_dir`. In contrast, the `fs` module is used for file system operations like creating, removing, renaming, and copying files and directories. While env helps access system-related information, `fs` is responsible for managing files and directories on the system. Let's break down the functions shown in Listing 17.12 step by step:

- **`current_dir`**
  The function `current_dir` retrieves the current working directory of the program and assigns it to the variable `curr_path`.
- **`create_dir`**
  The `create_dir` function creates a new directory at the specified path (`D:\rust1`) and prints the result of the operation. This function will return an `Err` if the directory already exists.
- **`create_dir_all`**
  The `create_dir_all` creates a directory structure, including any intermediate directories needed, at the specified path and prints the result. If any intermediate directories exist, they will remain as is and will not return an `Err`.
- **`remove_dir`**
  The `remove_dir` removes the directory at the specified path if the directory is empty.
- **`remove_dir_all`**
  The `remove_dir_all` will forcefully remove the specified directory even if the directory contains files and subdirectories.
- **`remove_file`**
  The function `remove_file`, as the name suggests, deletes a specific file indicated by a given path. The `rename` function renames a file. In the code, `prev` is the file for which we want to change the name and `new` is the new name of the file. Make sure you

provide the full name, including the extension, when renaming a file to avoid errors or unexpected behaviors.

- **copy**
  The final function is the `copy` function. This function copies the contents of a file at the source path (`D:\new1.txt`) to a new file at the destination path (`D:\new2.txt`).

17.3 Regular Expressions Basics

A *regular expression (regex)* is a search pattern that defines a specific text structure we aim to locate within an input text. This process involves two inputs: the *input text* and the *regex pattern*. The output is the instances of the search pattern found in the input text. Regexes are commonly used for tasks such as searching and replacing text, extracting specific information like email addresses or phone numbers, and solving text-processing problems programmatically.

This section provides a concise introduction on regexes in Rust. Since this book focuses on Rust, this discussion is brief and to the point. While many programming languages include regex functionality in their standard libraries, Rust opts to exclude it to maintain a lean standard library. Instead, Rust relies on external crates, such as the popular `regex` crate, for regex functionality. In this section, you'll learn how to use the `regex` crate for working with regexes in Rust.

17

17.3.1 Basic Methods

In this section, we'll explore the fundamental methods for defining regexes and capture patterns within strings using regexes. Consider the code shown in Listing 17.13.

```
use regex::Regex;
fn main() {
    let re = Regex::new(r"[prt]ain").unwrap();
    let text = "rrrain spain none";

    println!("The text has a match {:?}", re.is_match(text));
    println!("The text has a match {:?}", re.find(text));
}
```

Listing 17.13 Creating a Regex and Finding Its Matches in an Input String

The `Regex::new` function creates an instance of `regex`, which is a specialized type for handling regexes. The constructor function returns a `Result`; therefore, we use `unwrap` to handle potential errors if the `regex` syntax is invalid. The `r` before the `regex` string indicates a raw string in Rust, allowing you to write the pattern without escaping backslashes. The text inside the double quotes is a regular expression. Many kinds of constructs can be made in regular expressions, and the most fundamental is called a *character class*, which is indicated by the square brackets. The character class tells the regex engine to match only one out of several characters, specifically the characters included inside the square brackets. In this example, we are telling to match in the input text at the start, one of the three letters of either p, r, or t, followed by mandatory letters of ain.

Next, we define an input string ("rrrain spain none") to search for matches. The is_match function checks whether the regex matches any part of the input string, returning true or false. To find the exact location of the first match, the find function returns the start and end indices.

The functions shown in Listing 17.13 demonstrate if there's a match and the location of the first match. However, they don't provide the specific text that has been matched. The crate provides a convenient iterator for matching an expression repeatedly against a search string to find successive non-overlapping matches. For instance, consider the code shown in Listing 17.14.

```
use regex::Regex;
fn main() {
    let re = Regex::new(r"[prt]ain").unwrap();
    let text = "rrrain spain none";
    for cap in re.captures_iter(text) {
        println!("match: {:?}", &cap[0]);
    }
}
```

Listing 17.14 Repeatedly Matching an Expression in Text

The re.captures_iter method generates an iterator over all matches of the regex in the input text. Each match is represented as an instance of type captures. Inside the loop, &cap[0] accesses the full match (the entire portion of the text that satisfies the regex). This information is printed for each match found. The result in this case would be rain and pain. The first two letters of r in the first word of the input string do not match because after the r the regex demands the letters ain, which are absent. The last r matches because we have the letters of ain afterwards. The first letter in the second word does not match any of the characters in our character class and is therefore not part of the match. The letter p in the same word, followed by s matches because letter p is in the character class and is followed by the mandatory letters of ain.

The character class is quite useful and can be used for checking variations in spelling. For instance, consider Listing 17.15.

```
use regex::Regex;
fn main() {
    let re = Regex::new(r"gr[ae]y").unwrap();
    let text = "gray grey graye";
```

```
    for cap in re.captures_iter(text) {
        println!("match: {:?}", &cap[0]);
    }
}
```

Listing 17.15 Character Class for Checking Different Spellings

The spelling will be correct if we have letters of g and r followed by either a or e followed by mandatory letter of y. This expression will only return the valid spellings, which are gray and grey in this case.

17.3.2 Dot and Character Ranges

The dot (.) in a regex will match a single character including letters and digits. Dot should not be used inside a character class. For instance, let's change the regex in Listing 17.14 by adding a dot after the character class, as in `let re = Regex::new(r"[prt].ain").unwrap();`. This regex will now match when we have either one p, r, or t at the start followed by any character and then the mandatory letters of ain. For the input string shown earlier in Listing 17.14, this regex will match rrain. The extra r is matched due to the dot.

Similar to the character classes are *character ranges*, which check whether characters in a certain range are part of the text or not. For instance, the code shown in Listing 17.16 will match any word that starts from a lowercase character followed by mandatory letters of ain.


```
use regex::Regex;
fn main() {
    let re = Regex::new(r"[a-z]ain").unwrap();
    let text = "main pain tain rain but not 0ain";
    ...
}
```

Listing 17.16 Character Ranges in a Regex

The small a followed by a dash and then a z means "all the characters from a to z." This regex will match the words of main, pain, tain, and rain. The last word did not match because the start of the word does not have a lowercase letter character.

You can manipulate multiple ranges inside the square brackets, as in the following example:

```
let re = Regex::new(r"[^a-zA-Z]ain").unwrap();
```

This regex will now match all the words that start from either a lowercase or uppercase letters followed by ain.

You can also exclude certain ranges with a special character: the caret (^). For instance, consider the code shown in Listing 17.17.

```
use regex::Regex;
fn main() {
    let re = Regex::new(r"[^a-z]ain").unwrap();
    let text = "main pain tain rain but not 0ain";
}
```

Listing 17.17 Excluding Character Ranges

In this case, our regex does not want any lowercase characters from a to z at the start, and the word must also have the mandatory ain at the end. This match is only true for the word of 0ain in the input string.

Some shorthands exist for frequently used character classes. The most commonly used are \w and \d. The \w is a shorthand for character class including all upper and lowercase characters, digits, and underscore. The \d is a shorthand for all digits from 0 to 9. Listing 17.18 shows an example for \d shorthand.

```
fn main() {
    let re = Regex::new(r"\d\d\d\d\d\d").unwrap();
    let text = "My phone number is 816030 and the second phone number is 
816694";
}
```

Listing 17.18 Example Using a Shorthand

The regex in this case will match any number that contains 6 digits. The two telephone numbers in the input text will match in this case. As discussed earlier, a caret before the shorthand characters signifies a negation.

17.3.3 Starting and Ending Anchors

Anchors in regular expressions are special characters that define the position of a match within a string, rather than matching specific characters. These characters help you control where a pattern should start or end, making them essential for performing precise searches at the beginning, end, or within specific parts of a string.

The caret (^) when used outside the square brackets in a regex indicates the start of the input text. Let's look at the example shown in Listing 17.19.

```
use regex::Regex;
fn main() {
    let re = Regex::new(r"^aba").unwrap();
    let text = "aba aba bc";
```

```
    for cap in re.captures_iter(text) {
        println!("match: {:?}", &cap[0]);
    }
}
```

Listing 17.19 Example of Starting Anchor

The caret in this case will match the start of the input text before any character in the input text followed by the letters of aba. The first instance of aba will therefore match. The second instance of the aba will not match since it is not at the start of the input text.

Similar to the starting anchor, you can use the ending anchor, which is a dollar sign ($). This anchor will match at the end of the input text. For instance, consider the following regex and the input text:

```
let re = Regex::new(r"bc$").unwrap();
let text = "aba abc bc";
```

This regex will return a match of bc. The ending anchor ($) will match with the ending of the text, and then from the ending of the text, the last element should be c, and the second last should be b. These conditions are satisfied by the ending bc in the input text. The starting and ending anchors do not match any characters themselves; they are just telling the regex engine to match at the start and the end of the input text.

You can use both the starting and ending anchors at the same time. Consider the following regex:

```
let re = Regex::new(r"\d\d$").unwrap();
```

This regex will match any input text that contains exactly 2 digits.

17.3.4 Word Boundaries

Word boundaries in regular expressions match positions where a word begins or ends, without including any actual characters. Word boundaries are represented by \b.

Word boundaries in regular expressions will match at three key positions:

- Before the first character of the input text, if there are no characters before it.
- After the last character in the input text, if there are no characters after it.
- Between two characters, if the first one is a word character (like a letter or digit) and the second one is not (like a space or punctuation mark).

In simple words, a word boundary will match at the start or end of a string or between a word character and a non-word character. Consider the following regex and an input string:

```
let re = Regex::new(r"\b\w").unwrap();
let text = "Hi my name is nouman";
```

In this input string, the \b will match at the start before any character. Like the caret (^) and dollar ($) anchors, word boundaries (\b) do not match any specific characters. Instead, they match positions in the text. Specifically, a word boundary matches at the start or end of the string, or at any position where a word character (like a letter or digit) is adjacent to a non-word character (like a space or punctuation mark).

In the string, the \b (word boundary) will first match at the start of the string. Then, the regex expects a word character, so it will match the letter H in the word Hi. Next, the boundary will match the position between the two words Hi and my, specifically after the letter "I" because we have a word character before this position (the I) and a non-word character (a space) after it. However, the pattern will not match at the position after I because the following character is a space, not a word character.

Then, the boundary will match the position just before the letter m in my because it satisfies the conditions: there is a word character before it (the space before m) and a word character after it (the m itself). Finally, the pattern will also match the position after the letter m since it is followed by another word character.

In summary, the word boundary \b matches at the beginning of each word, allowing the regex to identify the starting positions of words in the string.

17.3.5 Quantifiers, Repetitions, and Capturing Groups

Quantifiers are used to denote repetitions. There are three quantifiers that are commonly used: the question mark (?), a plus (+), and a star (*). The question mark (?) is used for indicating 0 or 1 times repetition, the plus (+) is used to indicate 1 or more times and a star (8) to indicate 0 or more times of repetition.

We'll take a closer look at each type of quantifier in the following sections, followed by a look at a few relevant scenarios.

The Question Mark (?) Quantifier

Consider the following regex along with a string:

```
let re = Regex::new(r"a?aa").unwrap();
let text = "aa aaa";
```

The regex means that we need to have the 0 or 1 instance of letter a at the beginning followed by two mandatory letters of a. The question mark (?) after the letter a means that the letter a at the start is optional. In the given string, it will return two matches. The first word will match because the first a is optional, and it is not present in the word. The second word will also match because the first optional a is present in it.

Let's consider one more regex and some string:

```
let re = Regex::new(r"ba?").unwrap();
let text = "a ba b";
```

This regex means that at the start we need to have a mandatory letter of b followed by an optional a. In the input text, the first word will not match because it does not have the mandatory letter b. The second word of ba will match, the next word containing only letter b will also match.

The question mark (?) quantifier can be used to check for the file names with some constraints. For instance, consider the following regex and the sample text:

```
let re = Regex::new(r"\w?\w?\w?.rs").unwrap();
let text = "fil.rs t1.rs file.rs";
```

This regex will match any text that contains the mandatory .rs at the end. Note that only the last three characters of the file will be matched. In the input text, the first two file names will match completely while the third file name will have only a match of ile.rs.

The plus (+) Quantifier

The plus (+) is used to indicate 1 or more times. Consider the following regex along with some string:


```
let re = Regex::new(r"a+").unwrap();
let text = "a aa aaa baab bab";
```

In this case, the regex means that we need to have 1 or more occurrences of letter a in the input text for a successful match. This regex will match the first word containing a single a, the second word containing double aa, and the third word containing triple aaa. For the word baab, it will match only the pair of a's. In the last word, it will match only the single a.

The next example will return a file name that may contain any number of characters but must have the .gif extension:

```
let re = Regex::new(r"\w+\.gif").unwrap();
let text = "image1.gif and background.gif";
```

From the sample text, the regex will return only the file names of image1.gif and background.gif.

The star (*) Quantifier

The star (*) quantifier indicates 0 or more repetitions. Consider the following regex:

```
let re = Regex::new(r"ab*").unwrap();
let text = "a ab abbbbb";
```

The regex will match all three words because all the three words satisfy the regex, which demands the letter a is mandatory followed by 0 or more times of letter b.

Limited Repetitions

Sometimes, we don't want an unlimited number of repetitions, but instead, we want a limited number of repetitions. Limited repetitions are indicated inside curly brackets containing the least and greatest number of repetitions. Consider the following example:

```
let re = Regex::new(r"\w{3,5}").unwrap();
let text = "Welcome, I think you're happy";
```

This regex will match any text containing 3 to 5 characters, and in the text, it will match welco, think, you, are, and happy. The word welcome has partially matched because, for the words whose length is greater than 5 characters, only their first 5 characters will match. To enforce matching of only complete words, you can use the word boundary \b at the start and at the end of the regex. For instance, the following regex will now match only complete words that are 3 or more characters and 5 or less characters:

```
let re = Regex::new(r"\w{3,5}").unwrap();
```

A nice use case for limited repetitions is for limiting the number of digits in the fraction and whole number part of a number. Consider the following regex and sample string:

```
let re = Regex::new(r"\b\d{1,3}\.\d{1,3}\b").unwrap();
let text = "921.583 0.0 1456.25";
```

This regex in the example will match those numbers that are between 1 and 3 digits in its whole and fraction parts. The word boundary character (\b) will ensure that the whole number matches. The backslash before the dot (.) ensures that the dot character is used in its literal meaning. In the given string, the regex will match only the first two numbers.

Finally, note that the syntax \d{2} will mean fixed repetitions of exactly 2 digits.

Capturing Groups

Sometimes, the pattern you want to define is a bit more complex. Therefore, breaking it down into its individual components can really help for the sake of simplification. Groups can be created by placing parts of a regular expression inside parentheses to create groups or parts within a regular expression.

Consider a case for detecting the dates, which are written in the form of a year (4 digits), followed by a month (2 digits), and finally the day (2 digits). Let's write a regex for each part of a date (the year, day, and month) using capturing groups. Consider the following regex and string:

```
let re = Regex::new(r"(\d{4})-(\d{2})-(\d{2})").unwrap();
let text = "2012-03-14, 2013-01-01 and 2014-07-05";
```

The regex in the first parentheses will match the year. The year contains exactly 4 digits, so inside the parentheses, we use limited repetitions of exactly 4 digits. Next, we have a mandatory dash (-), followed by another group, which will contain a few digits for representing the month. This group is followed by a dash (-). Finally, we have another group for representing the day, which also must have fixed 2 digits. This regex will only return the valid dates from the input text.

One of the benefits of capturing groups is that the individual captures corresponding to the groups can also be accessed. The code shown in Listing 17.20 illustrates how this process can be performed.

```
fn main() {
    let re = Regex::new(r"(\d{4})-(\d{2})-(\d{2})").unwrap();
    let text = "2012-03-14, 2013-01-01 and 2014-07-05";

    for cap in re.captures_iter(text) {
        println!(
            "Month: {} Day: {} Year: {}, the whole: {}",
            &cap[2], &cap[3], &cap[1], &cap[0]
        );
    }
}
```

Listing 17.20 Accessing the Individual Groups

The individual groups are in the remaining indexes variable cap.

> **Note**
> We've left out many details intentionally for the sake of brevity since this book focuses on Rust and not on regular expressions. Nonetheless, we hope we've covered enough for a quick start to building simple regular expressions.

17.4 String Literals

Effective text processing relies on a strong understanding of string handling. Working with files and directories involves manipulating textual data, whether reading from files, constructing paths, or formatting outputs. Mastery of string literals is essential for these operations. This section covers string literals in Rust, providing important context for the text-related operations discussed throughout this chapter.

17.4.1 Working with Raw String Literals

Sometimes, you may encounter many double quotes inside a string. For instance, consider the following string:

```
let str = "The main said \" Hello world \"";
```

To include double quotes (") as part of a string, a backslash (\) is used to escape them. This step ensures that quotes are treated as literal characters within the string. However, when a string contains numerous double quotes, escaping each one is cumbersome and error prone. Missing an escape character may lead to an invalid string.

Rust offers raw string literals to address such scenarios. Raw string literals allow you to define strings without requiring escape characters. In fact, escape sequences within the body of a raw string are not processed at all, which makes handling complex strings much simpler and much less error prone.

A raw string is defined by putting an `r` before the string. Let's change the variable `str` to a raw string slice:

```
let str = r#"The main said " Hello world ""#;
```

The `r` before the string indicates that the string should be treated as a raw string, and the hashes (#) at the start and the end serve as a starting string marker and an ending string marker, respectively. These string markers indicate the start and end of a string in the presence of double quotes (included in the string). The double quotes are now part of the text, and we don't need backslashes before the double quotes. Printing the `str` will result in the following output:

```
The main said " Hello world "
```

If the string does not contain the double quotes, you can skip adding the starting and ending string markers. For instance, consider the following strings with different characters but no double quotes:

```
let str = r"The main said _Hello world_ \n \t ' ";
```

Printing the `str` in this case will display the following output:

```
The main said _Hello world_ \n \t '
```

In this context the double quotes indicate the end of a string, and to escape its special meaning, you must use hashes at the start and at the end.

17.4.2 Parsing JSON

JavaScript Object Notation (JSON) is a widely used data format that represents structured information as key-value pairs. This data format is commonly used for data exchanges between systems due to its simplicity and readability. An important use case of string literals is the processing of JSON strings. Consider the code shown in Listing 17.21.

```
fn main() {
    let jason_str = "{
        \"name\": \"Micheal\",
        \"age\": 40,
        \"sex\": Male
    }";
}
```

Listing 17.21 JSON String

This JSON string contains many double quotes, and to escape each of them, we used a backslash before each. These escape characters will ensure to get a correct output string. Writing JSON strings in this way is often error prone because of the difficulties of correctly placing the backslashes.

Let's now redefine the same string using string literals. Consider the revised version shown in Listing 17.22.


```
fn main() {
    let jason_str1 = r#"{
        "name": "Micheal",
        "age": 40,
        "sex": Male
    }"#;
}
```

Listing 17.22 Revised JSON String

This revised version eliminates the need for backslashes with each and every double quote. Moreover, it's more readable and easier to write and manage.

17.4.3 Using a Hash within a String

Using a hash (#) inside a string is useful when working with raw string literals that contain special characters, such as embedded code snippets, regular expressions, or configuration files where hash symbols are part of the syntax. Processing the hash inside a string literal is a bit tricky. Consider the following code:

```
let str = r#"Hello"# World!"#;    // Error
```

This line will not compile because the starting string marker (which is a single hash), matches the hash after the double quotes at the end of the word `Hello`. The hash at the end of the string is therefore not considered as the ending string marker. The code will compile if we change the starting marker to double hashes. Let's look at the correct version of the code:

```
let str = r##"Hello"# World!"##;
```

This code now compiles because the starting string marker of (##) matches at the end of the string.

In summary, the number of hashes at the start (which indicates the starting string marker) should be the same as the number of hashes at the end of the string. Moreover, the hash markers should not be included inside the string. If they are included, you'll need to add more hashes so that they become unique.

17.5 Practice Exercises

In this section, we'll provide some practice questions to help reinforce your learning. The exercise questions cover almost all the concepts introduced in the chapter. Solutions to each exercise will follow in Section 17.6.

1. **Matching telephone numbers without using \d shorthand**
   Write a regex to match the two telephone numbers in the provided string but without using the shorthand character class `\d`. Modify the regex in the following code to match the phone numbers 816030 and 816694.

```
use regex::Regex;
fn main() {
    let re = Regex::new(/*Your regex here*/r"^\d......").unwrap();
    let text = "My phone number is 816030 and the second phone number is
816694";

    for cap in re.captures_iter(text) {
        println!("match: {:?}", &cap[0]);
    }
}
```

2. **Understanding the question mark quantifier**
   Write a regex using the question mark (?) quantifier to find files with `.rs` extensions that may optionally start with an underscore (_). Given a string of filenames, the regex should identify only valid matches based on this pattern. Use the following code template.

```
fn main() {
    let re = Regex::new(r"// Add your regex here").unwrap();
    let text = "file.rs _file.rs test.rs _test.rs";
    for cap in re.find_iter(text) {
        println!("{}", cap.as_str());
    }
}
```

3. **Using the plus quantifier for word matching**
   Design a regex with the plus quantifier (+) to extract all words from a sentence that have one or more occurrences of the letter e in the middle of the word. Use the following code template.

```
fn main() {
    let re = Regex::new(r"// Add your regex here").unwrap();
    let text = "eager beaver sees three elephants";
    for cap in re.find_iter(text) {
        println!("{}", cap.as_str());
    }
}
```

4. **Build and inspect file paths**
   Write a program to dynamically build a file path, check its existence, and retrieve its metadata if it exists. The individual tasks to complete are provided in the following code template.

```
use std::path::{Path, PathBuf};
fn main() {
    let mut path = PathBuf::new();
    let components = vec!["C:", "Users", "Public", "example.txt"];
    /* Task 1: Add code to dynamically build the path using the components
               vector.
       Task 2: If the path exists as a file then display its metadata.
               Otherwise display that the path does not exists as a file */
}
```

5. **Simplify JSON representation with raw literals**
   Refactor the given JSON string to use a raw string literal. Update the json_data variable in the following code to correctly use raw string syntax while keeping the content valid.

```
fn main() {
    // Original JSON string with escaped quotes
    let json_data = "{
        \"user\": \"Alice\",
        \"details\": {
            \"age\": 29,
```

```
            \"city\": \"Wonderland\"
        },
        \"active\": true
    }";

    // Refactor the JSON string to use a raw string literal
    let json_data_raw = // Add your code here;
    println!("Original: {}", json_data);
    println!("Refactored: {}", json_data_raw);
}
```

6. **Guess the correct output**
   Consider the following code. For each print statement, try to guess the output at the terminal.

```
fn main() {
    let string1 = r#"""#;
    let string2 = r#""""""""#;
    let string3 = r#" He asked,"Is rust awesome?"""#;
    println!("{}", string1); // guess the output
    println!("{}", string2); // guess the output
    println!("{}", string3); // guess the output
}
```

17.6 Solutions

This section provides the code solutions for the practice exercises in Section 17.5. The code is largely self-explanatory; however, we have included comments and additional explanations wherever necessary to enhance understanding.

1. **Matching telephone numbers without using \d shorthand**

```
use regex::Regex;
fn main() {
    let re = Regex::new(r"^\d.....").unwrap();
    let text = "My phone number is 816030 and the second phone number is
816694";

    for cap in re.captures_iter(text) {
        println!("match: {:?}", &cap[0]);
    }
}
```

2. **Understanding the question mark quantifier**

```
fn main() {
    let re = Regex::new(r"_?\w+\.rs").unwrap();
    let text = "file.rs _file.rs test.rs _test.rs";

    for cap in re.find_iter(text) {
        println!("{}", cap.as_str());
    }
}
```

3. **Using the plus quantifier for word matching**

```
fn main() {
    let re = Regex::new(r"\b\w*e+\w*\b").unwrap();
    let text = "eager beaver sees three elephants";
    for cap in re.find_iter(text) {
        println!("{}", cap.as_str());
    }
}
```

4. **Build and inspect file paths**

```
use std::path::{Path, PathBuf};
fn main() {
    let components = vec!["C:", "Users", "Public", "example.txt"];
    let mut path = PathBuf::new();
    // Task 1:
    for component in components {
        path.push(component);
    }
    println!("Constructed Path: {:?}", path);

    // Task 2:
    if path.is_file() {
        println!("The path exists as a file.");
        let data = path.metadata().unwrap();
        println!("type {:?}", data.file_type());
        println!("length {:?}", data.len());
        println!("Permissions {:?}", data.permissions());
        println!("Modified {:?}", data.modified());
        println!("Created {:?}", data.created());
    } else {
        println!("The path does not exist as a file.");
    }
}
```

5. **Simplify JSON representation with raw literals**

```
fn main() {
    // Original JSON string with escaped quotes
    let json_data = "{
        \"user\": \"Alice\",
        \"details\": {
            \"age\": 29,
            \"city\": \"Wonderland\"
        },
        \"active\": true
    }";

    // Refactored JSON string using raw string literal
    let json_data_raw = r#"{
        "user": "Alice",
        "details": {
            "age": 29,
            "city": "Wonderland"
        },
        "active": true
    }"#;

    println!("Original: {}", json_data);
    println!("Refactored: {}", json_data_raw);
}
```

6. **Guess the correct output**

```
"

""""""

 He asked,"Is rust awesome?""
```

17.7 Summary

This chapter focused on text processing, file handling, and directory management in Rust, providing you with the essential skills for interacting with data in applications. We began with an overview of basic file handling, teaching fundamental operations for reading from files and writing to files. This chapter then explored directory- and path-related functions, demonstrating the effective navigation and manipulation of file systems. Additionally, we covered the basics of regular expressions, which are essential for text processing, along with details on repetitions, quantifiers, and capturing groups to perform advanced text manipulations. Our discussion also included string literals and their manipulation, equipping you with practical tools for handling various kinds of textual data. A comprehensive understanding of these topics is crucial for building applications that efficiently process text and manage file systems.

As we conclude our exploration of text processing and directory handling, in the next chapter, we turn our attention to addressing some practical, real-life problems and their effective implementations in Rust. You'll apply the concepts you've learned to real-world scenarios.

Chapter 18
Practical Problems

Real-world problems require practical solutions. In this final chapter, you'll tackle a series of challenges to put your newfound Rust skills to the test.

To finish our journey with Rust, let's walk through real-life problems that require sophisticated data structure solutions. This chapter presents a variety of scenarios, such as search results with word groupings, product popularity analysis, and highest stock price calculations. Other problems include searching for employees with no meetings, calculating the longest non-stop working hours, and suggesting items. Advanced examples like binary search trees (BSTs) for range queries, fetching top products, and efficient storage and retrieval are also covered. These practical examples illustrate how you can apply Rust's data structures to solve complex problems effectively. In this chapter, our problems progressively increase in difficulty.

18.1 Problem 1: Search Results with Word Groupings

Consider an online store that offers a wide range of items for sale, all of which are searchable. To enhance the user experience and customer satisfaction, the store wants to implement a functionality that displays accurate search results even if a user misspells a word.

The solution involves organizing the words describing the items into sets of anagrams. Anagrams are words that contain exactly the same letters but in a different order. This approach assumes that a misspelled word will likely have the same letters as the correct word, albeit in the wrong order.

18.1.1 Solution Setup

Let's explore how to set up this solution. We assume a list of words describing various items are given to us in the `main` function, and our task is to group together words that are anagrams of each other. To achieve this task, we'll utilize `HashMaps` and loops as key tools in our implementation.

Consider a vector in `main`, as shown in Listing 18.1, which contains some words. Some of the words are spelled correctly, and some are not.

```
fn main() {
    let words = vec![
        "The".to_string(),
        "teh".to_string(),
        "het".to_string(),
        "stupid".to_string(),
        "studpi".to_string(),
        "apple".to_string(),
        "appel".to_string(),
    ];
}
```

Listing 18.1 Vector of Words in main

Our job is to implement a functionality that displays accurate search results even if a user misspells a word. We'll implement this feature using a function called word_grouping. The input to this function will be a vector of strings containing words, and the output will be a grouping of these words into separate vectors. A skeleton of the function follows:

```
fn word_grouping(words_list: Vec<String>) -> Vec<Vec<String>> {}
```

18.1.2 Implementation

To code the problem, we'll use HashMaps. The keys of the HashMap will be strings of digits representing the frequencies of letters in a word, and the value will store all the words that have the same letter frequencies. For instance, for the word apple, we'll have the following key:

10001000000100020000000000

The first digit in the string corresponds to the frequency of letter a, the second digit to letter b and so on. The value corresponding to this key is apple. If for some reason, we receive a different spelling for the same word of apple, we'll still obtain the same key or digit string, and therefore, it will be added to the same key. Let's add the definition of the HashMap to the function definition, in the following way:

```
fn word_grouping(words_list: Vec<String>) -> Vec<Vec<String>> {
    let mut word_hash = HashMap::new();
}
```

Next, we'll compute the frequencies of the individual letters corresponding to each word. Listing 18.2 shows how this task is achieved through code.

```
use std::collections::HashMap;
fn word_grouping(words_list: Vec<String>) -> Vec<Vec<String>> {
    let mut word_hash = HashMap::new();
    let mut char_freq = vec![0; 26];

    for current_word in words_list {
        for c in current_word.to_lowercase().chars() {
            char_freq[(c as u32 - 'a' as u32) as usize] += 1;
        }

        let key: String = char_freq
            .into_iter()
            .map(|i| i.to_string())
            .collect::<String>();
        word_hash
            .entry(key)
            .or_insert(Vec::new())
            .push(current_word);
        char_freq = vec![0; 26];
    }
    word_hash.into_iter().map(|(_, v)| v).collect()
}
```

Listing 18.2 Computing the Digit String for Each Word

We'll first create a vector containing 26 entries all initialized from zero. Each entry will correspond to one letter of the alphabet. Next, we iterate through all the words in the input `words_list`. Then, we iterate through all the letters of a word. The function `lower_case` converts all the letters of the word to lowercase, and the `chars` function returns the individual characters. Let's focus on the following line inside the loop:

```
char_freq[(c as u32 - 'a' as u32) as usize] += 1;
```

This line updates the frequency for the current character. The character `c as u32` grabs the American Standard Code for Information Interchange (ASCII) value corresponding to the current character. We then subtract the ASCII value of letter `a` from it and use the result as an index to update the frequency for that letter. For example, if the current character is `d`, its ASCII value is `100`. Subtracting the ASCII value of the letter `a` (`97`) gives us `3`. Thus, at index `3`, we'll increment its value by `1`. Keep in mind that indexes in Rust start from `0`, so the index for `d` will be `3`.

Once the inner loop ends, the `char_freq` array contains the frequencies of all the letters in a word. We next use this value as a key for the `HashMap`. To use this value as a key, convert it into a `String` using the following lines:

```
let key: String = char_freq
    .into_iter()
    .map(|i| i.to_string())
    .collect::<String>();
```

Then, the following code inserts the word corresponding to the computed key:

```
word_hash
    .entry(key)
    .or_insert(Vec::new())
    .push(current_word);
```

The `or_insert` method ensures that, if no entry exists, then by default an empty vector is inserted. The `push` function adds the value to the vector corresponding to the same key.

Finally, we reinitialize the `char_freq` variable to all zeros before processing the next word in the `word_list` during the next iteration of the loop. This step is important to ensure that the frequencies from the previous word are not carried over and thus interfere with the processing of the current word. Next, the following line ignores the key part and collects all the values part in a vector:

```
word_hash.into_iter().map(|(_, v)| v).collect()
```

Let's add some code to our `main` function shown earlier in Listing 18.1 to test the function. The updated code is shown in Listing 18.3.

```
fn main() {
    ...
    let input_word = String::from("teh");
    let grouping = word_grouping(words);

    for i in grouping.into_iter() {
        if i.contains(&input_word) {
            println!("The group of the word is {:?}", i);
        }
    }
}
```

Listing 18.3 Code for Testing the Function word_grouping

This code defines `input_word` as `"teh"` and calls the `word_grouping` function to group words with the same character frequency pattern. It then iterates through the grouped words, checking if any group contains `input_word`. If the group includes the word, it prints the group of words to the console. This process helps identify and display the group of words sharing the same character frequencies as "teh". Executing the code will result in the following output:

```
The group of the word is ["The", "teh", "het"]
```

18.2 Problem 2: Product Popularity

Consider a business scenario where a company offers a wide range of products. For each product, the business collects data to calculate a popularity score over time. This score is derived from customer feedback, likes, dislikes, and reviews. The scores are updated weekly and appended to a list containing the previous weeks' scores.

Our business wants to determine whether the popularity of each product is fluctuating or increasing/decreasing over time. Additionally, they want to identify and categorize products that are consistently gaining or losing popularity, enabling them to implement appropriate strategies to address these trends.

18.2.1 Solution Setup

The business stores its product-related information in a HashMap for efficient storage. The keys of the HashMap are the product names, and the values are vectors containing the popularity scores. A sample of the company HashMap is shown in Listing 18.4.

18

```
use std::collections::HashMap;

   fn main() {
       let mut products = HashMap::new();
       products.insert("Product 1", vec![1,2,2,3]);
       products.insert("Product 2", vec![4,5,6,3,4]);
       products.insert("Product 3",vec![8,8,7,6,5,4,4,1] );
}
```

Listing 18.4 A Sample of Company's Data Given in main

Our task now is to implement the required functionality. We'll define a function of popularity_analysis, which will analyze a certain product popularity. The input to the function will be a popularity vector representing a product's scores. The output will be a Boolean value that is true if the popularity is consistently increasing or decreasing and false otherwise. The updated signature function follows:

```
fn popularity_analysis(scores: Vec<i32>) -> bool {}
```

18.2.2 Implementation

Inside the function, we'll declare two Boolean variables to track whether the values are increasing or decreasing. Their default values will be set to true, as shown in Listing 18.5.

```
fn popularity_analysis(scores: Vec<i32>) -> bool {
    let mut increasing = true;
    let mut decreasing = true;
}
```

Listing 18.5 Definition of popularity_analysis

Next, we'll iterate through all the values in the vector corresponding to the product. For the popularity to be considered increasing, the values must follow an ascending order. In other words, the value at a certain index should be less than the next value, and this condition must hold true for all values. If at any point this condition is not met, and a value is found to be greater, the product's popularity is not increasing. Let's check this functionality using an if statement, as shown in Listing 18.6, which adds code to the function definition shown in Listing 18.5.

```
fn popularity_analysis(scores: Vec<i32>) -> bool {
    let mut increasing = true;
    let mut decreasing = true;
    for i in 0..scores.len()-1 {
        if scores[i] > scores[i+1]{
            increasing = false;
        }
        if scores[i] < scores[i+1]{
                decreasing = false;
        }
    }
    return increasing || decreasing;
}
```

Listing 18.6 Updated Definition of popularity_analysis with Check for Ascending Popularity

The condition in the if statement means that, if the next value in the vector is greater than the current value, popularity is not increasing. In this case, we set the increasing variable to false.

Similarly, for the popularity to be decreasing, the values must follow a descending order. In other words, a value at a certain index should be greater than the next value, and this condition must hold true for all values. If at any point this condition is not met, and a value is found to be smaller, popularity is not decreasing steadily.

Let's add this condition. In this case, we'll set the decreasing variable to false. Listing 18.7 shows the updated code with the check for descending order.

```
fn popularity_analysis(scores: Vec<i32>) -> bool{
    let mut increasing = true;
    let mut decreasing = true;
    for i in 0..scores.len()-1 {
        if scores[i] > scores[i+1] {
            increasing = false;
        }
        if scores[i] < scores[i+1] {
            decreasing = false;
        }
    }
    return increasing || decreasing;
}
```

Listing 18.7 Updated Definition of popularity_analysis with Check for Descending Popularity

By the end of the loop, if either `increasing` or `decreasing` is set to `false`, this value indicates that the popularity is fluctuating. If both remain `true`, it signifies that the popularity is either consistently increasing or decreasing. The function eliminates the need for an additional check for the fluctuating case.

In summary, the function will return a `true` value when either `increasing` or `decreasing` is `true`, and in all other cases, it will return `false`. This behavior aligns perfectly with the requirement because our business is specifically interested in identifying products whose popularity is consistently either increasing or decreasing.

Now, let's test the function in `main`. The code shown in Listing 18.8 calls the function and displays its results.

```
fn main() {
    let mut products = HashMap::new();
    products.insert("Product 1", vec![1,2,2,3]);
    products.insert("Product 2", vec![4,5,6,3,4]);
    products.insert("Product 3",vec![8,8,7,6,5,4,4,1] );
    for (product_id, popularity) in products {
        if popularity_analysis(popularity) {
            println!("{} popularity is increasing or decreasing", product_id);
        }
        else{
            println!("{} popularity is fluctuating", product_id);
        }
    }
}
```

Listing 18.8 Code in main for Testing the Function

Executing the code in `main` results in the following output:

```
Product 1 popularity is increasing or decreasing
Product 3 popularity is increasing or decreasing
Product 2 popularity is Fluctuating
```

The products are not shown in the order they were created because the order was not enforced in the `HashMaps`. The popularity of the first product is increasing or decreasing. If we look at its respective vector containing the value of `[1, 2, 2, 3]`, indeed it is the case, since all the value are in ascending order. For product 2, the values are fluctuating because the vector `[4, 5, 6, 3, 4]` contain values that are neither in ascending nor in descending order. Finally, for product 3, the popularity is decreasing.

18.3 Problem 3: Highest Stock Price

A business dealing with various types of stocks maintains weekly records of their values. They are interested in implementing a feature to quickly retrieve the highest value a stock has reached in any given week. Additionally, the company wants to sequentially analyze past weeks to determine the highest stock value up to that point. Given the large amount of data and diverse stock types, they need a solution that provides this information with minimal delay. To better understand the problem, consider the diagram shown in Figure 18.1.

	Stock Value	Highest Stock Value
Week 6	145	145
Week 5	140	140
Week 4	22	120
Week 3	120	120
Week 2	80	80
Week 1	55	55

Figure 18.1 Stock Values and Highest Stock Prices Week-Wise

The data shown in Figure 18.1 corresponds to weekly stock prices. On the left side, we have stock prices, and on the right side, we have the highest stock recorded price. Week 1 represents the oldest record, and Week 6 represents the most recent. For each week, we record the highest stock value observed so far. In Week 1, the value of the stock is 55, which is also the highest recorded so far. In Week 2, the price increases to 80, making it the new highest value. In Week 3, the price increases further, setting a new record highest stock value. In Week 4, the price does not exceed 120, and so the highest value remains unchanged at 120. This approach provides a clear way to track the maximum stock value week by week.

18.3.1 Solution Setup

To solve this problem, we'll use a data structure called a *max stack*. A max stack functions as a regular stack but also keeps track of the maximum element at all times, allowing this information to be retrieved in little to no time. A max stack can be implemented in various ways, but we'll design it using a struct containing two stacks. Let's explore how these two stacks will work together.

The two stacks will be referred to as the `main_stack` and the `max_stack`. The `main_stack` will work like an ordinary stack while the `max_stack` will keep track of the maximum element. For the first element, we'll push it into the `main_stack`. The element will be checked against the top of the `max_stack`. If `main_stack` is greater, then `main_stack` will be pushed; otherwise, a copy of the top of the `max_stack` will be pushed in the `max_stack`. By default, the top will be set to a very low value so that the first value is always being pushed. This process will be repeated for all values. This approach ensures that `max_stack` represents the maximum value in the stack at any point of time.

We'll then call a simple `pop` operation to pop out elements from both stacks. The top of the `max_stack` will still represent the maximum value in the stack. To grab the record of the maximum value three weeks ago, we'll call pop three times and then display the top of the `max_stack`. In this way, we can retrieve the maximum value in the stack in little to no time. Now that we understand the problem, let's implement its solution.

18

18.3.2 Implementation

We'll start by defining our `main` struct, which will contain the two stacks. Listing 18.9 shows the definition of the struct.

```
struct MaxStack {
    main_stack: Vec<i32>,
    max_stack: Vec<i32>,
}
```

Listing 18.9 Definition of the MaxStack Struct

Next, we'll add an implementation block for the `MaxStack` containing a `new` constructor function for initializing the two stacks. The function will have no inputs and will return an instance of `Self`. Listing 18.10 shows the definition of this new function.

```
impl MaxStack {
    fn new() -> Self {
        MaxStack {
```

```
            main_stack: Vec::new(),
            max_stack: Vec::new(),
        }
    }
}
```

Listing 18.10 New Constructor Function for the MaxStack

The function returns an instance of MaxStack with two empty vectors, representing the two stacks.

Now, we'll implement the push and pop methods. The push method will accept a &mut self and a value that we intend to add to the two stacks. The value is simply pushed into the main_stack without checking against any condition. However, to add value to the max_stack, we'll check the value against the top of the max_stack. Listing 18.11 shows the code for the push function.

```
impl MaxStack {
    ...
    fn push(&mut self, value: i32) {
        self.main_stack.push(value);
        if !self.max_stack.is_empty() && self.max_stack.last().unwrap() > &
value {
            self.max_stack.push(*self.max_stack.last().unwrap());
        } else {
            self.max_stack.push(value);
        }
    }
}
```

Listing 18.11 Definition of the push Method

In this case, the push method pushes a copy of the top of the max_stack if the max_stack is not empty and the top of the stack is greater in value than the value that we intend to push. The last method results in the last element inserted in the vector and therefore represents the top of the stack.

First In, Last Out

Vectors in Rust follow the first in, last out (FILO) order, which is the same principle used by stacks. This order means that elements are added and removed in reverse order, with the most recently added element being the first to be removed. In the else part of the code, we are sure that the current value passed into the function is greater than the top of the stack and therefore needs to be inserted at the top of the max_stack.

Next, we'll add the pop method. The input to the function is again a &mut self. The pop simply removes the top of the stack elements from both the stacks. Listing 18.12 shows the code for the pop method.

```
impl MaxStack {
    ...
    fn pop(&mut self) {
        self.main_stack.pop();
        self.max_stack.pop();
    }
}
```

Listing 18.12 Definition of the pop Method

Finally, we'll add one more method that will simply return the maximum value in the stack at any point of time. Listing 18.13 shows the definition of this function.

```
impl MaxStack {
    ...
    fn max_value(&self) -> i32 {
        *self.max_stack.last().unwrap()
    }
}
```

Listing 18.13 Definition of the max_value Method

The max_value method returns the top value of the max_stack.

Let's now test the code in main. Consider the code shown in Listing 18.14.

```
fn main() {
    let mut stack = MaxStack::new();
    stack.push(55);
    stack.push(80);
    stack.push(120);
    stack.push(99);
    stack.push(22);
    stack.push(140);
    stack.push(145);

    print!("Maximum value of stock: ");
    println!("{:}", stack.max_value());

    println!("After going one week back");
    print!("Maximum value of stock: ");
    stack.pop();
```

```
    println!("{:}", stack.max_value());
}
```

Listing 18.14 Code in main for Testing the Functionality

In this case, first we created an instance of the MaxStack and then added some stock prices to it using the pop method. The added stocks are [55, 80, 120, 99, 22, 140, 145], which represents the weekly stock prices. Next, we retrieve the maximum stock price by calling the max_value method. This method should return 145. Next, we look at the highest stock price going one week back. This step should return 140. Executing the program in shown Listing 18.14 should produce the following output:

```
Maximum value of stock: 145
After going one week back
Maximum value of stock: 140
```

This result confirms the correctness of our code.

18.4 Problem 4: Identify Time Slots

Consider a boss who relies on two secretaries for assistance. Secretaries perform various activities and have their own schedules of meetings. Some of their meetings may therefore overlap with one another. The boss must identify the time slots during which both secretaries are simultaneously busy in meetings. By determining these overlapping periods, the boss can easily identify the time slots where at least one secretary is free to assign tasks more effectively.

To achieve this goal, we must analyze schedules and pinpoint the overlapping meeting times of the two secretaries ultimately to streamline task delegation and optimize the workflow.

18.4.1 Solution Setup

We assume that the meeting schedule for the two secretaries is given to us in the form of vectors defined in main. Each meeting is associated with the start and end times which are also given in the form of vectors. These vectors are shown in Listing 18.15.

```
fn main() {
    let meetings_sec_a: Vec<Vec<i32>> = vec![vec![13, 15], vec![15, 16],
vec![7, 9]];
    let meetings_sec_b: Vec<Vec<i32>> = vec![vec![14, 15], vec![5, 10]];
}
```

Listing 18.15 Vectors Representing the Meeting Schedules of Each Secretary

Each employee meeting schedule is given by vectors and the entries in the vectors are showing the time slots in which the meeting will be held. The first value of the time slot represents the starting time, and the second value represents the ending time of the meeting. The times in this case are in 24-hour system. For example, the first time slot of Secretary A, containing a value of 13 means 1 pm in the afternoon. Note that the vectors in Listing 18.15 are two dimension vectors, which means each element of the outer vector is itself a vector. This is important because the inner vectors represent individual meetings with start and end times, allowing you to organize and manage meeting schedules more effectively. By using a two-dimensional structure, you can easily store and access multiple meeting time intervals in a structured way, making it easier to perform operations like finding overlapping meetings or sorting meetings by time. To access the starting or ending time information for a specific meeting, we'll use the syntax of couple of square brackets. For instance, to access the starting time of the second meeting of Secretary A, we'll use the following syntax:

```
meetings_sec_a[1][0]
```

The index value in the first square bracket corresponds to the meeting number and the index in the second square bracket corresponds to either the starting or ending times.

We'll implement the required functionality using a dedicated function named `overlapping_meetings`. The function will accept the meeting schedules of the two secretaries given in the form of vectors and will return a vector containing the overlapping time slots. Following is the basic definition of the function:

18

```
fn overlapping_meetings(meetings_a: Vec<Vec<i32>>, meetings_b: Vec<Vec<i32>>) ->
Vec<Vec<i32>> {}
```

The function's logic involves iterating through all the meetings scheduled for Secretary A and comparing each of them, one by one, with all the meetings of Secretary B. This process helps determine if there are any overlapping time slots between the two secretaries' schedules. The important part of this logic is to determine the conditions for an overlap. Let's explain it through some visuals, as shown in Figure 18.2.

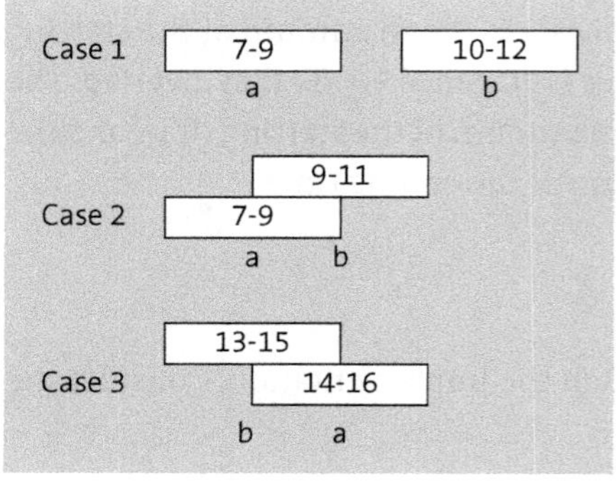

Figure 18.2 Understanding the Overlapping of Meeting Slots Represented by Time Intervals

The rectangles represent the meetings, and the values inside them are the starting and ending times of the meeting. The letters a and b represent the two secretaries. A rectangle with letter a beneath it means that the time slot represents the meeting of Secretary A. We have at least three possible cases:

- Case 1: We do not have any overlap.
- Case 2: We have an overlap of one hour with Secretary A's meeting starting first.
- Case 3: We have an overlap of one hour, but Secretary B's meeting starts first followed by the meeting of Secretary A.

The key condition for two meetings to overlap is that neither meeting should end before the other begins, which can be verified by checking if the maximum of the starting times of the two meetings is less than the minimum of their ending times. This step ensures that the meetings share a common time interval. The overlapping interval is defined by the difference of the two values. This condition is shown in Figure 18.3.

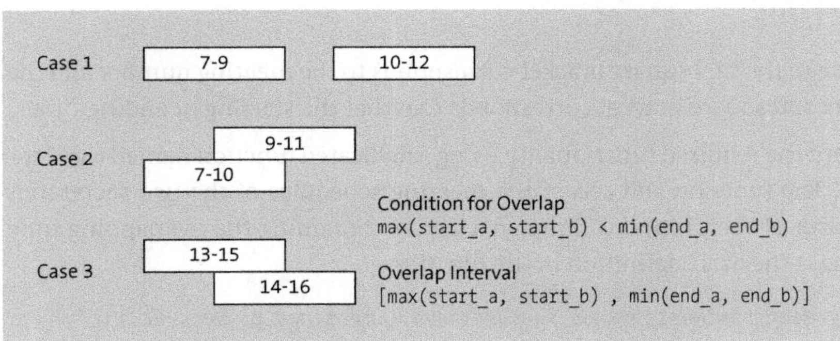

Figure 18.3 Overlapping of Meeting Slots

The `start_a` and `start_b` are the beginning times of the meeting for Secretary A and Secretary B, respectively. In the same way, `end_a` and `end_b` are the end times of the meetings. In case 1, the maximum of the two starting times, i.e., 7 and 10, is 10, and the minimum of the ending times, i.e., 9 and 12, is 9. Since 10 is not less than 9, the two meetings do not overlap. However, in case 2 and case 3, there is an overlap because the condition remains true for them. For instance, in case 2, the maximum of the starting times is 9, and the minimum of the ending times is 10. Since 9 < 10, they overlap. The lower end of the overlap interval is given by the maximum of the starting time, and the upper end of the interval is given by the minimum of the ending time.

18.4.2 Implementation

Now that we understand the logic, let's go through the implementation. Consider the code shown in Listing 18.16.

```
use std::cmp;
fn overlapping_meetings(meetings_a: Vec<Vec<i32>>, meetings_b: Vec<Vec<i32>>)
-> Vec<Vec<i32>> {
    let mut intersection: Vec<Vec<i32>> = Vec::new();
    for i in 0..meetings_a.len() {
        for j in 0..meetings_b.len() {
            let (start_a, start_b) = (meetings_a[i][0], meetings_b[j][0]);
            let (end_a, end_b) = (meetings_a[i][1], meetings_b[j][1]);
            let overlap_status = overlap(start_a, start_b, end_a, end_b);
            if overlap_status != None {
                intersection.push(overlap_status.unwrap());
            }
        }
    }
    intersection
}

fn overlap(start_a: i32, start_b: i32, end_a: i32, end_b: i32) ->
Option<Vec<i32>> {
    let mut intersection_time: Vec<i32> = Vec::new();
    if cmp::max(start_a, start_b) < cmp::min(end_a, end_b) {
        intersection_time.push(cmp::max(start_a, start_b));
        intersection_time.push(cmp::min(end_a, end_b));
        Some(intersection_time)
    } else {
        None
    }
}
```

Listing 18.16 Implementation of overlapping_meetings

The function starts with a couple of loops. The outer loop iterates through the meetings of Secretary A, and the inner loop iterates through the meetings of Secretary B. Inside the loop, we grab the starting and ending times for the meetings of the secretaries and pass this information to an `overlap` function, which implements the logic shown in Figure 18.3. If an overlap exists, it is returned inside the `Some` variant, and in the case of no overlap, a `None` is returned. When the call to the `overlap` function completes in the `overlapping_meetings`, we'll check whether an overlap exists. In the case of an overlap, we add the overlapping period to the vector of `intersection`.

Let's check the functionality of `main` shown in Listing 18.16. The updated code is shown in Listing 18.17.

18

```
fn main() {
    let meetings_sec_a: Vec<Vec<i32>> = vec![vec![13, 15], vec![15, 16],
vec![7, 9]];
    let meetings_sec_b: Vec<Vec<i32>> = vec![vec![14, 15], vec![5, 10]];

    let intersection = overlapping_meetings(meetings_sec_a, meetings_sec_b);
    println!("The overlapping timings are {:?}", intersection);
}
```

Listing 18.17 Updated Code in main for Testing the Required Functionality

Executing the code in `main` results in the following output:

```
The overlapping timings are [[14, 15], [7, 9]]
```

We have two overlapping time slots in the schedules. For instance, the first meeting of Secretary A overlaps with the first meeting of Secretary B from 14:00 to 15:00. Similarly, the last meeting of Secretary A overlaps with the second meeting of Secretary B from 7:00 to 9:00.

This overlapping information enables the boss to identify time slots where at least one secretary is free, facilitating better task assignment and scheduling. As a final note, the three cases shown earlier in Figure 18.2 covers all the remaining cases, and you don't need to explicitly handle the remaining possible cases.

18.5 Problem 5: Item Suggestions

An online shopping business recently held a lucky drawing with thousands of participants. The winners received a $50 gift card, which can be used to purchase items from the online store. However, the business has set a restriction—customers can buy a maximum of two products using the gift card.

To assist customers, we want to suggest possible pairs of products that have a combined price of exactly $50, helping them avoid out-of-pocket expenses. Our goal is to identify and recommend as many such product combinations as possible. To achieve this goal, we'll use a list of products that the customer is likely to purchase, which includes items from their wish lists and products based on their previous purchases. Our task is to return all possible pairs of products whose combined cost equals $50.

18.5.1 Solution Setup

In `main`, we are given a vector that contains the prices of the products. We'll next pass the vector to a function that will implement the required functionality. The code in `main` is shown in Listing 18.18.

```
fn main() {
    let product = vec![11, 30, 55, 34, 45, 10, 19, 20, 60, 5, 23];
    let suggestions = product_suggestions(product, 50);
    println!("{:?}", suggestions);
}
```

Listing 18.18 Sample Data and Code Given to Us in main

The `product_suggestion` function will take a vector of product prices and the value of the gift card as inputs. The function will return a two-dimensional vector, where each entry represents a pair of product prices that add up to the total value of the gift card. The following code is the signature of the function:

```
fn product_suggestions(product_prices: Vec<i32>, amount: i32) -> Vec<Vec<i32>>
{}
```

18.5.2 Implementation

One approach to implement the required functionality is to select the first value and iterate through the remaining values to check if it can be paired with any of them. We then repeat this process for the second value, pairing it with the remaining elements in the list. This method, however, is inefficient because it requires two nested loops. The outer loop selects one value at a time, while the inner loop checks for possible pairings with the remaining values in the list.

18

We'll use a more efficient solution involving the hash set with a single loop. The loop will be used to iterate through all the values, and if the value can be paired up with any value that is already in the `HashSet`, then we'll add it to the list of possible pairs. In any other case, we'll add it to the `HashSet`. This scenario may seem a bit hard to comprehend right now, but let's write it out in code, and then things will become easy to comprehend. Consider the code shown in Listing 18.19.

```
use std::collections::HashSet;
fn product_suggestions(product_prices: Vec<i32>, amount: i32) -> Vec<Vec<i32>>
{
    let mut prices_hash = HashSet::new();
    let mut offers = Vec::new();
    for i in product_prices {
        let diff = amount - i;
        if prices_hash.get(&diff).is_none() {
            prices_hash.insert(i);
        } else {
            offers.push(vec![i, diff]);
        }
    }
```

```
    offers
}
```

Listing 18.19 Definition of product_suggestion Function

First, we define a `price_hash`, which is a `HashSet`, and a vector, which will store the final offers. Next, we iterate through the vector `product_prices`.

During each iteration, we compute the difference between the current product price values from the gift amount, which is given by variable `amount`. The `HashSet` initially does not contain any value. We can check if the difference value exists in the `HashSet` or not. If we are unable to find a value equal to the difference in the `HashSet`, then we'll add it to the `HashSet`. The `get` method needs a reference to the value, and calling `is_none` will return true if there is no value found in the `HashSet` after calling the `get` function. If the difference value does exist in the `HashSet`, we add the value to the offers along with the difference because the difference value does exist in the `HashSet`.

Let's execute the code for the sample data vector in `main`, as shown earlier in Listing 18.18. The vector is repeated here, for convenience:

```
vec![11, 30, 55, 34, 45, 10, 19, 20, 60, 5, 23];
```

The details of the iterative processing are shown in Table 18.1.

Iteration	currect_price	Diff	prices_hash	offers
1	11	50 - 11 = 39	{11}	[]
2	30	50 - 30 = 20	{11, 30}	[]
3	55	50 - 55 = -5	{11, 30, 55}	[]
4	34	50 - 34 = 16	{11, 30, 55, 34}	[]
5	45	50 - 45 = 5	{11, 30, 55, 34, 45}	[]
6	10	50 - 10 = 40	{11, 30, 55, 34, 45, 10}	[]
7	19	50 - 19 = 31	{11, 30, 55, 34, 45, 10, 19}	[]
8	20	50 - 20 = 30	{11, 30, 55, 34, 45, 10, 19}	[[20, 30]]
9	60	50 - 60 = -10	{11, 30, 55, 34, 45, 10, 19, 60}	[[20, 30]]
10	5	50 - 5 = 45	{11, 30, 55, 34, 45, 10, 19, 60}	[[20, 30], [5,45]]
11	23	50 - 23 = 27	{11, 30, 55, 34, 45, 10, 19, 60, 23}	[[20, 30], [5,45]]

Table 18.1 Dry Run of the Code from Listing 18.19

At the start, the `prices_hash` is empty, and the `offers` list is also empty. For the first price of 11, the difference between 50 and 11 is 39. Since 39 is not in `prices_hash`, 11 is added to `prices_hash`. For the next value of 30, the difference is 20. Since 20 is also not in the `prices_hash`, 30 is added to `prices_hash`. For the next values of 55, 34, 45, 10 and 19, their differences are also absent in the `prices_hash`; therefore they are also inserted in the `prices_hash`.

Next, 20 is processed. The difference between 50 and 20 is 30. This time, 30 is already in `prices_hash`, which means the pair [20, 30] sums up to 50. The pair [20, 30] is added to `offers`. The difference of the next value of 60 is absent, and therefore, 60 is added to the `prices_hash`. The value 5 is next processed since its difference, which is 45, is already in `prices_hash`; therefore, the pair [5, 45] is added to `offers`. Finally, 23 is also added to `prices_hash`. By the end, `prices_hash` contains all the processed prices, and `offers` contains the prices pairs that add up to 50.

The solution does not require two loops and uses `HashSet`, which offers extremely fast lookups for values. Note that the solution is based on the assumption that the values will not repeat. For repeating values, the solution may be a bit different.

18.6 Problem 6: Items in Range Using Binary Search Trees

A retail business specializing in electronics and gadgets wants to enhance its online shopping experience by allowing customers to quickly search for products within a specific price range. With thousands of products in its catalog, the business faces the challenge of managing and retrieving product price information efficiently. When a customer searches for items within a certain price range, the system must quickly filter and display the relevant products without delay.

18

To achieve this goal, we propose implementing a *binary search tree (BST)* to store and manage the prices of all products. Each node of the BST will represent a product price, enabling the system to maintain a sorted structure. This organization allows for efficient insertion, deletion, and, most importantly, quick range-based searches. When a customer specifies a price range, the BST can swiftly locate the starting point and traverse the tree to collect all products within the desired range. This approach significantly reduces search time compared to scanning through an unsorted list or using a linear search, ensuring fast and responsive customer interactions. By utilizing a BST, the business can provide a seamless shopping experience, even as the product catalog continues to expand.

18.6.1 Solution Setup

Before starting our implementation, you must understand what a BST is and how it stores values. A tree is a collection of entities called nodes. Nodes are connected by edges. Each node contains a value or data. The first node of the tree is called the root. Figure 18.4 shows an example of a tree.

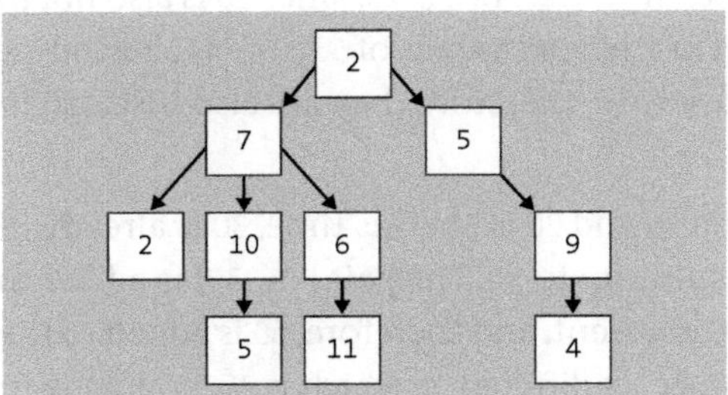

Figure 18.4 An Example Tree

The root node is the topmost node in a tree structure. It is unique in that no other node points to it, distinguishing it from other nodes that have parent nodes pointing to them. The node at the top of the diagram shown in Figure 18.4, containing the value 2, is the root node. The root node acts as the entry point for traversing, searching, or modifying the tree, with all other nodes descending from it either directly or through intermediate nodes.

In a tree structure, each node may have child nodes, which are the nodes directly connected and pointed to by that node. For example, if we consider a root node shown in Figure 18.4, the nodes with values 7 and 5 are its child nodes. Nodes that have no child nodes are referred to as *leaf nodes*.

A tree shares some conceptual similarities with a linked list but with a key difference in structure. While a linked list arranges elements in a straight, linear sequence, a tree organizes elements hierarchically. Each node in a tree, except the root, is pointed to by a single parent node. However, unlike linked lists where each node points to a single next node, a tree node can point to multiple child nodes, creating a branching structure. This branching arrangement allows trees to efficiently represent hierarchical data and support faster search, insertion, and deletion operations.

A BST is a special type of tree where each node can have no more than 2 children. The children are referred to as the left child and the right child. The tree part that is connected with the left child is called left subtree, and the tree part connected with the right child is called the right subtree. The essential requirement in the BST is that the left subtree of a node contains only nodes with values less than the node's value. In the same way, the right subtree of a node contains only nodes with values greater than the node's value. Moreover, the left and right subtrees must also be BSTs themselves.

During insertion in a BST, the first value becomes the tree root. Subsequent insertion proceeds according to the following order:

1. Starting from the root of the tree, the value is checked against a current node. If the new value is less than the current node's value, move to the left child.

2. If the new value is greater than the current node's value, move to the right child.
3. Continue this comparison and move left or right depending on the comparison until a null position (empty spot) is found in the tree.
4. Make a new node and place the value inside it.

Let's apply the procedure on the list of values:

```
[9, 6, 14, 20, 1, 30, 8, 17, 5]
```

The first value in the list will become the root of the BST. The next value of 6 will be compared with the value at the root. Since it is less than 9, it is moved to the left subtree. As there are no nodes in the left subtree, we'll make a new node and insert it to the left of root containing the value 6. Following the same procedure, the next value of 14 will be inserted to the right of the root. Repeating this operation for all the values in the list, we'll obtain the BST shown in Figure 18.5.

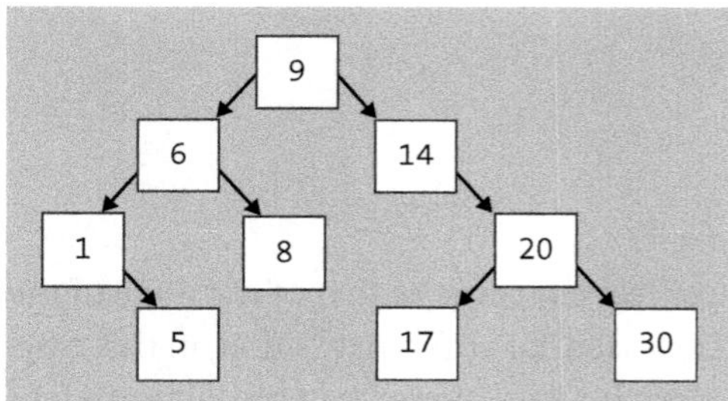

Figure 18.5 BST Construction

Once the construction of the tree is completed for each node, all the nodes in its left subtree are less than its value, and all the nodes in its right subtree are greater than its value. This facilitates the checking of some interval since we won't be looking for all the data at all times. For instance, say we want to retrieve values within the range of 6 to 15. We can define the following three simple conditions for investigating each node starting from the root:

- If value is >= 6 and <=15, add the value to the result.
- If value is >= 6, add the value to the result and investigate the left subtree for the three conditions again.
- If value is <= 15, add the value to the result and investigate the right subtree for the three conditions again.

When the value is greater than 6, we investigated the left subtree because of the chance that the left subtree may contain values that are less than the current value but greater than the value 6. Recall that all values in the left subtree are less than the values in the node.

Finally, we'll only check in the right subtree if the value of the node is less than or equal to 15, again due to the property of the BST, which ensures that all the values in the right subtree will always be greater than the value of the node.

These three conditions will be checked recursively for each and every node. If the value at the node is in range, then we'll add that value to the result. Applying the three conditions on the tree shown in Figure 18.5 will result in values of [9, 6, 8, 14].

18.6.2 Basic Data Structure

We'll start the implementation of the BST by defining a node first. Listing 18.20 shows the definition of this node.

```
#[derive(Clone)]
struct Node {
 val: i32,
 left: Option<Box<Node>>,
 right: Option<Box<Node>>
}
```

Listing 18.20 Definition of a Node of a Tree

The definition shown in Listing 18.20 is more or less the same as that of the node of the linked list defined in Chapter 11, Section 11.1. The essential difference is that, in this case, a Node contains two pointers called left and right for keeping track of the left and right children of a node. Box pointers are used in the definition of a Node as there is always a single pointer pointing to each Node in a tree structure.

Next, we'll add functionality to the Node by defining functions and methods in an implementation block for a Node. First, we add the new function, which will create a new Node, as shown in Listing 18.21.

```
impl Node {
    fn new(value: i32) -> Self {
        Node {
            val: value,
            left: None,
            right: None,
        }
    }
}
```

Listing 18.21 Definition of the new Constructor Function for Creating a New Node

The left and right pointers are initially set to None, and the value is initialized to the value passed in.

Next, we'll add the insert method, as shown in Listing 18.22.

```
impl Node {
    ...
    fn insert(&mut self, value: i32) {
        if value > self.val {
            match self.right {
                None => self.right = Some(Box::new(Node::new(value))),
                Some(ref mut node) => node.insert(value),
            }
        } else {
            match self.left {
                None => self.left = Some(Box::new(Node::new(value))),
                Some(ref mut node) => node.insert(value),
            }
        }
    }
}
```

Listing 18.22 Definition of the insert Method

The insert method will add a new value to a BST. The process begins by checking if the new value is greater than the current node's value. If it is, the function looks at the right child of the current node.

If the right child is empty (None), a new node is created with the given value, and this node becomes the right child. If the right child is not empty (Some), the function calls itself on the right child. This behavior is an example of recursion, where the same process is repeated for the right subtree. Note that the syntax Some(ref mut node) is used to borrow and modify a value. More specifically, the ref keyword ensures that the node is used as a reference and its ownership is not transferred.

If the new value is less than or equal to the current node's value, the function follows a similar process with the left child. If the left child is empty, a new node is created and placed as the left child. If the left child is not empty, the function calls itself recursively on the left child, repeating the process.

The method will keep on checking the value with the left child and right child and will go down the hierarchy by recursively calling itself until it finds a suitable place for insertion. This recursive approach allows the method to keep drilling down into the tree until it finds the correct position to place the new value, ensuring that the BST structure is maintained.

For keeping track of the root of the tree, we'll define a wrapper struct called BinarySearchTree containing a single field of root with type of Node. Listing 18.23 shows the definition of the wrapper BinarySearchTree.

18

```
struct BinarySearchTree {
    root: Node,
}
```

Listing 18.23 Definition of the BinarySearchTree Struct

This code will help us manage the BST using the root information. Specifically, the `root` field holds the reference to the root node of the tree, which is the entry point for accessing all the other nodes in the BST. By storing this reference, the struct allows us to perform various operations, such as inserting, searching, or deleting nodes.

18.6.3 Implementation

To search for items in range, we'll define a function called `products_in_range`. This function will essentially implement the desired functionality. The definition of the function is shown in Listing 18.24.

```
fn products_in_range(root: Node, low: i32, high: i32) -> Vec<i32> {
    let mut output: Vec<i32> = Vec::new();
    // recursively checks all the nodes for the three conditions outlined in
    // Section 1.7.2 output
}
```

Listing 18.24 Definition of the function products_in_range

This function will recursively check each node for the three conditions outlined in Section 18.6.1. To enable the function to accomplish this task, we must introduce another function called `traversal`, which will be called within the function. Listing 18.25 is the definition of the traversal function and essentially implements the logic described in Section 18.6.1.

```
fn traversal(node: Option<Box<Node>>, low: i32, high: i32, mut output: &mut
Vec<i32>) {
    if !node.is_none() {
        if node.as_ref().unwrap().val >= low && node.as_ref().unwrap().val <=
high {
            output.push(node.as_ref().unwrap().val);
        }
        if node.as_ref().unwrap().val >= low {
            traversal(node.as_ref().unwrap().left.clone(), low, high, &mut
output);
        }
        if node.as_ref().unwrap().val <= high {
            traversal(node.as_ref().unwrap().right.clone(), low, high, &mut
output);
```

```
        }
    }
}
```

Listing 18.25 Definition of the traversal Function

The `traversal` function visits each node and collects values that lie within a given range, specified by `low` and `high`. The function takes four arguments, namely, the current node, the range (`low` and `high`), and a mutable reference to a vector (`output`) where the matching values will be stored.

The function first checks if the current `node` is not empty. If the `node` exists, it retrieves the value from the `node`. If this value is within the given range (greater than or equal to `low` and less than or equal to `high`), it adds the value to the `output` vector. Next, the function checks if the current node's value is greater than or equal to `low`. If it is, more values may exist within the range in the left subtree, so the function calls itself recursively on the left child of the current node. Similarly, if the current node's value is less than or equal to `high`, the right subtree may also contain values within the range, so the function calls itself recursively on the right child. This process continues recursively for each `node`, visiting only the parts of the tree that could have values within the range. In this way, the function efficiently collects the required values in the output vector.

You can now update the definition of the function `products_in_range` by making a call to the `traversal` function. Listing 18.26 shows the updated definition of the function.


```
fn products_in_range(root: Node, low: i32, high: i32) -> Vec<i32> {
    let mut output: Vec<i32> = Vec::new();
    traversal(Some(Box::new(root)), low, high, &mut output);
    output
}
```

Listing 18.26 Updated Definition of the products_in_range

Let's use the implementation in `main` now. Assuming a list of products given by a vector, we'll first make a BST tree and then populate it from the values in the vector. Finally, we'll call the `products_in_range` function. Listing 18.27 shows the code in the `main` function.

```
fn main() {
    let product_prices = vec![9, 6, 14, 20, 1, 30, 8, 17, 5];
    let mut bst = BinarySearchTree {
        root: Node::new(product_prices[0]),
    };

    for i in 1..product_prices.len() {
        bst.root.insert(product_prices[i]);
```

```
    }

    let result = products_in_range(bst.root, 6, 15);
    println!("{:?}", result);
}
```

Listing 18.27 Code in main for Testing the Functionality

Executing the code shown in Listing 18.27 produces the following output: `[9, 6, 8, 14]` This result confirms the correctness of our solution.

18.7 Problem 7: Fetching Top Products

A business has operations in multiple countries. In each country, it maintains a list of top products. The country-based lists are stored in the form of linked lists that contain information regarding the rankings of the products. The business wants to combine all these lists into one consolidated list so that it can determine the top products across all countries. For this purpose, we want to combine all the individual linked lists into a singly linked list with the ranks sorted. Figure 18.6 shows this problem visually.

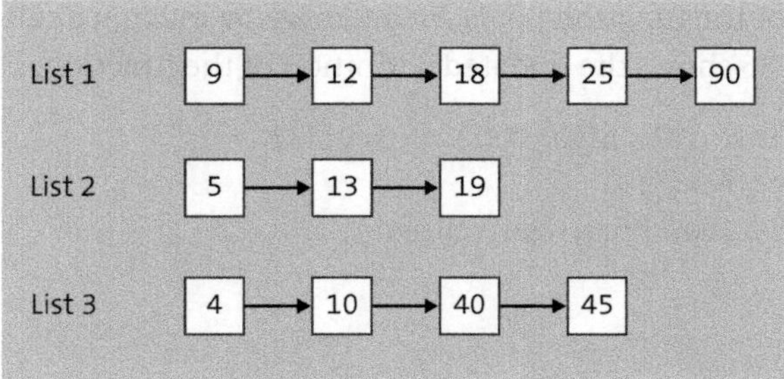

Figure 18.6 Description of the Problem

The three lists shown in Figure 18.6 correspond to the top ranked products in three different countries. The higher the rank value, the better the rank of the product is. The lists of individual countries are given in descending order. Moreover, each of these lists is stored as a linked list data structure, where each node is associated with a value that represents its rank and a pointer to the next node in the list. Our job is to combine all these lists into one consolidated list containing the ranks in descending order. For the sake of simplicity, the product names are ignored, which we can separately store in a `HashMap` or within the same data structure by modifying a node definition.

18.7.1 Solution Setup

As we'll be using the link list in the implementation, so we'll start with the code for the link list that we defined earlier in Chapter 11, Section 11.1. The code is shown again in Listing 18.28.

```
#[derive(Debug)]
struct Linklist<T: std::fmt::Debug> {
    head: pointer<T>,
}
#[derive(Debug)]
struct Node<T> {
    element: T,
    next: pointer<T>,
}
type pointer<T> = Option<Box<Node<T>>>;
impl<T: std::fmt::Debug> Linklist<T> {
    fn create_empty_list() -> Linklist<T> {
        Linklist { head: None }
    }
    fn add(&mut self, element: T) {
        let previous_head = self.head.take();
        let new_head = Box::new(Node {
            element: element,
            next: previous_head,
        });
        self.head = Some(new_head);
    }
    fn remove(&mut self) -> Option<T> {
        let previous_head = self.head.take();
        match previous_head {
            Some(old_head) => {
                self.head = old_head.next;
                Some(old_head.element)
            }
            None => None,
        }
    }
    fn peek(&self) -> Option<&T> {
        match &self.head {
            Some(H) => Some(&H.element),
            None => None,
        }
    }
```

```
    fn printing(&self) {
        let mut list_traversal = &self.head;
        println!();

        while true {
            match list_traversal {
                Some(Node) => {
                    print!("{:?} ", Node.element);
                    list_traversal = &list_traversal.as_ref().unwrap().next;
                }
                None => break,
            }
        }
    }
}
```

Listing 18.28 Basic Link List Implementation Covered in Chapter 11

More or less the same code that we developed previously, let's quickly recap this scenario.

First, we have a Node definition that contains a couple of fields consisting of an element and the next pointer. The pointer is a recursive type and is defined as an Optional Box Node. The wrapper struct of Linklist contains information about the head of the list. The new function in the implementation of Linklist creates an empty list. The same implementation also contains methods of add, remove, peek, and printing.

The code for main is shown in Listing 18.29.

```
fn main() {
    let mut list1 = Linklist::create_empty_list();
    list1.add(45);
    list1.add(40);
    list1.add(35);
    list1.add(23);
    list1.add(11);

    let mut list2 = Linklist::create_empty_list();
    list2.add(60);
    list2.add(44);

    let mut list3 = Linklist::create_empty_list();
    list3.add(85);
    list3.add(20);
    list3.add(15);
}
```

Listing 18.29 Code in main Containing the Country Wise Lists of Ranking

The add method adds values at the front. The values are inserted (highest values first) to ensure that the lists contain values in ascending order.

18.7.2 Implementation

To implement the required functionality, we'll define a function called sorting_lists. This function will accept a vector of Linklists and return a single consolidated list that combines all the individual Linklist. The following code is the signature of the function:

```
fn sorting_lists(vec_list: &mut Vec<Linklist<i32>>) -> Linklist<i32> {}
```

The logic in this function iteratively compares only the heads of the lists. During each iteration, the minimum of the three heads of the lists will be deleted from its respective list and will be added to the combined list. During computation, if a list becomes empty, it will not take part in the computation. If a singly linked list is left during the computation, the remaining elements in the list will be simply added to the combined list. Figure 18.7 shows the process visually.

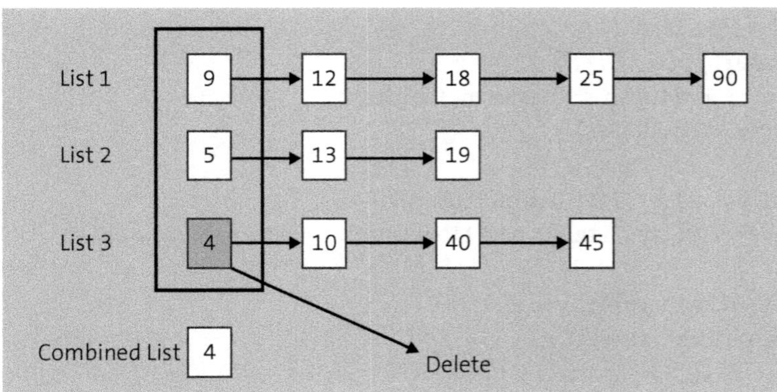

Figure 18.7 Illustrating the Logic inside the sorting_lists Function

In the first iteration, the heads of the three lists containing the values of 9, 5, and 4 are compared, and the smallest value of 4 is deleted from list 3 and added the final combined list. Next, the heads of the three lists will be compared again, and the smallest value now will be added to the combined list. This process is shown in Figure 18.8.

The value 5, which happens to the smallest of the three heads, is added to the combined lists. In this way, the algorithm will populate the combined list until there are no more elements or nodes left. Note that the values are in descending order in the final list and not in ascending order. We'll take care of this issue soon.

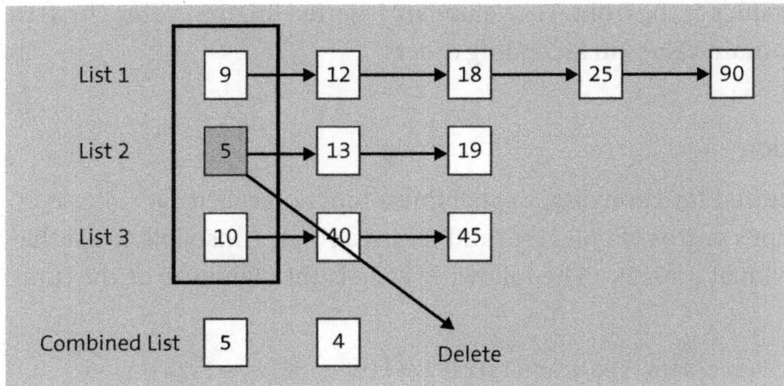

Figure 18.8 Second Iteration for Populating the Computing the Combined List

Listing 18.30 shows the code for the logic shown in Figure 18.7 and Figure 18.8.

```
fn sorting_lists(vec_list: &mut Vec<Linklist<i32>>) -> Linklist<i32> {
    let mut sorted_list: Linklist<i32> = Linklist::create_empty_list();
    let mut values: Vec<i32> = Vec::new();
    while true {
        let values = vec_list
            .into_iter()
            .map(|x| x.head.as_ref().unwrap().element)
            .collect::<Vec<i32>>();

        let min_val = *values.iter().min().unwrap();
        let min_index = values.iter().position(|x| *x == min_val).unwrap();

        sorted_list.add(min_val);
        vec_list[min_index].remove();

        if vec_list[min_index].head.is_none() {
            vec_list.remove(min_index);
        }
        if vec_list.len() == 0 {
            break;
        }
    }
    sorted_list
}
```

Listing 18.30 Definition of sorting_lists Function

The function first creates an empty `Linklist` named `sorted_list` that will contain the final result. The variable `values` will be used during computations. Next, we have a `while` loop. During each iteration, we first obtain the heads of the three lists. The `element` parts of the three heads are stored in the values. Next, we compute the minimum of the heads and determine the lists where this value happens. The minimum value is next added to the sorted_list. A remove operation is next carried out on the list containing the head with the minimum value. Finally, if all the three lists become empty due to successive removals of heads, we'll break out of the while loop. The function ends by returning the sorted_list.

Let's call the implementation in main. Starting from the main function shown earlier in Listing 18.29, the updated code is shown in Listing 18.31.

```
fn main() {
    ...
    let mut result = sorting_lists(&mut vec![list1, list2, list3]);
    result.printing();
}
```

Listing 18.31 Updated Code from Listing 18.29 in main

Executing the code will produce the following output:

```
85 60 45 44 40 35 23 20 15 11
```

Notice how the ranks are displayed in one combined list but in descending order. However, according to the problem description, we need to display the ranks in ascending order. To properly display the elements, we'll add another method to the Linklist implementation called reverse.

18

The logic of the reverse function is centered on three variables called previous, current, and next. When the algorithm starts, the previous variable is set to null, and the current variable is the first node or the head, while next is the next node after the head. During each iteration, the algorithm will first update the pointer of the current node to that of the previous node. This step will be followed by incrementing the previous, current, and next by one node. Figure 18.9 shows the process for the first node.

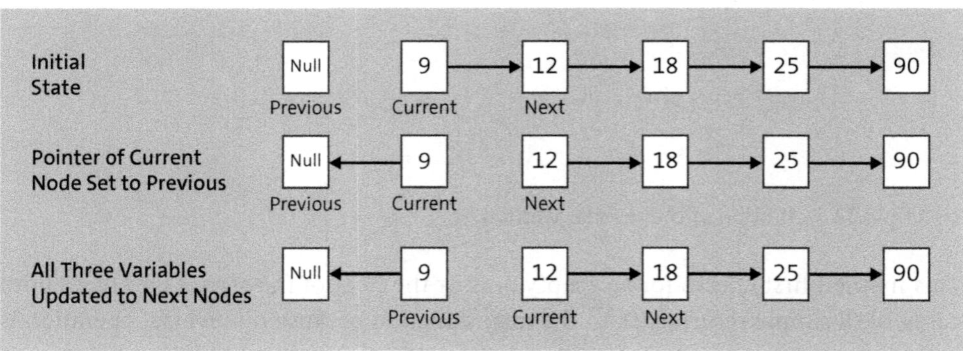

Figure 18.9 Logic of the reverse Method for the First Node

The same process is repeated for each and every node. The process will stop when the current node becomes null or empty. At this stage, the previous will become the new head of the list. Figure 18.10 shows the process for the second node in the list.

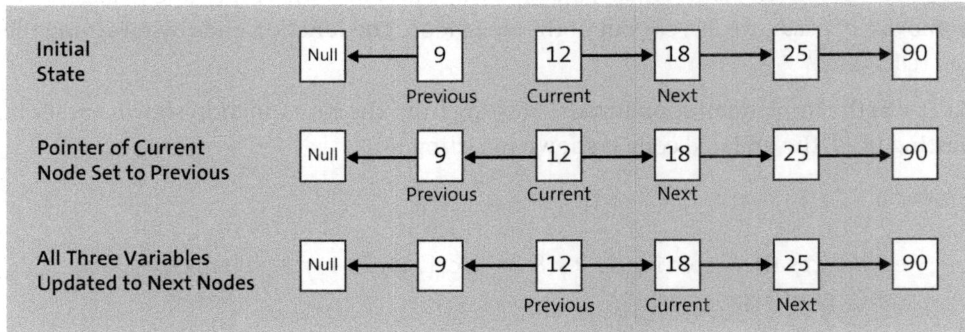

Figure 18.10 Logic of the reverse Method for the Second Node

Listing 18.32 shows the implementation of the reverse method.

```
impl<T: std::fmt::Debug> Linklist<T> {
    ...
    fn reverse(&mut self) {
        if self.head.is_none() || self.head.as_ref().unwrap().next.is_none() {
            return;
        }

        let mut previous = None;
        let mut current_node = self.head.take();
        while current_node.is_some() {
            let next = current_node.as_mut().unwrap().next.take();
            current_node.as_mut().unwrap().next = previous.take();
            previous = current_node.take();
            current_node = next;
        }

        self.head = previous.take();
    }
}
```

Listing 18.32 Definition of the reverse Method

This method first checks for an empty head or the next of the head is empty. In both cases, we'll simply return back to the main function because no reversal operation is needed. Next, let's code the logic. The previous is first set to None, and the current node is set to the head of the list. Following this step, we iterate until the current_node has some value. Inside the loop, we initialize variable `next` as the `next` of the `current_node`. Next, we update the pointer of the `current_node` to that of the `previous` node. This step is followed by updating the `previous` node to that of the `current_node`. Finally, the `current_node` becomes the next. At the end of the loop, we'll finally set the head of the list as the `previous`.

Let's update the main function from Listing 18.31 and call the `reverse` method for obtaining the correct result. The updated code is given in Listing 18.33.

```
fn main() {
    let mut result = sorting_lists(&mut vec![list1, list2, list3]);
    result.reverse();
    result.printing();
}
```

Listing 18.33 Updated main from Listing 18.36

Executing the code produces the following output:

```
11 15 20 23 35 40 44 45 60 85
```

This output conforms to the requirements set in the problem description.

18.8 Problem 8: Effective Storage and Retrieval

Consider a business case where we are involved in designing a module for a search engine that will efficiently store and fetch words. This module will act as a dictionary with insert and search functionalities. Ideally, we want a search functionality with blazingly fast speeds since speed is extremely critical to the overall success of the system.

To achieve speed in insertion and searching, the business has decided that they will use a *trie data structure*. The trie data structure is also known as digital tree or prefix tree. It is a type of *k-ary search tree*, which is used for locating specific keys from within a set. These keys are most often *strings*, with links between nodes defined not by the entire key, but by individual *characters*.

18.8.1 Solution Setup

Let's assume we have three words *The*, *This*, and *Any* that we want to store inside a tree. The diagram shown in Figure 18.11 illustrates how the tree looks when these words are inserted into the tree.

The first word is *The*. The letters of the word will be processed individually and inserted one after the other in the tree. When the second word of *This* comes, the first letter will be checked at the root. The root has a child starting with the same letter of *T*, so no new letter will be added. The same procedure will be repeated for all the letters of the word.

In other words, if the letters happen to be children of the current node, no new nodes will be inserted. Following this approach, a new node for letter *h* is not created because *T* has a *h* as a child. The next letter *I* is not a child of *h*; therefore, a new node is created. Finally, *s* is also not an existing child of *I*, and therefore, a new node of *s* is also created.

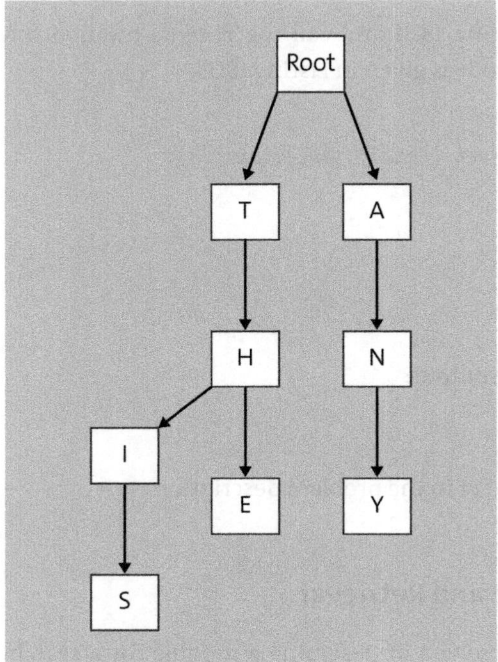

Figure 18.11 The Trie Data Structure after Inserting the Three Words

In this way, words that start with the same letters will follow the same path from the root, and only the parts of the words that start to diverge will be added. This approach allows for the quick searching of words throughout the entire tree. The third word will be inserted as new nodes because their prefixes haven't matched with any of the already existing words.

18.8.2 Basic Data Structure

To make our solution faster and more efficient, we'll use `HashMaps` in defining the nodes of the tree. Let's look at the definition of a node.

Each node will have two fields, `children` and `is_word`. The `children` will keep track of the nodes that are children of the node, and the `is_word` is a Boolean value indicating whether or not the node corresponds to a word. The `children` will be a `HashMap` so we can quickly search whether a letter happens to be a node child or not. Listing 18.34 shows the definition of a `Node`.

```
use std::collections::HashMap;
#[derive(Default, Debug, PartialEq, Eq, Clone)]
struct Node {
    children: HashMap<char, Node>,
    is_word: bool,
}
```

Listing 18.34 Definition of a Node

Let's examine visually how the data structure will look for some of the nodes. Figure 18.12 shows how the nodes will be defined for the letters *T*, *h*, and *e*.

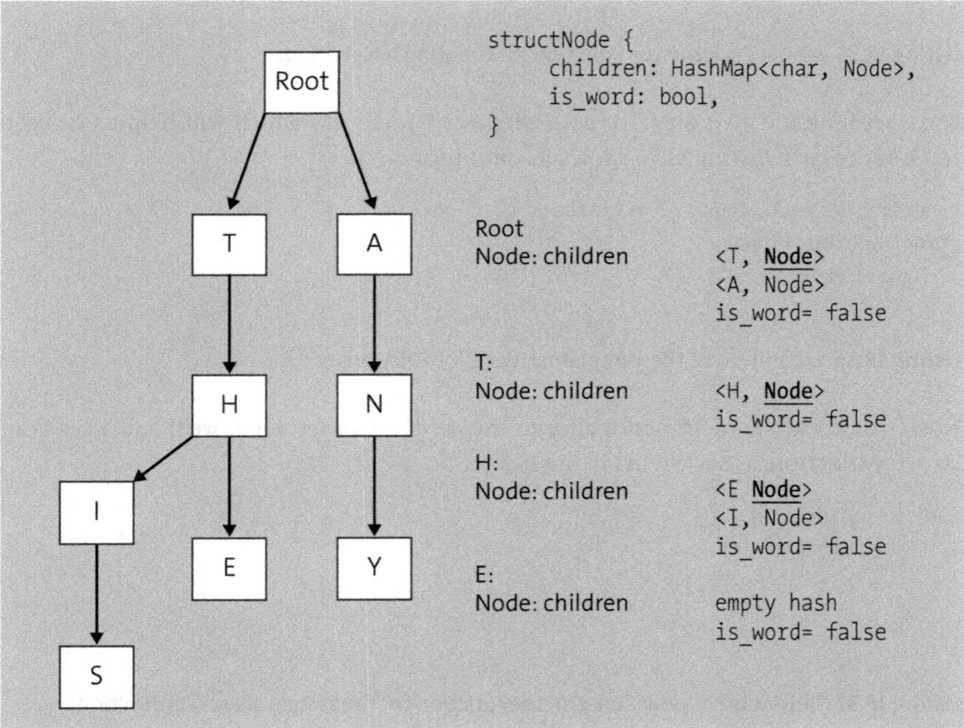

Figure 18.12 Definition of Some of the Nodes in the Tree

In the top right, we have the definition of the `Node`. At the root level, the `Node` for the child field will store two values in its `HashMap` corresponding to its two children: one for letter `T` and another one for letter `A`. Each entry contains the respective letter and another `Node`. The `is_word` field for the root is false since it does not correspond to any word. The node corresponding to `T` has its child information in the `HashMap` associated with its children field. The `is_word` is again false since `Node T` also does not correspond to a valid word. The remaining `Nodes` are defined in the same way.

18

18.8.3 Implementation

Now that you understand the basic data structure, let's add some functionality to Node. Listing 18.35 shows the code for new constructor function.

```
impl Node {
    fn new() -> Self {
        Node {
            is_word: false,
            children: HashMap::new(),
        }
    }
}
```

Listing 18.35 A new Constructor Function for Creating a New Node

Next, we'll define a wrapper struct called WordDictionary which will help us to keep track of the root. Listing 18.36 shows its definition.

```
#[derive(Default, Debug, PartialEq, Eq, Clone)]
struct WordDictionary {
    root: Node,
}
```

Listing 18.36 Definition of the Wrapper struct WordDictionary

Now, we can add some functionality to the WordDictionary. First, we'll add a new constructor function, as shown in Listing 18.37.

```
impl WordDictionary {
    fn new() -> Self {
        Self::default()
    }
}
```

Listing 18.37 Definition of new Constructor Function for Creating a New WordDictionary

The call to default will set the field to their default values.

Our next step is to look at the insert method. This method will take one word at a time and insert it into the tree or WordDictionary. The function will implement the same logic that we described earlier in Section 18.8.1. Listing 18.38 shows the definition of the insert method.

```
impl WordDictionary {
    ...
    fn insert(&mut self, word: &String) {
        let mut current = &mut self.root;
```

```
        for w in word.chars() {
            current = current.children.entry(w).or_insert(Node::new());
        }

        if !current.is_word {
            current.is_word = true;
        }
    }
}
```

Listing 18.38 Definition of the insert Method

This method accepts `&mut self` and a reference string to a word for which we want to make entries in the tree. Inside the function we first create a reference to the root of the tree. Next, we iterate through each letter of the word. We insert each letter into the `HashMap` as a key (corresponding to the `children` field of a node), and the value part will be a new `Node`. The `or_insert` part will add a default value corresponding to the key if no value already exists.

When the statement `current.children.entry(w).or_insert(Node::new());` executes, it will return the value part from the `HashMap`, which is a `Node`. This `Node` is next stored in the variable `current`. As a result, the `current` variable has the value part of the entry in it, corresponding to the specific key part mentioned by the letter `w` in this case.

An important point to remember is that `HashMaps` do not allow duplicate keys. If the same key appears while inserting letters of another word, a new entry will not be created. Instead, the value part, which is a `Node`, will be returned. Thus, for new words, keys will not be updated as long as the prefix remains the same. Once the loop ends, the word insertion is complete. At this point, you must set `is_word` to `true` for the last processed letter since now the data makes up a complete word. 18

Now, let's look at the implementation of the `search` method. Given a particular word, the method will start from the root and determine if the letters of the word follow a path from a root to a certain leaf of the tree. Listing 18.39 shows the definition of this method.

```
impl WordDictionary {
    ...
    fn search(&self, word: &String) -> bool {
        let mut current = &self.root;
        for w in word.chars() {
            if current.children.get(&w).is_some() {
                current = current.children.get(&w).unwrap();
            } else {
                return false;
            }
```

```
        }
        current.is_word
    }
}
```

Listing 18.39 Definition of the search Method

Starting from the root Node, the method checks each character of the word iteratively in the HashMap associated with the children field of the Node. If a key corresponding to a letter exists for a certain Node in the HashMap, we update the variable current to that Node. If at any stage, the key is not found, we'll return a false, meaning that the word does not exist. When the loop ends, we return the is_word field of the respective Node. In summary, if the word exactly matches some other word, then is_word will be true, and if the word matches partially with some existing word, then false is returned.

Let's add some code in main to test the required functionality. Consider the code shown in Listing 18.40.

```
fn main() {
    let words = vec![
        "the", "a", "there", "answer", "any", "by", "bye", "their", "abc",
    ].into_iter().map(|x| String::from(x)).collect::<Vec<String>>();

    let mut d = WordDictionary::new();

    for i in 0..words.len() {
        d.insert(&words[i]);
    }
    println!(
        "Word 'there' in the dictionary: {}",
        d.search(&"there".to_string())
    );
}
```

Listing 18.40 Code in main for Testing the Required Functionality

In this example, first, the vector of raw strings is converted to owned strings. Then, a new WordDictionary is initialized, and the words are inserted into it. Finally, we search for the word there in the WordDictionary. When you execute the code, the result should be the following output:

```
Word 'there' in the dictionary: true
```

This result confirms the validity of our code.

18.9 Problem 9: Most Recently Used Product

A business is interested in knowing the products that have been purchased most recently by a customer. They will use this information in their promotional and marketing campaigns.

To store this information, we need a suitable data structure where they can easily generate lists of some predetermined number of most recently purchased products. The business is also critical about the speed of retrieving this information from the data structure.

Let's describe the problem in a bit more detail. Consider the list of products to be [1 2 3 4 5 4 6 8 9] where each number represents a certain product. The company is only interested in storing information about the four most recently purchased items or products. Figure 18.13 shows how the business wants to keep this information in a list.

Product Purchases [1 2 3 4 5 4 6 8 9]

Product 1 is Purchased	1			
Product 2 is Purchased	1	2		
Product 3 is Purchased	1	2	3	
Product 4 is Purchased	1	2	3	4

Figure 18.13 Visual Description of the Problem

When product 1 is purchased, we store it in the list. Next product 2 is being purchased, so it is also stored at the end of the list. In the same way, the next two products are also stored at the end, respectively. After purchase of product 4, the list is at full capacity. The most recently purchased product is at the end of the list while the least recently purchased or the oldest purchased product is at the beginning of the list.

Now, when a customer purchases a new product (product 5, in this case), we'll delete the oldest item in the list, which will create an empty space in the list. Next, we'll push all the elements to the left by one place and will insert the newly purchased item at the end. Figure 18.14 shows the updated list after the purchase of product 5.

Next, when product 4 is purchased again, we'll move it to the end of the list also. The revised position of the list is shown in Figure 18.15.

When an existing product is purchased again, we'll just move it to the end, and the deletion step will not occur. Similarly, the purchases of all the remaining items will be maintained in the list.

Product Purchases [1 2 3 4 5 4 6 8 9]				
Delete Least Recently Purchased		2	3	4
Push All Elements to the Left	2	3	4	
Insert New Purchase at the End	2	3	4	5

Figure 18.14 List after Purchase of Product 5

After Product 4 is Purchased	2	3	5	4

Figure 18.15 List after Purchase of Product 4

Using this strategy, the least recently purchased item is always at the beginning, and the most recently purchased item is at the end.

18.9.1 Solution Setup

The business requires that the solution data structure should be designed keeping in view the speed aspect. Moreover, we also want to keep track of the first and last element of the list of values. For speed, we'll use the `HashMaps`, and for keeping track of the first and last element, we'll use the doubly linked list, where we have explicit information about the first and the last element of the list (the head and the tail, respectively).

Let's first look at the solution, which will make the implementation quite easy. Assume the same list of products that we have just seen, that is, [1 2 3 4 5 4 6 8 9]. When the first product is purchased, we'll make a node of the doubly linked list and store it as a node. Since we only have one node, both the head and tail will point to it. Moreover, we'll also make an entry into a `HashMap` where the key will be the value of the node, and the value will be basically a pointer to the node. This whole process is shown in Figure 18.16.

When the next few products are purchased (i.e., products 2, 3, and 4), we'll make entries at the tail of the list, and we'll also add an entry in the `HashMap`. The updated state of the `HashMap` and the doubly linked list is shown in Figure 18.17 after purchase of products 2, 3, and 4.

When product 5 is purchased, we'll delete the head and insert the new product at the tail. The entries in the `HashMap` will also be updated accordingly. The new state of the `HashMap` and the doubly linked list is shown in Figure 18.18.

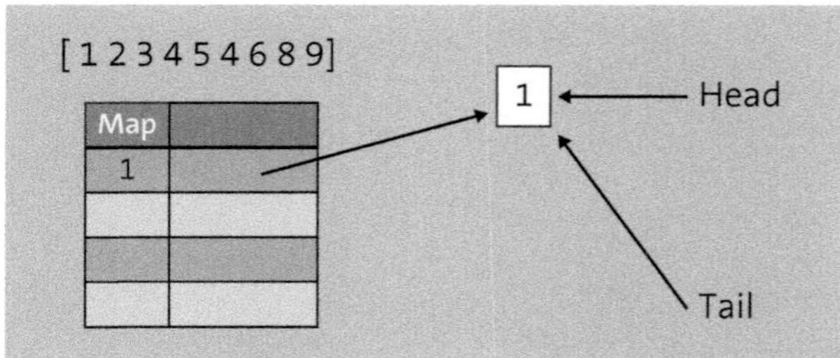

Figure 18.16 State of HashMap and Doubly Linked List after Purchase of Product 1

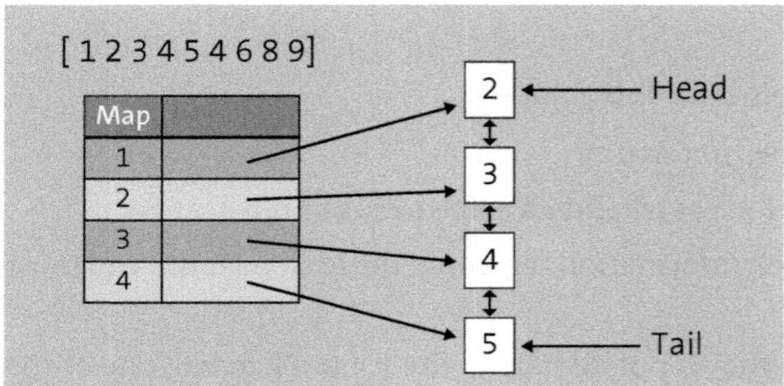

Figure 18.17 State of HashMap and Doubly Linked List after Purchase of Products 2, 3, and 4

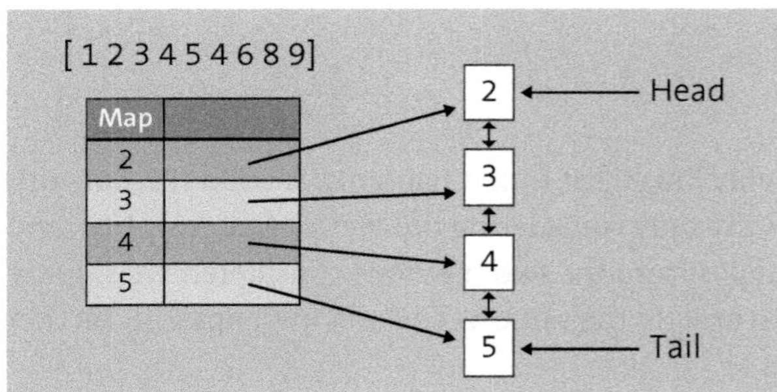

Figure 18.18 State of HashMap and Doubly Linked List after Purchase of Product 5

Next, when product 4 is purchased for the second time, the node with value 4 will be moved to the tail, and the node with value 5 will replace the node with value 4. In this case, the map will not change, and only the doubly linked list will change. The updated state is shown in Figure 18.19.

Note that we don't need to replace the values in the map because the order inside the map doesn't matter. Moreover, the map allows us to determine whether a specific value exists without scanning all entries, significantly improving lookup efficiency when compared against searching through a list. The process for the purchase of the remaining products will be performed in the same way.

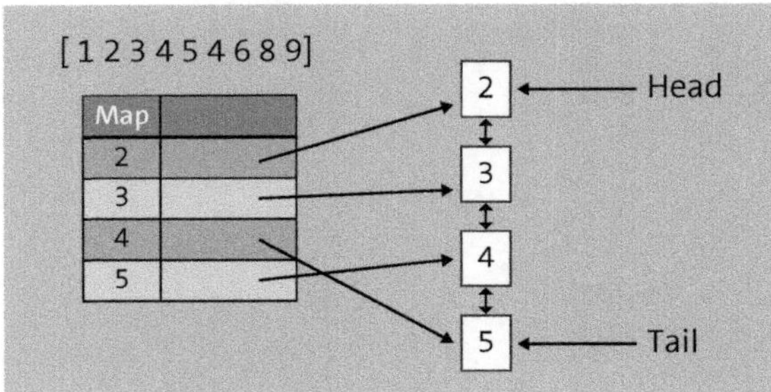

Figure 18.19 State of HashMap and Doubly Linked List after Purchase of Product 4 for the Second Time

Let's summarize the important points for this solution:

- The order of the map entries does not matter.
- The map is just used to quickly look at whether a value exists or not.
- The head and tail provide explicit information regarding the least recently and most recently purchased items.
- When a new item is purchased that is not already in the list, we'll delete one of the nodes and insert a new node at the tail.
- If the item that is being purchased is already in the list, then we'll just move it to the end without updating the map.

18.9.2 Basic Data Structure

We'll start with the code of the doubly linked list from Chapter 11, Section 11.2, but with minor changes in the code. First, we're only considering the methods of `push_back` and `remove_front`. Second, the return types from the two methods considered are slightly changed. The rest of the details are exactly the same as covered in Chapter 11, Section 11.1. The code is shown in Listing 18.41.

```
use std::cell::RefCell;
use std::collections::HashMap;
use std::rc::Rc;
#[derive(Debug)]
struct Node {
    prod_id: i32,
    prev: Link,
    next: Link,
}

impl Node {
    fn new(elem: i32) -> Rc<RefCell<Self>> {
        Rc::new(RefCell::new(Node {
```

```
            prod_id: elem,
            prev: None,
            next: None,
        }))
    }
}

type Link = Option<Rc<RefCell<Node>>>;
#[derive(Default, Debug)]
struct DoublyLinkList {
    head: Link,
    tail: Link,
}

impl DoublyLinkList {
    fn new() -> DoublyLinkList {
        DoublyLinkList {
            head: None,
            tail: None,
        }
    }
    pub fn push_back(&mut self, elem: i32) -> Link {
        let new_tail = Node::new(elem);
        match self.tail.take() {
            Some(old_tail) => {
                old_tail.borrow_mut().next = Some(new_tail.clone());
                new_tail.borrow_mut().prev = Some(old_tail);
                self.tail = Some(new_tail);
            }
            None => {
                self.head = Some(new_tail.clone());
                self.tail = Some(new_tail);
            }
        }
        self.tail.clone()
    }

    pub fn remove_front(&mut self) -> Option<Link> {
        self.head.take().map(|old_head| {
            match old_head.borrow_mut().next.take() {
                Some(new_head) => {
                    new_head.borrow_mut().prev.take();
                    self.head = Some(new_head);
                    self.head.clone()
                }
```

18

```
                None => {
                    self.tail.take();
                    None
                }
            }
        })
    }
}
```

Listing 18.41 Doubly Linked List Implementation

The code has already been covered in Chapter 11, Section 11.1, so we'll skip its explanation.

Next, we'll define the data structure explained in Section 18.9.1. Listing 18.42 shows the definition of the required data structure.

```
#[derive(Debug)]
struct MRPProduct {
    map: HashMap<i32, Rc<RefCell<Node>>>,
    product_list: DoublyLinkList,
    size: i32,
    capacity: i32,
}
```

Listing 18.42 Definition of the MRPProduct Containing the Required Data Structure

The `MRPProduct` struct contains a `HashMap` where the product numbers serve as the keys, and the values of the HashMap are pointers to the respective nodes in the doubly linked list. The `capacity` defines the maximum number of elements that can be stored in the `HashMap`.

Next, we'll add some functionality to `MRPProducts`. As usual, we'll first add the `new` constructor function for creating a new instance of an `MRPProduct`. Listing 18.43 shows the definition of the new constructor.

```
impl MRPProduct {
    fn new(capacity: i32) -> Self {
        Self {
            map: HashMap::new(),
            product_list: DoublyLinkList::new(),
            size: 0,
            capacity: capacity,
        }
    }
}
```

Listing 18.43 Definition of new Constructor Function for MRPProduct

18.9.3 Implementation

Next, we'll define a purchase method, the core function in the implementation. It will update the doubly linked list and the HashMap based on a new purchased product. The following is the signature of the method:

```
impl MRPProduct {
    ...
    fn purchase(&mut self, prod_id: i32) {}
}
```

Inside the method, we first check if the HashMap already contains the prod_id passed in. This check can be performed using the get function. If the value already exists, then we'll move it to the tail of the list. Listing 18.44 contains the partial code for the method.

```
impl MRPProduct {
    ...
    fn purchase(&mut self, prod_id: i32) {
        if let Some(node) = self.map.get(&prod_id) {
            self.item_list.move_to_tail(node);
        }
    }
}
```

Listing 18.44 Partial Code for the purchase Method

We currently do not have a method that moves a node to the tail of the list, and we need to implement this method. The method move_to_tail will have the following signature:

```
impl DoublyLinkList {
    ...
    fn move_to_tail(&mut self, node: &Rc<RefCell<Node>>) {}
}
```

The input to the method is &mut self and a reference to the node that we want to move to the tail.

Next, we'll grab the information of the next and prev nodes of a given node. We would also want to own those nodes. Listing 18.45 shows how these tasks can be performed.

```
impl DoublyLinkList {
    ...
    fn move_to_tail(&mut self, node: &Rc<RefCell<Node>>) {
        let prev = node.borrow().prev.as_ref().map(|a| Rc::clone(a));
        let next = node.borrow().next.as_ref().map(|a| Rc::clone(a));
    }
}
```

Listing 18.45 Partial Implementation of move_to_tail

To grab the previous node's information, we'll call the `borrow` on a node and access its `prev` field. The map function maps the node from a reference to a node to an owned copy using a `clone` method. Thus, we can change its contents. In the same way, we can enable access to the `next` of the node using the `clone` method.

There are now four cases for the `prev` and `next`:

1. **Case 1: No previous and next nodes**
   The first case is when the `prev` and `next` are both `None`. This can happen if we currently have no product information, and the `node` passed into the method is the only `node` in the list. In this case we do not need to do anything.
2. **Case 2: `Some` previous node, no next node**
   The second case is when there is `Some` previous `node` but there is no `next` node. This will happen when the `node` passed in is already at the `tail` since `next` of the `tail` is always empty. This again suggests that we do not need to do anything because the `node` is already at the `tail`.
3. **Case 3: No previous node, `Some` next node**
   The third case is when there is no previous `node` but there is `Some next node`. This suggests that the `node` we want to move is at the `head` of the list since the `head` has no previous node. Let's explain this case visually in some detail. The case is visually represented in Figure 18.20.

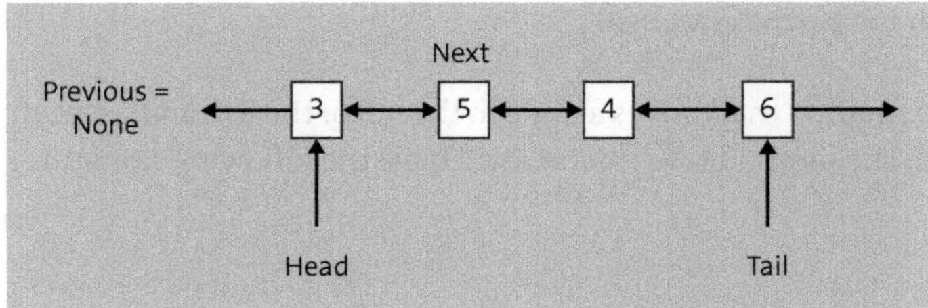

Figure 18.20 Case 3: When the Node That Needs to Be Moved to the Tail Is Head of the List

We'll first update the `next` of the old `head` to that of `None`. Then, the `Next` node previous pointer is set to `None`, and finally, the `head` of the list updated to that of the `Next` node. This process is shown in Figure 18.21.

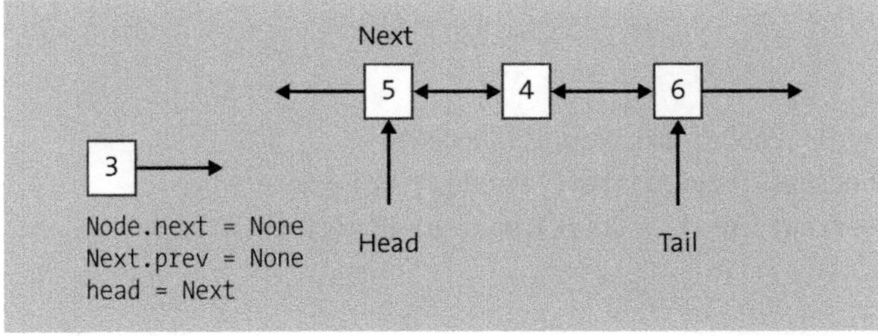

Figure 18.21 Updating of Head in Case 3

The removed head, which we refer to as node in the diagram, needs to be installed at the tail. Figure 18.22 shows the necessary steps to follow in this regard.

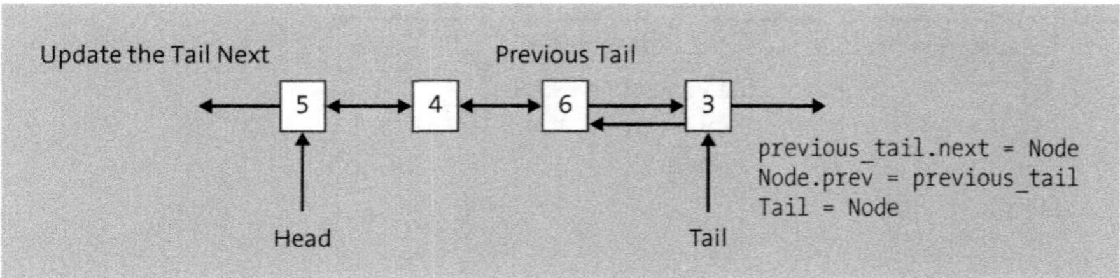

Figure 18.22 Insertion of the Removed Head at the Tail of the List

First, we'll set the previous `tail` next field to point to the node (old head). The `previous` of the node needs to point to the previous `tail`; finally we'll update the tail to that of the node.

4. **Case 4: `previous` and `next` point to nodes**
   The fourth case arises when both the `previous` and `next` nodes to the node are passed into the function. This situation is shown in Figure 18.23.

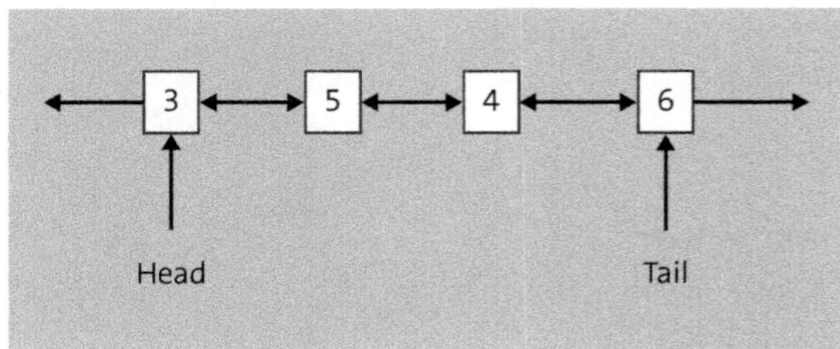

Figure 18.23 Case 4: When the previous and next Fields Are Pointing to Nodes

In this case, we'll first remove the node from the list and set the points of the `previous` and `next` nodes accordingly. The process is shown in Figure 18.24.

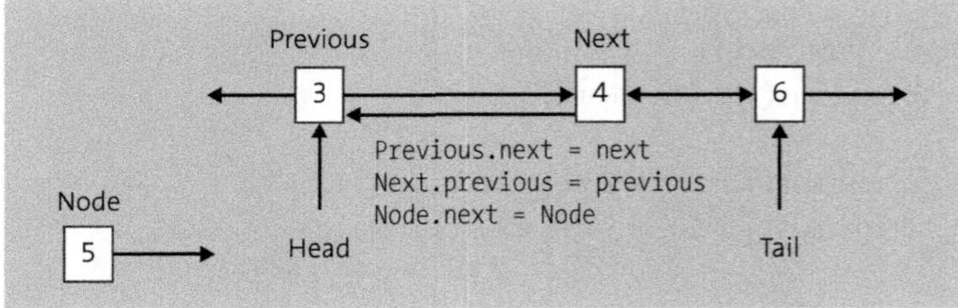

Figure 18.24 Updating of the List after Removing the Node

Next, we'll move the node the end of the list and will update the `tail` and the pointers accordingly. Figure 18.25 shows the necessary updates.

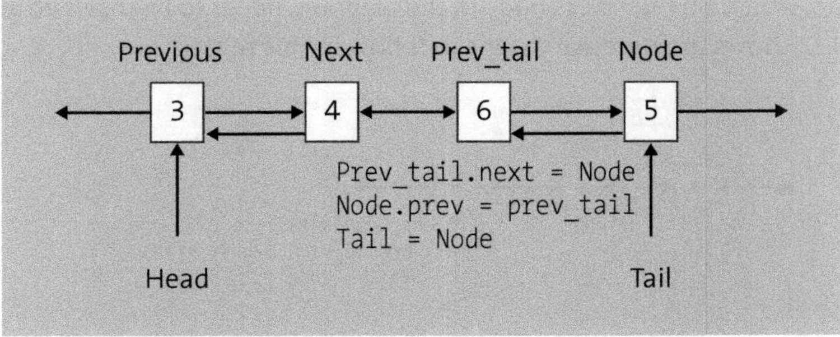

Figure 18.25 Insertion of the Removed Node at the Tail

Now that you understand the four cases, the implementation is quite easy to comprehend. Listing 18.46 shows the complete implementation of the function `move_to_tail`.

```
impl DoublyLinkList {
    ...
    fn move_to_tail(&mut self, node: &Rc<RefCell<Node>>) {
        let prev = node.borrow().prev.as_ref().map(|a| Rc::clone(a));
        let next = node.borrow().next.as_ref().map(|a| Rc::clone(a));
        match (prev, next) {
            (None, None) => {}
            (Some(_), None) => {}
            (None, Some(next)) => {
                node.borrow_mut().next = None;
                next.borrow_mut().prev = None;
                self.head = Some(next.clone());

                let prev_tail = self.tail.as_ref().unwrap();
                prev_tail.borrow_mut().next = Some(node.clone());
                node.borrow_mut().prev = Some(prev_tail.clone());
                self.tail = Some(node.clone());
            }
            (Some(prev), Some(next)) => {
                node.borrow_mut().next = None;

                prev.borrow_mut().next = Some(next.clone());
                next.borrow_mut().prev = Some(prev.clone());

                let prev_tail = self.tail.as_ref().unwrap();
                prev_tail.borrow_mut().next = Some(node.clone());
                node.borrow_mut().prev = Some(prev_tail.clone());
                self.tail = Some(node.clone());
            }
        }
```

```
    }
}
```

Listing 18.46 Complete Implementation of the Method move_to_tail

The implementation of `move_to_tail` follows a `match` statement to handle the four cases systematically. When both `prev` and `next` are `None` (case 1) or when `next` is `None` (case 2), the function does nothing since the node is either the only one in the list or already at the tail. In case 3, where there is no `prev` but a `next`, the function updates the head, disconnects the node from the front, and reattaches it at the tail. Finally, in case 4, where both `prev` and `next` exist, the function first detaches the node from its current position by updating the adjacent nodes' pointers before placing it at the tail. The method is based around the four cases explained in detail using visuals.

Now that the function of `move_to_tail` is complete, we'll complete the core method of `purchase`, which was incomplete in Listing 18.44.

In the `purchase` method, if a `prod_id` already exists in the `HashMap`, we move it to the end by calling the `move_to_tail` method. If the node is not in the `HashMap`, then we'll first check if the list has some space for adding the new product. If the list has no space, then we'll remove the `head` from the list and then proceed with inserting the new element at the `tail`. Listing 18.47 shows the corresponding code.

```
impl MRPProduct {
    ...
    fn purchase(&mut self, prod_id: i32) {
        if let Some(node) = self.map.get(&prod_id) {
            self.product_list.move_to_tail(node);
        } else {
            if self.size >= self.capacity {
                let prev_head = self.product_list.remove_front().unwrap();
                self.map.remove(&prev_head.unwrap().borrow().prod_id);
            }
            let node = self.product_list.push_back(prod_id).unwrap();
            self.map.insert(prod_id, node);
            self.size += 1;
        }
    }
}
```

Listing 18.47 Completed Definition of purchase Method

If there is no capacity, we remove the `head` using the `remove_front` method and also remove the entry from the `HashMap`. Finally, we simply add the new `node` at the `tail` using the `push_back` method and update the `size`.

We also need to add one more final method of `print` to `MRPProduct` implementation for printing the list. This method will essentially traverse all products in the list and print out the product number. Listing 18.48 shows the implementation of this method.

```
impl MRPProduct {
    ...
    fn print(&self) {
        let mut traversal = self.product_list.head.clone();
        while !traversal.is_none() {
            let temp = traversal.clone().unwrap();
            print!("{} ", temp.borrow().prod_id);
            traversal = temp.borrow().next.clone();
        }
        println!("");
    }
}
```

Listing 18.48 Definition of the print Method

The code for the intended functionality is now complete. To test this functionality, we'll add some code in `main`. We'll first create a new instance of `MRPProduct` with a capacity of size `3` and then add a few products to the list. Listing 18.49 shows the code in `main`.

```
fn main() {
    let mut products_list = MRPProduct::new(3);
    products_list.purchase(10);
    products_list.print();
    products_list.purchase(15);
    products_list.print();
    products_list.purchase(20);
    products_list.print();
    products_list.purchase(25);
    products_list.print();
    products_list.purchase(20);
    products_list.print();
}
```

Listing 18.49 Code in main for Testing the Required Functionality

Executing the code produces the following output:

```
10
10 15
10 15 20
15 20 25
15 25 20
```

Initially, the list is empty. So we inserted the first three purchased products of 10, 15, and 20. The most recently purchased is always at the end (extreme right) of the list. At this stage, the list is at full capacity. The purchase of the new product 25 leads to the deletion of the least recently purchased product of 10 from the beginning of the list. Finally, the purchase of product 20 brings the same product to the end position.

18.10 Problem 10: Displaying Participants in an Online Meeting

Let's say our business uses software for conducting online meetings with its employees. Our company wants to keep track and display the list of participants attending the meeting using a suitable data structure. The company wants the names of attendees in a meeting to be displayed in alphabetical order. The attendees can join or leave a meeting at random, so the underlying data structure should allow for easy updating, accordingly. After analyzing the problem, the design team has agreed that the name of the attendees should be stored using a BST.

Once the names are stored, the company is further interested in a special mode of display called "Gallery Mode." This mode displays the names of the participants in paginated form (i.e., divided into pages that can be scrolled down). Only ten participants can be shown on one page.

Our job is to use the BST containing the names of participants and implement the pagination feature. Whenever our function is called, the names of the next ten participants should be returned in alphabetical order.

Figure 18.26 shows the BST corresponding to some list of participants.

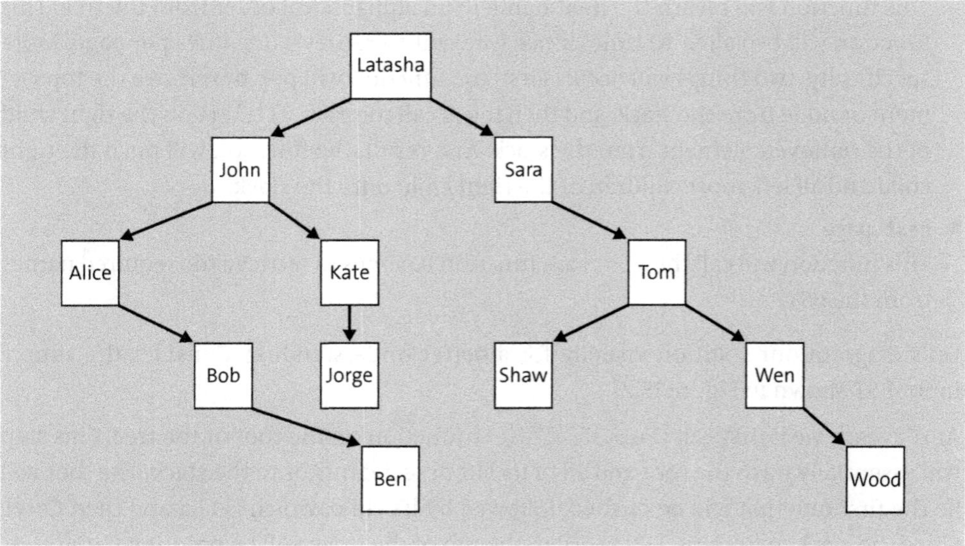

Figure 18.26 BST for the List of Participants

The details of the BST were covered earlier in Section 18.6.1. A minor difference in this case is that the values are names of participants and not numbers. A comparison of two names will be made based on their place in alphabetical order. A letter that comes first is less than a letter that comes later. If the first letters of two names are the same, then we'll compare their second letters and so on.

The essential function of pagination will take the BST as an input and will produce the first ten names from the tree as an output in alphabetical order. For the BST shown in Figure 18.26, the first call to the pagination will produce the following list of names:

```
[Alice, Bob, Ben, John, Jorge, Kate, Latasha, Sara, Shawn, Tom]
```

The second call to the function will produce the remaining names:

```
[Wen, Wood]
```

Because in a BST, values are stored in an order, retrieving the values in alphabetical form should be easy. Let's see how this solution can be created.

18.10.1 Solution Setup

Our proposed solution will consist of three key functions:

- **push_all_left**
  This function, when called for a certain node, will push the node and all of its left descendants onto the stack. The function will only terminate when there are no more left nodes to push onto the stack.
- **next_name**
  This function will return the next name in an alphabetical order from the tree. This function will be called 10 times since we want to retrieve 10 names per page. More specifically, two things will occur: First, the function will pop or remove the top element or node from the stack, and then it will call the `push_all_left` on the right child of the removed element from the stack. As a result, the function will push the right child and all left-most children of the right child onto the stack.
- **next_page**
  This function will call the `next_name` function 10 times to retrieve the required names from the BST.

Let's diagram our solution visually for a better understanding. Consider the simple input BST shown in Figure 18.27.

At the start, we'll first call the `push_all_left` function on the root of the tree. This step will essentially push the root and all of its left descendants onto the stack. The root will be the first one that will be pushed, followed by its child, which is Elia and then Caryl. Since the stack maintains a FILO order, the top of the stack will be pointing to Caryl, as shown in Figure 18.28.

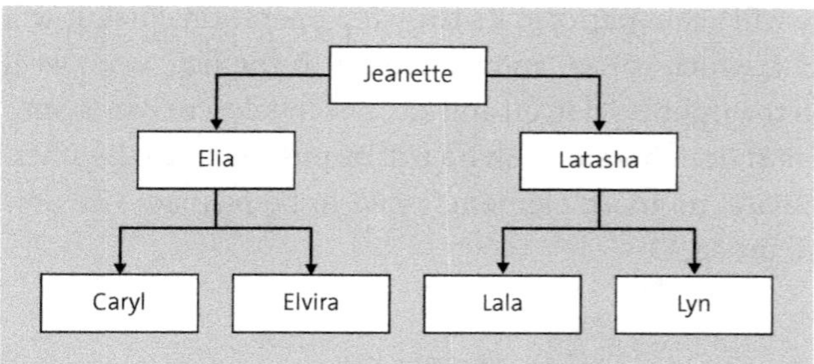

Figure 18.27 A Sample BST Tree

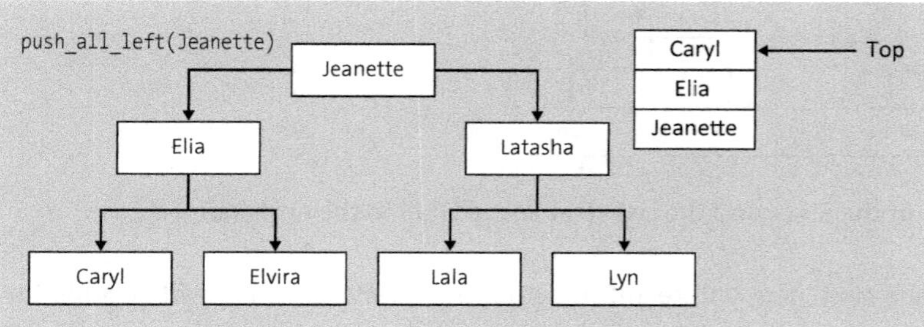

Figure 18.28 Function of push_all_left on Jeanette

Following this step, we'll make a call to the `next_page` function, which will further call the `next_name` function `10` times to retrieve the names of 10 participants in an alphabetical order.

The `next_name` function performs two tasks. First it will pop the top-most element from the stack. As a result, Caryl will be popped from the stack shown in Figure 18.28. Once popped, we'll add Caryl to a resultant list. Next, the function will move to the right child of the popped element and will call the `push_all_left` on that node. Since there is no right child of Caryl in this case, nothing else will be pushed onto the stack. If a right child existed, we would push the right child onto the stack, along with all its left descendants. The state of the stack is shown in Figure 18.29.

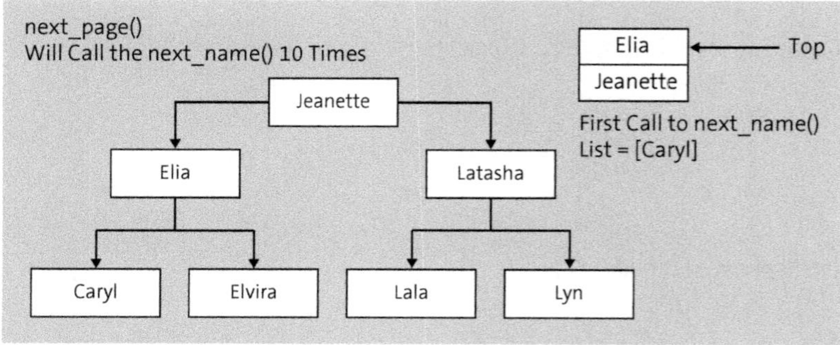

Figure 18.29 State of the Stack and the List after First Call to the next_name

The second call to the next_name will again perform its two-step operation. First, it will pop the top of the stack element, which is Elia, and will add it to the list. Next, we'll move to the right child and push the right child itself and all of its left descendants onto the stack. The right child of Elia is Elvira; therefore, Elvira will be pushed onto the stack. Since there is no left child of Elivira, no more elements need to be pushed. The new shape of the stack is shown in Figure 18.30.

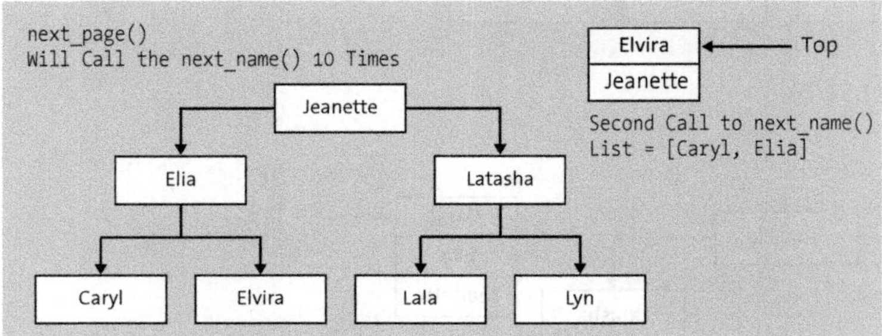

Figure 18.30 State of the Stack and the List after Second Call to the next_name

The third call to the next_name will result in a popping of Elvira, which will be added to the list. Since there is no right child, no new element will be pushed. The new stack will only contain the root now. The fourth call will pop Jeanette out, who will be added to the list. Next, the left descendants of the right child will be pushed onto the stack. The process will continue in this manner until the first 10 names are returned and there are no more elements to return.

18.10.2 Basic Data Structure

Now that we understand the problem and the solution in detail, the solution is easy to implement.

First, we'll define the BST and its relevant implementation. Listing 18.50 contains the relevant code.

```
use std::cell::RefCell;
use std::rc::Rc;
#[derive(Debug, Default, PartialEq, Eq, Clone)]
struct Node {
    val: String,
    left: Link,
    right: Link,
}
type Link = Option<Rc<RefCell<Node>>>;
impl Node {
    fn new(val: String) -> Self {
        Node {
```

```
            val,
            left: None,
            right: None,
        }
    }
    fn insert(&mut self, val: String) {
        if val > self.val {
            match &self.right {
                None => self.right =
Some(Rc::new(RefCell::new(Self::new(val)))),
                Some(node) => node.borrow_mut().insert(val.to_string()),
            }
        } else {
            match &self.left {
                None => self.left =
Some(Rc::new(RefCell::new(Self::new(val)))),
                Some(node) => node.borrow_mut().insert(val.to_string()),
            }
        }
    }
}
```

Listing 18.50 Definition of BST and Its Relevant Methods

In Section 18.6.2, we implemented the BST using a box pointer. We can use that implementation as well. However, in this problem, we'll consider the BST implementation using Rc and RefCell smart pointers. This approach allows multiple parts of a program to share ownership over nodes (provided by Rc) while still enabling interior mutability (provided by RefCell). This approach is especially useful when you need to modify nodes while keeping multiple references to them, which is not possible with Box alone. The code shown in Listing 18.50 also defines a new constructor function and an insert method for inserting a new value in the BST.

Next, we define a wrapper struct for constructing a new BST along with its relevant functions and methods. Listing 18.51 shows the code for a wrapper struct named BinarySearchTree.

```
#[derive(Debug, Default, PartialEq, Eq)]
struct BinarySearchTree {
    root: Node,
}
impl BinarySearchTree {
    fn new(val: String) -> Self {
        BinarySearchTree {
            root: Node::new(val.to_string()),
        }
```

```
    }
    fn insert(&mut self, val: String) {
        self.root.insert(val.to_string());
    }
}
```

Listing 18.51 Definition of the binarySearchTree and Its Implementation

The new constructor function initializes an empty BST, and the insert method inserts the first element or the root into the BST.

Finally, to implement the required functionality, we'll define another struct called DisplayLobby to contain a stack of nodes, as shown in Listing 18.52.

```
struct DisplayLobby {
    stack: Vec<Rc<RefCell<Node>>>,
}
```

Listing 18.52 Definition of DisplayLobby

18.10.3 Implementation

We'll implement the required functionality discussed in Section 18.10.1 inside an implementation block for DisplayLobby.

First, we'll add a push_all_left method to the implementation block. Listing 18.53 shows the code for the method.

```
impl DisplayLobby {
    ...
    fn push_all_left(mut p: Option<Rc<RefCell<Node>>>, stack: &mut
Vec<Rc<RefCell<Node>>>) {
        while let Some(link) = p.clone() {
            stack.push(p.clone().unwrap());
            p = link.borrow().left.clone();
        }
    }
}
```

Listing 18.53 Definition of push_all_left

This function keeps on iterating and pushing elements to the stack until some existing node in the left child of the tree arises.

Next, let's look at the next_name method. As pointed out earlier, this method does two things. First, it pops the top element from the stack. Then, it calls the push_all_left method on the right child of the node. the code shown in Listing 18.54 shows an implementation of this function.

```
impl DisplayLobby {
    ...
    fn next_name(&mut self) -> String {
        let node = self.stack.pop().unwrap();
        let name = &node.borrow().val;
        let mut next_node = node.borrow().right.clone();
        Self::push_all_left(next_node, &mut self.stack);
        name.to_string()
    }
}
```

Listing 18.54 Definition of the new_name Method

The name returned from the method will be added to the list of resultant names.

Next, the `next_page` method returns the names of `10` participants in an alphabetical order. This method will essentially call the `next_name` 10 times. Listing 18.55 shows its implementation.

```
impl DisplayLobby {
    ...
    fn next_page(&mut self) -> Vec<String> {
        let mut resultant_names: Vec<String> = Vec::new();
        for i in 0..10 {
            if !self.stack.is_empty() {
                resultant_names.push(self.next_name());
            } else {
                break;
            }
        }
        resultant_names
    }
}
```

Listing 18.55 Definition of next_page Method

The method calls the `next_name` as long as the stack is not empty, in this case, 10 times.

Finally, we'll add the implementation of the `new` constructor function, which will set up elements in the stack from the left descendants of the root. The code shown in Listing 18.56 illustrates the implementation of the function.

```
impl DisplayLobby {
    fn new(root: Option<Rc<RefCell<Node>>>) -> Self {
        let mut stack = Vec::new();
        Self::push_all_left(root.clone(), &mut stack);
```

```
        DisplayLobby { stack }
    }
}
```

Listing 18.56 Definition of the new Constructor Function

Now that our code is complete, let's test it in `main`. We'll first create a vector containing some names and populate a BST from these names. Next, we'll call the `next_page` a few times to see the list of participants. Listing 18.57 contains the code in `main`.

```
fn main() {
    let mut bst = BinarySearchTree::new("Jeanette".to_string());
    let names: Vec<String> = vec![
        "Latasha",
        "Elvira",
        "Caryl",
        "Antoinette",
        "Cassie",
        "Charity",
        "Lyn",
        "Lia",
        "Anya",
        "Albert",
        "Cherlyn",
        "Lala",
        "Kandice",
        "Iliana",
        "Nouman",
        "Azam",
    ]
    .into_iter()
    .map(String::from)
    .collect();
    for name in names.into_iter() {
        bst.insert(name.to_string());
    }
    let mut display = DisplayLobby::new(Some(Rc::new(RefCell::new(bst.root))));
    println!("Participants on first page: {:?}", display.next_page());
    println!("Participants on second page: {:?}", display.next_page());
}
```

Listing 18.57 Code in main for Testing the Implementation

The code shown in Listing 18.57 generates the following output:

```
Participants on first page: ["Albert", "Antoinette", "Anya", "Azam", "Caryl",
"Cassie", "Charity", "Cherlyn", "Elvira", "Iliana"]
Participants on second page: ["Jeanette", "Kandice", "Lala", "Latasha", "Lia",
"Lyn", "Nouman"]
```

The output confirms the correctness of the implementation.

18.11 Summary

This chapter focused on practical problems that require sophisticated data structure solutions in Rust. In particular, we covered some real-life scenarios to illustrate the application of various data structures to solve complex issues efficiently. The problems we presented emphasized critical thinking and practical skills in analyzing data, managing information, and optimizing processes. Through these exercises, you gained hands-on experience with Rust's enhanced capabilities in areas such as data organization, data retrieval, and performance optimization. This chapter reinforced the strategies and techniques we explored throughout this book, which we hope has equipped you with the knowledge you need to implement effective solutions in your own projects.

The Author

Dr. Nouman Azam is an associate professor of computer science at the National University of Computer and Emerging Sciences. He also teaches online programming courses about Rust and MATLAB and reaches a community of more than 50,000 students.

He received his PhD in computer science from the University of Regina in Canada. Prior to that, he completed his MSc in computer software engineering from the National University of Sciences and Technology, Pakistan, and his BSc in computer sciences from the National University of Computer and Emerging Sciences, Pakistan.

Dr. Azam's research interests include game theory, rough sets, conflict analysis, and group decision-making.

Index

N

O

P

T

U

V

W

Z

- Your all in one guide to modern C++
- Work with the C++ language, from basic syntax and functions to more advanced features such as pointers and macros
- Learn to use the standard library and containers

Torsten T. Will

C++

The Comprehensive Guide

If you need to know C++, look no further! This comprehensive guide has everything you need to master the modern C++23 language, from syntax fundamentals to advanced development concepts. Follow practical code examples as you learn object-oriented programming, work with standard library containers, program concurrent applications, and more. Don't just learn how to code—learn how to code better with expert tips and guidance on the rules of compact, secure, and efficient code.

1089 pages, pub. 10/2024
E-Book: $64.99 | **Print:** $69.95 | **Bundle:** $79.99

www.rheinwerk-computing.com/5927